T0137700

Lecture Notes in Computer Science 14742

Founding Editors

Gerhard Goos
Juris Hartmanis

The series Lecture Notes in Computer Science (LNCS), including its subseries Lecture Notes in Artificial Intelligence (LNAI) and Lecture Notes in Bioinformatics (LNBI), has established itself as a medium for the publication of new developments in computer science and information technology research, teaching, and education.

LNCS enjoys close cooperation with the computer science R & D community, the series counts many renowned academics among its volume editors and paper authors, and collaborates with prestigious societies. Its mission is to serve this international community by providing an invaluable service, mainly focused on the publication of conference and workshop proceedings and postproceedings. LNCS commenced publication in 1973.

Lecture Notes in Computer Science 14742

Founding Editors

Gerhard Goos
Juris Hartmanis

Editorial Board Members

Elisa Bertino, *Purdue University, West Lafayette, IN, USA*
Wen Gao, *Peking University, Beijing, China*
Bernhard Steffen, *TU Dortmund University, Dortmund, Germany*
Moti Yung, *Columbia University, New York, NY, USA*

The series Lecture Notes in Computer Science (LNCS), including its subseries Lecture Notes in Artificial Intelligence (LNAI) and Lecture Notes in Bioinformatics (LNBI), has established itself as a medium for the publication of new developments in computer science and information technology research, teaching, and education.

LNCS enjoys close cooperation with the computer science R & D community, the series counts many renowned academics among its volume editors and paper authors, and collaborates with prestigious societies. Its mission is to serve this international community by providing an invaluable service, mainly focused on the publication of conference and workshop proceedings and postproceedings. LNCS commenced publication in 1973.

Bistra Dilkina

Editor

Integration of Constraint Programming, Artificial Intelligence, and Operations Research

21st International Conference, CPAIOR 2024
Uppsala, Sweden, May 28–31, 2024
Proceedings, Part I

 Springer

Editor
Bistra Dilkina 🆔
University of Southern California
Los Angeles, CA, USA

ISSN 0302-9743   ISSN 1611-3349 (electronic)
Lecture Notes in Computer Science
ISBN 978-3-031-60596-3   ISBN 978-3-031-60597-0 (eBook)
https://doi.org/10.1007/978-3-031-60597-0

This Springer imprint is published by the registered company Springer Nature Switzerland AG
The registered company address is: Gewerbestrasse 11, 6330 Cham, Switzerland

If disposing of this product, please recycle the paper.

Preface

This book constitutes the proceedings of the 21st International Conference on the Integration of Constraint Programming, Artificial Intelligence, and Operations Research (CPAIOR 2024). The conference was held as an in-person event in Uppsala, Sweden at the Uppsala University campus, May 28–31, 2024.

The conference received a total of 104 paper submissions of original unpublished work, which were reviewed by at least three Program Committee members in a single-blind process. The reviewing phase was followed by an author response period and an extensive discussion period carried out by the Program Committee. At the end of the review period, 42 regular papers were accepted for presentation during the conference and were published in this volume. In addition, the conference received 17 extended abstract submissions, containing either original unpublished work or a summary of work, which were reviewed for appropriateness for the conference, and 12 abstracts were accepted for a short presentation and a poster at the conference. Of the regular papers accepted to the conference, the paper "Assessing Group Fairness with Social Welfare Optimization" by Violet Chen, John Hooker and Derek Leben was selected for the Best Paper Award, and the paper "Probabilistic Lookahead Strong Branching via a Stochastic Abstract Branching Model" by Gioni Mexi, Somayeh Shamsi, Mathieu Besançon and Pierre Le Bodic was selected for the Best Student Paper Award. The selection was completed by a 3-member Best Paper Committee.

The conference program also included three keynote talks. Elina Rönnberg (Linköping University, Sweden) gave a keynote on "Decomposition to Tackle Large-Scale Discrete Optimisation Problems". Giacomo Nannicini (University of Southern California, USA) gave a keynote on "Optimization and Machine Learning on Quantum Computers: Yes/No/Maybe?". Tias Guns (KU Leuven, Belgium) gave a keynote on "Decision-Focused Learning: Foundations, State of the Art, Benchmarks and Future Opportunities". The conference program also included a Master Class on "Quantum Computing for CP, AI, and OR, and vice-versa" with invited talks by Carleton Coffrin (Los Alamos National Laboratory, USA), Ashley Montanaro (University of Bristol, UK), Tamás Terlaky (Lehigh University, USA), Harsha Nagarajan (Los Alamos National Laboratory, USA), Andreas Bärtschi (Los Alamos National Laboratory, USA), Zachary Morrell (Los Alamos National Laboratory, USA), Xiaodi Wu (University of Maryland College Park, USA), and David Bernal Neira (Purdue University, USA).

The organization of this conference would not have been possible without the help of many individuals. We would like to thank the Program Committee members and external reviewers for their hard work. We are also very grateful to the Master Class chair Carleton Coffrin (Los Alamos National Laboratory, USA), Sponsorship chair Thiago Serra (Bucknell University, USA), the Publicity chair María Andreína Francisco Rodríguez (Uppsala University), Sweden and Diversity, Equity, and Inclusion (DEI) chair Amira Hijazi (Georgia Institute of Technology, USA). Special thanks go to the conference general chairs Pierre Flener, María Andreína Francisco Rodríguez, and Justin Pearson

(Uppsala University, Sweden), whose support has been instrumental in making this event a success.

Lastly, we want to thank all sponsors for their generous contributions. At the time of writing, these include Artificial Intelligence Journal (AIJ), OptalCP, Quantum Technologies Group at Carnegie Mellon University's Tepper School of Business, Coupa, Association for Constraint Programming, Google, Boeing, Uppsala municipality, nextmv, Gurobi, MERL, The Optimization Firm, SOAF, COSLING, INQA, Uppsala University.

April 2024   Bistra Dilkina

Organization

Program Chair

Bistra Dilkina — University of Southern California, USA

Conference Chairs

Pierre Flener — Uppsala University, Sweden
María Andreína Francisco Rodríguez — Uppsala University, Sweden
Justin Pearson — Uppsala University, Sweden

Master Class Chair

Carleton Coffrin — Los Alamos National Laboratory, USA

Program Committee

Deepak Ajwani — University College Dublin, Ireland
Aliaa Alnaggar — University of Toronto, Canada
J. Christopher Beck — University of Toronto, Canada
Nicolas Beldiceanu — IMT Atlantique (LS2N), France
Mathieu Besançon — Inria, Université Grenoble Alpes, France
Armin Biere — University of Freiburg, Germany
Christian Blum — Spanish National Research Council (CSIC), Spain
Merve Bodur — University of Edinburgh, UK
Quentin Cappart — Polytechnique Montréal, Canada
Carlos Cardonha — University of Connecticut, USA
Mats Carlsson — RISE Research Institutes of Sweden, Sweden
Margarita Castro — Pontificia Universidad Católica de Chile, Chile
Andre Augusto Cire — University of Toronto, Canada
Simon de Givry — INRAE, France
Mathijs De Weerdt — Delft University of Technology, Netherlands
Emir Demirović — Delft University of Technology, Netherlands
Guillaume Derval — University of Liège, Belgium

Bistra Dilkina (Chair)	University of Southern California, USA
Aaron Ferber	Cornell University, USA
Ambros Gleixner	HTW Berlin, Germany
Carla Gomes	Cornell University, USA
Oktay Gunluk	Cornell University, USA
Emmanuel Hebrard	LAAS, CNRS, France
John Hooker	Carnegie Mellon University, USA
Matti Järvisalo	University of Helsinki, Finland
Serdar Kadioglu	Brown University, USA
George Katsirelos	MIA Paris, INRAE, AgroParisTech, France
Joris Kinable	Amazon, USA
Zeynep Kiziltan	University of Bologna, Italy
T. K. Satish Kumar	University of Southern California, USA
Arnaud Lallouet	Huawei Technologies, France
Thi Thai Le	Zuse Institute Berlin, Germany
Pierre Le Bodic	Monash University, Australia
Christophe Lecoutre	CRIL, Université d'Artois, France
Jimmy Lee	Chinese University of Hong Kong, China
Michele Lombardi	DISI, University of Bologna, Italy
Pierre Lopez	LAAS-CNRS, Université de Toulouse, France
Leonardo Lozano	University of Cincinnati, USA
Arnaud Malapert	Université Côte d'Azur, CNRS, I3S, France
Ciaran McCreesh	University of Glasgow, UK
Laurent Michel	University of Connecticut, USA
Michael Morin	Université Laval, Canada
Nysret Musliu	TU Wien, Austria
Barry O'Sullivan	University College Cork, Ireland
Justin Pearson	Uppsala University, Sweden
Laurent Perron	Google France, France
Gilles Pesant	Polytechnique Montréal, Canada
Milena Petkovic	Zuse Institute Berlin, Germany
Claude-Guy Quimper	Université Laval, Canada
Jean-Charles Regin	University Nice-Sophia Antipolis, I3S, CNRS, France
Michael Römer	Bielefeld University, Germany
Elina Rönnberg	Linköping University, Sweden
Louis-Martin Rousseau	Polytechnique Montréal, Canada
Domenico Salvagnin	University of Padova, Italy
Pierre Schaus	UCLouvain, Belgium
Thomas Schiex	INRAE, France
Thiago Serra	Bucknell University, USA
Paul Shaw	IBM, France

Mohamed Siala	INSA Toulouse & LAAS-CNRS, France
Helmut Simonis	University College Cork, Ireland
Kostas Stergiou	University of Western Macedonia, Greece
K. Subramani	West Virginia University, USA
Guido Tack	Monash University, Australia
Kevin Tierney	Bielefeld University, Germany
Christian Tjandraatmadja	Google, USA
Michael Trick	Carnegie Mellon University, USA
Dimosthenis C. Tsouros	KU Leuven, Belgium
Willem-Jan van Hoeve	Carnegie Mellon University, USA
Hélène Verhaeghe	KU Leuven, Belgium
Thierry Vidal	École nationale d'Ingénieurs de Tarbes, France
Petr Vilím	OptalCP, Czech Republic
Mark Wallace	Monash University, Australia
Roland Yap	National University of Singapore, Singapore
Neil Yorke-Smith	Delft University of Technology, Netherlands
Tallys Yunes	University of Miami, USA

Additional Reviewers

Younes Aalian	Ian Gent
Mehmet Anil Akbay	Mohammed Ghannam
Valentin Antuori	Luca Giuliani
Elif Arslan	Bernhard Gstrein
Christian Artigues	Tias Guns
Federico Baldo	Matthias Horn
Arthur Bit-Monnot	Taoan Huang
Ignace Bleukx	Marie-José Huguet
Camille Bonnin	Gabriele Iommazzo
François Camelin	Tanuj Karia
Jonas Charfreitag	Olivier Lhomme
Alexandre Dubray	Haoming Li
Aloïs Duguet	Steve Malalel
Valentin Durante	Yuri Malitsky
Suhendry Effendy	Jayanta Mandi
Maaike Elgersma	Imko Marijnissen
James Fitzpatrick	Valentin Mayer-Eichberger
Mathias Fleury	Matthew McIlree
Maarten Flippo	Gioni Mexi
Marco Foschini	Yimeng Min
Matteo Francobaldi	Eleonora Misino
Nikolaus Frohner	Tobias Paxian

Felipe Pereira
Léon Planken
Zhongdi Qu
Jaume Reixach
Noah Schutte
Konstantin Sidorov
Anand Subramani

Ajdin Sumic
Fabio Tardivo
Charles Thomas
Kevin Tierney
Junhan Wen
Daniel Wetzel
Damien T. Wojtowicz

Contents – Part I

Contents – Part II

Online Optimization of a Dial-a-Ride Problem with the Integral Primal Simplex

Elahe Amiri[1,2,3]([✉]) [ID], Antoine Legrain[1,2,3] [ID], and Issmaïl El Hallaoui[1,3] [ID]

[1] Polytechnique Montréal, Montréal, QC H3T 1J4, Canada
{elahe.amiri,antoine.legrain,issmail.el-hallaoui}@polymtl.ca
[2] CIRRELT, Montréal, QC H3T 1J4, Canada
[3] GERAD, Montréal, QC H3T 2A7, Canada

Abstract. This paper focuses on developing a real-time dispatching system for a ride-sharing service. The primary goal is to address the dynamic dial-a-ride Problem, aiming to minimize waiting times while ensuring service quality by limiting ride duration. We introduce a rolling horizon-based framework, involving the division of the time horizon into small epochs, batching requests within each epoch, and re-optimizing the problem for the batch of requests. Unlike prior studies that restart optimization for each period from scratch, we leverage the strength of integral primal simplex to reuse effectively the previously computed solutions as a warm start, extending current routes with new incoming requests. Moreover, using integral primal methods allows us to provide an algorithm that is tractable in real-time and scales effectively to handle thousands of customers per hour. Experiments using historic taxi trips in New York City, involving up to 30,000 requests per hour, illustrate the efficacy and potential advantages of the method in effectively managing large-scale and dynamic scenarios.

Keywords: Real-time dial-a-ride · integral column generation · integral simplex using decomposition

1 Introduction

In recent years, the rise of Commercial ride-sourcing services like Uber and Lyft has revolutionized urban mobility by offering on-demand transportation through mobile applications. Despite the theoretical advantages of these services in mitigating congestion and pollution, their rapid expansion has introduced new challenges and worsened the situation. Promoting shared alternatives like ride-sharing can alleviate this issue by utilizing vehicles more efficiently. However, its implementation comes with various challenges including the *high degree of dynamism* and the *large amount of trip requests* to handle. Therefore, advanced optimization techniques are rarely used in practice which limits the potential benefits of these systems. Alonso-Mora et al. [1] was the first who demonstrate the potential of systematic ride-sharing, revealing that 98% of ride requests in

© The Author(s), under exclusive license to Springer Nature Switzerland AG 2024
B. Dilkina (Ed.): CPAIOR 2024, LNCS 14742, pp. 1–16, 2024.
https://doi.org/10.1007/978-3-031-60597-0_1

New York City (NYC) could be served using 15% of the taxi fleet, with an average waiting time of 2.8 min.

From a mathematical point of view, ride-sharing is modeled as a dial-a-ride Problem (DARP) which involves planning a set of vehicle routes to transport people between their origins and destinations. In these systems, dispatching decisions are made online without prior knowledge of trip requests, necessitating the solution of the DARP in a dynamic mode to adapt the dispatching plan in real-time. Traditional approaches, typically employ a rolling horizon strategy, attempting to batch requests to take advantage of the optimization [1,2]. However, these methods restart the optimization for each new batch without utilizing the previous solution. This paper introduces an innovative re-optimization framework, denoted as F-ICG, using the strength of primal methods to reuse previously computed solutions as a warm start. To be more precise, the F-ICG divides the time horizon into short periods, termed epochs, and frequently re-optimizes the static DARP for the requests batched within these epochs, employing the Integral Column Generation (ICG). Utilizing primal methods involves decomposing the problem into smaller subproblems, thereby simplifying their resolution by reducing density. Importantly, the solution consistently remains feasible, enhancing the real-time tractability of the algorithm.

The paper is organized as follows. Section 2 presents related work and summarizes our main contributions. Then, in Sect. 3 we describe the problem. Section 4 gives an overview of the solution approach. The computational results are reported in Sect. 5, and finally, Sect. 6 concludes the paper.

2   Related Work

Over the past few decades, DARPs have been widely studied and attracted lots of interest in operations research. Cordeau et al. [3] proposed a three-index formulation known as *Standard DARP* and conducted a comprehensive review of various formulations of the DARP since 2003. The reader is also referred to [4,5] for a thorough survey on the DARP, classification of the problem variants, the solution methodologies, and references to benchmark instances. Our focus lies specifically in the field of dynamic DARP, which has received comparatively less attention than its static counterpart.

Most of the studies in the literature on dynamic DARP have focused on heuristics to find fast solutions and metaheuristics for further optimization. The typical two-step approach involves assigning new requests with fast insertion heuristics initially and refining the solution during idle periods using another heuristic or metaheuristic. Regarding the first step, some insertion heuristics add the new request to the dispatching plan without relocating previously assigned ones [6], while some others allow for relocation of requests that are scheduled but not served yet [7]. To enhance solutions during idle times between request arrivals, many studies have suggested metaheuristics. For example, [8] enhanced solutions by reinserting unserved requests, while [6] introduced a regret heuristic, assessing the benefit of transferring requests between vehicles.

While most research has focused on heuristics and metaheuristics, a few studies have also developed exact solutions for the dynamic DARP. These methods are mainly based on re-optimizing the static version of the problem within a rolling horizon framework. In this regard, Bertsimas et al. [9] proposed a *backbone* algorithm and discussed methods to simplify optimization-based approaches for large-scale taxi routing problems without ride-sharing. Riley et al. [2] considered a large-scale ride-sharing system modeled as a DARP, which guarantees to serve all the requests. They utilize a rolling horizon strategy for request batching with the epoch length of 30 s and proposed a solution based on column generation (CG) to optimize operations. Their objective is to minimize the average waiting times of customers while constraining ride durations. A key difference of their work compared to prior studies, is their commitment to serve *all the requests*. In order to meet the real-time requirements, they employ an iterative pricing algorithm that generates routes of increasing lengths. They evaluate their algorithm on real data from NYC dataset [10] with up to 32,869 requests per hour and outperformed the achievements of [1] in terms of the waiting times. In a follow-up study [11], their approach was enhanced with a model predictive control for idle vehicle relocation, resulting in a significant 30% improvement in average waiting time compared to their previous work.

Note that the algorithms proposed in [2] and [11] are based on the branch-and-price algorithm which combines CG and branch-and-bound and is placed in the class of *dual fractional* methods. The methods in this class maintain optimality as well as the feasibility of the linear model and terminate when reaching integrality. Although these methods are generally effective, certain limitations can hinder their ability to provide timely solutions in real-time scenarios. Exploring a branch-and-bound tree to achieve integrality, and restarting optimization from scratch are two such limitations. Another class of algorithms for integer programs are *Integral Primal methods* that maintain feasibility (and integrality) throughout the process and terminate when optimality is achieved. These approaches offer advantages such as avoiding combinatorial branching to reach integrality, utilizing existing solutions as a warm start, and having a feasible integer solution available at any point, allowing early termination whenever the quality is satisfactory or the time is running out. Despite these advantages, an examination of state-of-the-art showed that exact primal methods had received very little attention in the literature compared to dual-fractional ones.

The first primal approach was introduced by [12] for Set Partitioning Problems (SPP), but was limited to small instances. Zaghrouti et al. [13] later introduced Integral Simplex Using Decomposition (ISUD) to tackle large SPPs. ISUD involves decomposing the problem into a Reduced Problem (RP) to improve the current solution, and a Complementary Problem (CP), to find descent directions qualified as integer. Several strategies have been proposed to improve the performance of ISUD such as adding primal cuts to eliminate fractional directions [14], the multi-phase strategy which solves the CP in a sequence of phases [13], and Zooming method which explores the neighbourhood around fractional directions to find an integer descent direction [15].

Recently, Tahir et al. [16] embedded primal ISUD in CG and introduced ICG to solve SPPs. Their computational results on large-size instances with up to 2000 constraints indicate that this method outperforms two popular CG-based heuristics. They also indicated that ICG can be adopted even on SPP with side constraints [17]. Messaoudi et al. [18] examined the performance of ICG on a realistic multi-attribute Rich Vehicle Routing Problem with the instances reaching 140 customers. The capability of ISUD and ICG algorithms in using warm start for re-optimization and their ability to provide solutions at any time makes them worth pursuing. We believe that such methods have to be evaluated in a real-time optimization framework. The research conducted in [2] serves as the foundational benchmark for our study and we share a common mathematical framework for modeling DARP, and utilize the same rolling horizon strategy for batching request. However, our approach diverges significantly in the methodologies employed for addressing the pricing subproblems and the Master Problem (MP). Our main contributions are summarized as follows:

1. We introduce the first implementation of a primal algorithm based on integral simplex for a Set Covering Problem within the context of the DARP.
2. We conduct a thorough assessment of primal-based methods within a real-time optimization framework.
3. Our approach demonstrates robust scalability, handling up to 30,000 requests per hour, and outperforms prior studies in terms of average waiting time.

3   Problem Statement

The notations presented in Table 1 are used to formulate the problem. We borrowed the model proposed in [2] which guarantees the service of all requests. The primary objective is the minimization of average waiting time while making sure that the ride duration and capacity limitations are met.

Our proposed strategy centers around the use of ICG, a methodology that combines a primal approach with CG. In this setup, variables are defined as the vehicle routes. The model itself is comprised of a MP, dedicated to selecting the optimal routes, and Pricing Subproblems, which are solved to generate feasible routes. A route $r \in R^v$ starts from the departure stop of the vehicle v, and visits a sequence of pickup and drop-off nodes while satisfying the following constraints to ensure feasibility:

1. Each request $i \in P$ should be picked up and dropped off by the same vehicle.
2. Each request $i \in P$ can only be visited after its desired pickup time e_i.
3. Travel time for the passengers of each request $i \in P$ should not deviate too much from the shortest possible time t_i. Constants α and β are considered to determine the permitted deviation from the shortest path as $\max\{\alpha t_i, \beta + t_i\}$
4. Each vehicle $v \in V$ is constrained by its specified carrying capacity Q_v.

Table 1. Definition of the parameters and variables

symbol	type	description
P	–	Set of pickup nodes; $P = \{1, \cdots, n\}$
D	–	Set of drop-off nodes; $D = \{n+1, \cdots, 2n\}$
V	–	Set of vehicles
R^v	–	Set of feasible routes for vehicle v
R	–	Set of all feasible routes ($R = \cup_{v \in V} R^v$)
c_r	real	Total waiting times of customers served along the route r
p_i	real	Penalty for not serving request i in the current period
e_i	real	Earliest possible pickup for request i
q_i	real	Number of passengers for request i
t_i	real	Shortest travel time from the origin of request i to its destination
Q_v	integer	Capacity of the vehicle v
a_i^r	binary	Parameter that equals to 1 if customer i is served by route r
y_r	binary	Variable that equals to 1 if route r is selected
z_i	binary	Variable that equals to 1 if request i is not served

The MP is formulated as an SPP presented by formulation (1)–(5). The objective (1) minimizes the waiting time of customers that are planned to be served and the penalties for the unserved customers. Constraints (2) ensure that variable z_i is set to 1 if its corresponding request $i \in P$ is not served by any of the selected routes. Constraints (3) ensure that exactly one route is assigned to each vehicle. And finally, constraints (4) and (5) restrict the domain of the decision variables.

$$Z_{MP}^* = \min \sum_{r \in R} c_r y_r + \sum_{i \in P} p_i z_i \tag{1}$$

$$\text{s.t.} \quad \left(\sum_{r \in R} y_r a_i^r\right) + z_i = 1 \qquad \forall i \in P \qquad (\pi_i) \tag{2}$$

$$\sum_{r \in R^v} y_r = 1 \qquad \forall v \in V \qquad (\sigma_v) \tag{3}$$

$$z_i \in \{0, 1\} \qquad \forall i \in P \tag{4}$$

$$y_r \in \{0, 1\} \qquad \forall r \in R \tag{5}$$

To generate feasible routes, the subproblems are modeled as a shortest-path problem with resource-constrained (SPPRC) aiming to minimize the reduced cost. Let π_i and σ_v be dual variables associated to constraints (2) and (3). The reduced cost associated with route $r \in R$ is calculated as $c_r - \sum_{i \in P} a_i^r \pi_i - \sigma_v$.

4   Solution Method

This section gives an overview of our real-time optimization approach for a large-scale ride-sharing system. Our solution method involves re-optimizing the static DARP with all known ride requests periodically. The stages of the overall approach are illustrated schematically in Fig. 1 and is described in this section.

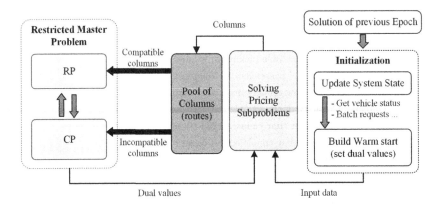

Fig. 1. Schematic representation of the stages in the overall approach

4.1   Online Re-optimization Architecture

Our re-optimization strategy is motivated by the work of [2,11] who suggested using a rolling horizon scheme. This involves dividing the time into small epochs, each lasting 30 s, and periodically updating the departure plan. Let ℓ denote the length of each epoch. Let define time intervals as $[0,\ell),[\ell,2\ell),\cdots$ where $[\tau\ell,(\tau+1)\ell)$ corresponds to epoch τ. For each epoch τ, inputs are computed as:

- Batch the requests arrived in epoch $\tau-1$ into a set P_τ
- Determine the departing stop and departure time of the vehicles based on the solution in $\tau-1$. It is defined as the first stop falling within the interval $[(\tau+1)\ell,\infty)$ if it exists. If a vehicle was idle at the current epoch, its departing stop would be its last stop and its earliest departure time would be $(\tau+1)\ell$.
- Determine the set of unserved requests up to $\tau-1$ and update penalties.
- Initialize the dual variables π_i and σ_v using the solutions from the preceding epoch $\tau-1$. For the new arrival requests, π_i is set to their respective penalties.

To calculate the penalties for unserved requests, the function proposed in [2] is employed: $p_c = \rho 2^{(T\ell-e_c)/10\ell}$, where e_c represents the earliest possible pickup time for request c, and ρ serves as the parameter intended to encourage the scheduling of a request in the earliest possible epoch. Once the inputs are determined, the static DARP is re-optimized and the system advances to the next epoch, iterating the entire process again.

4.2   Integral Column Generation

ICG is a sequential procedure comprising three stages that utilizes ISUD to solve the MP inside the CG method. The conventional CG involves two primary steps. Initially, the linear relaxation of the MP restricted to columns generated (called RMP) is solved. Subsequently, the subproblems are defined based on the duals associated with the constraints of the RMP and are solved to generate columns with negative reduced costs. In ICG, instead of solving the linear relaxation of the RMP, ISUD is directly applied to the RMP which is an integer problem. The goal is to produce a feasible (integer) solution at each iteration which is a key advantage in online applications. This approach introduces an additional level of decomposition, breaking down the RMP into an RP and a CP for resolution using ISUD.

Pricing Subproblems. As mentioned in Sect. 3, the subproblems are modeled as SPPRC which involves determining the shortest route within a network starting from a source node and ending at a sink node while meeting specific constraints related to a set of resources. We consider the following resources in our method:

– The capacity resource indicating the accumulated load in the vehicle
– The time resource specifies available travel time for onboard passengers based on the max permitted deviation from the shortest path.

To solve the subproblems, we implemented a modified version of the label setting algorithm proposed in [19]. The labeling method builds labels that are partial paths originating from the source node. It starts with an initial label at the source node, and through an extension function, recursively extends it to successor nodes while updating the reduced cost and resources consumption. The extension function also examines the path's feasibility based on resource limitations and eliminates labels related to infeasible paths. Dominated labels that cannot provide a Pareto-optimal path can also be identified and removed during the labeling process to speed up the algorithm. This procedure continues until all labels have been extended in all feasible ways or discarded.

We also use two acceleration techniques to speed up the process. The first, known as *truncated labeling* [19], involves retaining only a limited number of labels at each node for a potential extension. In our case, this limit is set to 15 labels. Another technique is to avoid extending labels to pick up nodes if their corresponding partial paths have already visited a drop-off node. Additionally, we have established a maximum limit on the number of pickups on the path, which significantly reduces the total number of labels generated during the process. For real-time optimization tractability, rather than solving the subproblems for each vehicle in every epoch, we selectively solve a subset of the subproblems, matching the number of requests. The selection process identifies the vehicle that can serve each request sooner based on departing time and location. This policy, along with other heuristics, enhances scalability in real-time optimization.

Master Problem. After solving the subproblems, generated routes are stored in a pool of columns. We applied ISUD to solve the RMP with some modifications to better scale in real-time. ISUD is one of the most promising primal methods that have been notably applied to SPPs, which adhere to the **Trubin theory** [20]. Trubin proved that every edge of the convex hull of the feasible set of an SPP corresponds to an edge of its linear relaxation. This implies that integer solutions (extreme points) can be reached through linear programming pivots in SPPs. To apply ISUD, the preliminary stage involves constructing the RP and the CP models. This necessitates the computation of the *Incompatibility Degree* for the generated columns.

Let (y^0, z^0) be the initial feasible integer solution (warm start), where $\mathcal{P}_y = supp(y^0)$ and $\mathcal{P}_z = supp(z^0)$ are index sets of positive value variables related to our initial solution. Each vehicle v is planned to serve a subset of requests $T_v \subseteq P$ in the current solution, referred to as a cluster. These clusters are disjoint (i.e., $\cap_{v \in V} T_v = \emptyset$). Each request $i \in P$ belongs to exactly one cluster or is not planned to be served (i.e., $i \in \mathcal{P}_z$). We define the compatibility concept as follows:

Definition 1. *A column y_r is deemed **compatible** concerning $\mathcal{P} = \mathcal{P}_y \cup \mathcal{P}_z$, if its corresponding route $r \in R^v$ encompasses all the requests of its cluster T_v as well as the union of one or more cluster, along with any requests of $\mathcal{P}_z$, without breaking existing clusters. Otherwise, it is declared incompatible and its **degree of incompatibility** is equal to the minimum number of times that current clusters must be subdivided to ensure compatibility of y_r with the resulting clusters.*

Reduced Problem. In a standard ISUD approach, the RP model is defined the same as the MP formulation considering the set of compatible columns. Pivoting on a compatible column with negative reduced cost leads to an improved integer solution. The RP is then redefined with the new basis, and repeated until no more such columns are found. Inspired from the multi-phase strategy developed in [13], we modify our RP model. Rather than solely focusing on compatible columns, we consider $R_\mathcal{C}$ includes both compatible columns and those incompatible ones with incompatibility degree of one. Using the prior solution as a warm start, coupled with the limited number of stops in the routes, resulted in fewer compatible columns during the RP construction. Consequently, this refinement helps to achieve a more balanced composition of the problem and reduces the number of CP iterations. The RP model is then defined as depicted in Eqs. (6)–(10). The objective function and constraints of the formulation are almost the same as the MP (formulation (1)–(5)); However, the feasible region is restricted to $R_\mathcal{C}$. Following each iteration of solving the RP, dual values are updated based on the linear relaxation of the RP solution. Subsequently, the reduced cost and incompatibility degree of the columns in the pool are updated. The RP is refined iteratively by adding eligible columns with negative reduced costs from the pool. The process continues until no further improvement is detected, then is the time to build the CP model.

$$Z_{RP}^* = \min \sum_{r \in R_C} c_r y_r + \sum_{i \in P} p_i z_i \tag{6}$$

$$\text{s.t.} \quad \left(\sum_{r \in R_C} y_r a_r^r \right) + z_i = 1 \qquad \forall i \in P \qquad (\pi_i) \tag{7}$$

$$\sum_{r \in R_C^v} y_r = 1 \qquad \forall v \in V \qquad (\sigma_v) \tag{8}$$

$$z_i \in \{0, 1\} \qquad \forall i \in P \tag{9}$$

$$y_r \in \{0, 1\} \qquad \forall r \in R_C \tag{10}$$

Complementary Problem. The CP searches for descent directions to enhance the solution. A descent direction is a subset of the incompatible column whose linear combination is compatible and has a negative reduced cost. To Build CP, we consider incompatible columns and formulate the CP by Eqs. (11)–(16). In this formulation, ν_r and μ_i serve as weight variables defining the linear combination of incompatible columns, while λ_r and γ_i define the linear combination of columns in the current solution.

$$Z_{CP}^* = \min_{\nu, \lambda} \sum_{r \in R_I} c_r \nu_r + \sum_{i \in P \backslash \mathcal{P}_z} p_i \mu_i - \sum_{r \in \mathcal{P}_y} c_r \lambda_r - \sum_{i \in \mathcal{P}_z} p_i \gamma_i \tag{11}$$

$$\text{s.t.} \quad \sum_{r \in R_I} \nu_r a_r^r + \mu_i - \sum_{r \in \mathcal{P}_y} \lambda_r a_r^r = 0 \qquad i \in P \backslash \mathcal{P}_z \tag{12}$$

$$\sum_{r \in R_I} \nu_r a_r^r - \sum_{r \in \mathcal{P}_y} \lambda_r a_r^r - \gamma_i = 0 \qquad \forall i \in \mathcal{P}_z \tag{13}$$

$$\sum_{r \in R_I} \omega_r \nu_r + \sum_{i \in P \backslash \mathcal{P}_z} \omega_i \mu_i = 1 \tag{14}$$

$$\nu_r \geq 0 \qquad \forall r \in R_I \tag{15}$$

$$\mu_i \geq 0 \qquad \forall i \in P \backslash \mathcal{P}_z \tag{16}$$

The objective (11) minimize the reduced cost. Constraints (12) and (13) ensure the compatibility of the direction, and constraint (14) is the normalization constraint that bounds the feasible region. As suggested in [21], we used the incompatibility degree of the columns as normalization weights ω to encourage finding integer directions. The density of the constraint coefficient matrix in the RP and the CP can be reduced by removing the redundant constraints during the pre-processing and considering the linearly independent rows. Consequently, both the CP and the RP are computationally faster than the original problem [13,21]. If the solution satisfies $Z_{CP}^* \geq 0$, or the CP is infeasible, then the current solution is proved to be optimal [13]. Otherwise, the solution of the CP form a descent direction $d = (\nu, \mu, \lambda, \gamma, 0)$ which is integer if it consists of pair-wise row-disjoint columns. In such cases, the solution can be refined by introducing variables y_r with $\nu_r > 0$ and variables z_i with $\mu_i > 0$ into the basis, while simultaneously removing variables y_r with $\lambda_r > 0$ and variables z_i with $\gamma_i > 0$ from the basis. After each iteration of solving the CP, the reduced costs of the columns in the pool are adjusted according to the dual values obtained from the CP. The main steps of our ICG approach is presented in Algorithm 1 (For each set (e.g., $\overline{V}$ and $\overline{R}$), using the symbol *overline* denotes the set within the current iteration).

Algorithm 1: ICG Algorithm for the DARP

---

0   build an initial feasible integer solution (y^0, z^0), initialize dual values π^0, σ^0

1    $R \leftarrow \emptyset$; $k \leftarrow 0$

2   **while** *true* **do**

    Step 1: Generate Columns

3        $\overline{V} \leftarrow$ Select a subset of vehicles

4   **for** *all vehicle $v \in \overline{V}$* **do**

5            $\overline{R}^v \leftarrow$ Solve the subproblem for vehicle v

6            $\overline{R} \leftarrow \overline{R} \cup \overline{R}^v$

7   **if** $\overline{R} = \emptyset$ **then**

8   **return** *current solution* (y^k, z^k)

9        $R \leftarrow R \cup \overline{R}$

10       $k \leftarrow k + 1$

    Step 2: Solving RMP

11       $(y^k, z^k), \pi^k, \sigma^k \leftarrow \texttt{ISUD}(y^{k-1}, z^{k-1}, R)$

Function *ISUD* (y^{k-1}, z^{k-1}, R)

12   **while** *an improved solution is found* **do**

13   **for** *all routes $r \in R$* **do**

14   Calculate Incompatibility Degree

15           $R_C \leftarrow$ Columns $r \in R$ with $deg^r \leq 1$ and negative reduced cost

16           $(y^k, z^k), \pi^k, \sigma^k \leftarrow$ Construct and solve RP based on R_C

17       $R_I \leftarrow$ Incompatible Columns in R

18       $d \leftarrow$ Construct and solve CP to find a descent direction

19   **if** $Z^*_{CP} < 0$ *and d is integer* **then**

20           $(y^k, z^k), \pi^k, \sigma^k \leftarrow$ update the current solution using d

21   Go to line 12

22   **else**

23   **return** *current solution* (y^k, z^k)

---

5   Experimental Results

Instance Description and Algorithmic Setting. We assessed the efficiency of our method using 24 instances from [11], derived from real data in NYC [10].

These instances cover a two-hour horizon each day during peak hours (7–9 AM) for two days per month, from July 2015 to June 2016. Instances are classified into three groups according to their customer count, varying from 19,276 to 59,820, and averaging 48,100 customers. In accordance with [2,11], Manhattan was divided into a grid consisting of cells of 200 square meters which serve as pickup or drop-off points. The reader is referred to [2,11] for additional details on instance creation. All experiments were carried out using the following parameters unless otherwise specified: $\alpha = 1.5$, $\beta = 240$ s, and $\rho = 420$ s, and $\ell = 30$, and a fleet of 2000 vehicles with capacity of 4, initially evenly distributed across all stop locations. We have implemented our algorithm in C++ language and utilized version 22.1 of the IBM ILOG CPLEX Optimization Studio to solve RPs and CPs. All tests were conducted on a Linux machine, using 16 CPU cores of 2.7 GHz and 32 GB of RAM.

Numerical Results. Table 2 provides an overview of the computational performance of our method across the 24 instances, with a focus on key performance metrics. The columns labeled WT/Req. and TD/Req. present the average waiting time and average trip delay, respectively, measured in minutes per request. The subsequent pair of columns display the number of ISUD iterations per epoch, and the average time spent for re-optimizing the problem during epochs. According to the table, as instance size grows, the algorithm tends to spend more time solving the subproblems, as well as conducting more iterations to re-optimize the solution using ISUD. This signifies that the algorithm allocates more time to re-optimize the problem, a trend that is evident in the increasing epoch run times. Furthermore, the average runtime per epoch stands at approximately 13 s, constituting 40% of the available time of 30 s.

Table 2. Computational results of F-ICG

Number of Customers	WT/Req.	TD/Req.	#ISUD Iter./epoch	Epoch Runtime	% in MP Time		% in Total Time	
					RP	CP	MP	SP
<40,000	1.69	1.25	11.93	1.51	32.3%	65.4%	65.6%	34.4%
40,000–50,000	2.43	2.07	23.14	11.63	30.9%	68.3%	46.4%	53.6%
50,000<	2.51	2.21	23.73	17.19	31.2%	68.0%	42.1%	57.9%
Total Average	**2.33**	**1.99**	**21.17**	**12.99**	**31.4%**	**67.5%**	**47.7%**	**52.3%**

The last four columns detail the runtime distribution across solution approach components. As per the table, approximately 47.7% of the total runtime is devoted to solving the MP, while the remaining 52.3% is utilized for generating routes through the solutions of the subproblems, indicating a balanced distribution of computational effort. Within the time dedicated to solving the MP, roughly 31.4% is dedicated to enhancing the solution through the RP, while the rest addresses the CP. This allocation strategy reflects a significant adjustment in our approach, incorporating a subset of incompatible columns with a degree of one in modelling the RP. This modification contributes to achieving a more balanced problem composition and efficiently manages the runtime budget. Consequently, it facilitates a balanced and effective treatment of both the RP and the CP during the re-optimization process.

Wait Times. Figure 2 illustrates the distribution of waiting times for all customers in a large-size instance (R40567). In this case, F-ICG achieves an average waiting time of 2.53 min with a standard deviation of 1.39.

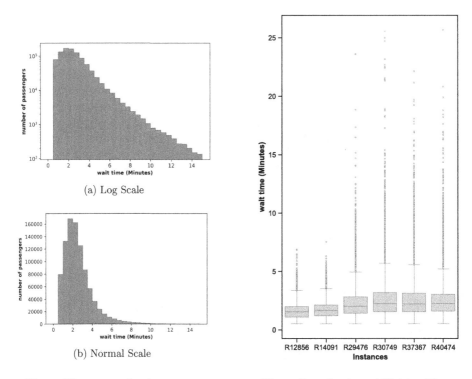

(a) Log Scale

(b) Normal Scale

Fig. 2. Histogram of waiting times

Fig. 3. Distribution of Waiting Times

Figure 3 depicts the distribution of waiting times across six instances, including two small-sized (R12858, R14091), two medium-sized (R34544, R30749), and two large-sized (R40474, R37367) instances. The figure suggests that the median wait time appears to remain relatively stable across all categories, with no significant increase, signifying that despite the instance size, the median service efficiency is maintained. Nevertheless, there is a noticeable increase in outliers, especially in the large-sized instances, pointing to a greater variability in the waiting times experienced by customers within these instances.

Comparison with Prior Studies. This section conducts a comparative analysis between our primal-based approach (F-ICG) and the CG-based approach in [2], denoted as M-RTRS. As mentioned in Sect. 2, M-RTRS utilizes a rolling horizon strategy for batching requests and applies the CG method to optimize the solution iteratively. An iterative algorithm that generates routes of increasing lengths is employed to address the pricing subproblems, thereby generating feasible paths. To ensure a fair comparison, we utilized the same instances and tried our best to replicate the simulation environment, although it remains uncertain if the algorithms are evaluated in exactly the same manner. The values presented in Table 3 correspond to those in [11].

Table 3. Average Waiting Times by Instance Size

Number of Customers	2000 vehicles			1800 vehicles
	M-RTRS	F-ICG	RF-ICG	RF-ICG
<40,000	2.22	1.69	1.29	1.33
40,000–50,000	3.83	2.43	1.72	1.91
50,000<	3.78	2.51	1.87	2.19

Notably, the study's outcomes underscore the superior performance of the F-ICG algorithm compared to M-RTRS, showcasing a substantial 33% reduction in average waiting time, lowering it from 3.78 to 2.51 min, particularly in large instances. The last two columns, consider a scenario wherein idle vehicles (i.e., the vehicles whose recent stop occurred in the preceding epoch and have no scheduled visit in the current epoch) relocate to their initial locations, a simplistic approach to reposition vehicles without employing complicated methods. Termed RF-ICG, this strategy is applied to both 2000 and 1800 vehicles. As shown in the table, we managed to employ 10% fewer vehicles while still surpassing M-RTRS results by an impressive 40%. These results clearly suggest the advantages of using primal approaches for re-optimization, emphasizing the effectiveness of F-ICG and the benefits of strategic assumptions to achieve significant improvements in average waiting times and resource utilization.

Figure 5 displays the total optimization time and the time required to solve the MP across sequential epochs in the optimization process for a large-size instance (R40567). We emphasize that our re-optimization process is a time-constrained process, where a limit of 30 s is set for batching requests and resolving the problem. This figure may reflect the algorithm's responsiveness to varying problem complexities or sizes over time, where certain epochs require more computational resources, potentially due to more complex routing configurations or larger data batches.

Figure 4 illustrates the relationship between the number of customers and the overall time dedicated to optimization per instance. Here, the total optimization time, accounted for over an hour to provide a point of comparison to Fig. 12

in [2]. While [2] reported an approximate total processing time of 800 s for larger instances, our findings indicate an average of around 1800 s similar cases. Despite the apparent increase in total run time in comparison to their findings, it is critical to note that our re-optimization times consistently remain within the 30-s limit, a fact that is also corroborated by Fig. 5. The allocation of additional optimization time has been employed to enhance the quality of the optimization outcomes. As a result, our methodology has led to markedly better performance regarding waiting times, demonstrating the effective application of the available optimization time to refine the solutions.

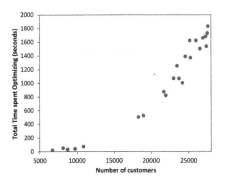

Fig. 4. Optimization Times

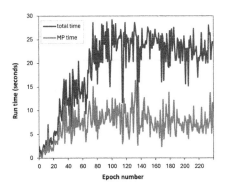

Fig. 5. Optimization Times per Epoch

Impact of the Fleet Size. Figure 6 evaluates the impact of fleet size on the waiting time and trip delay, considering the assumption of repositioning idle vehicles as outlined. The investigation specifically focuses on the consequences of reducing the number of vehicles from 3000 to 1500 across varying customer counts.

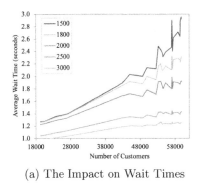

(a) The Impact on Wait Times

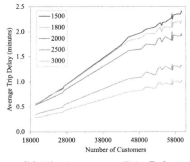

(b) The impact on Trip Delay

Fig. 6. Impact of the Fleet Size on the Average Waiting Times and Average Trip Delay across all instances

As anticipated, a reduction in the number of vehicles correlates with an increase in both average waiting time and trip delay. Nevertheless, even with a fleet size of 1500, the average waiting time remains below 3 min, and the average trip delay stays below 2 min. This suggests the method's potential to alleviate congestion in large cities while maintaining customer satisfaction.

6   Conclusion

This paper introduces F-ICG, to address the dynamic DARP within the context of a large-scale ride-sharing system. F-ICG utilizes a rolling horizon-based framework and incorporates a primal-based algorithm embedded with CG to re-optimize the dispatching plan. Employing the integral primal methods allows us to provide a tractable algorithm in real-time and scale effectively to handle thousands of customers per hour. The evaluation of F-ICG on instances from the NYC dataset, reveals that adopting primal approaches for re-optimization can considerably improve the average waiting time. Results showcase the superiority of the F-ICG, demonstrating a reduction of about 33% across all tests compared to prior methods. Additionally, using the strategy of returning idle vehicles to their initial point, resulted in an additional 18% improvement. Given that the underlying mathematical model is an SPP, the proposed re-optimization framework can be adapted to a variety of dynamic operation management problems, promising significant operational benefits. Future research will be devoted to enhancing the performance of the method by using machine learning techniques to adjust the parameters of the CP in ISUD.

References

1. Alonso-Mora, J., Samaranayake, S., Wallar, A., Frazzoli, E., Rus, D.: On-demand high-capacity ride-sharing via dynamic trip-vehicle assignment. Proc. Natl. Acad. Sci. **114**(3), 462–467 (2017). https://doi.org/10.1073/pnas.1611675114
2. Riley, C., Legrain, A., Van Hentenryck, P.: Column generation for real-time ride-sharing operations. In: Rousseau, L.-M., Stergiou, K. (eds.) CPAIOR 2019. LNCS, vol. 11494, pp. 472–487. Springer, Cham (2019). https://doi.org/10.1007/978-3-030-19212-9_31
3. Cordeau, J.-F., Laporte, G.: The dial-a-ride problem: models and algorithms. Ann. Oper. Res. **153**(1), 29–46 (2007). https://doi.org/10.1007/s10479-007-0170-8
4. Ho, S.C., Szeto, W.Y., Kuo, Y.-H., Leung, J.M.Y., Petering, M., Tou, T.W.H.: A survey of dial-a-ride problems: literature review and recent developments. Transp. Res. Part B Methodol. **111**, 395–421 (2018). https://doi.org/10.1016/j.trb.2018.02.001
5. Molenbruch, Y., Braekers, K., Caris, A.: Typology and literature review for dial-a-ride problems. Ann. Oper. Res. **259**, 295–325 (2017). https://doi.org/10.1007/s10479-017-2525-0
6. Lois, A., Ziliaskopoulos, A.: Online algorithm for dynamic dial a ride problem and its metrics. Transp. Res. Procedia **24**, 377–384 (2017). https://doi.org/10.1016/j.trpro.2017.05.097

7. Luo, Y., Schonfeld, P.: Online rejected-reinsertion heuristics for dynamic multive-hicle dial-a-ride problem. Transp. Res. Rec. **2218**(1), 59–67 (2011)
8. Carotenuto, P., Martis, F.: A double dynamic fast algorithm to solve multi-vehicle dial a ride problem. Transp. Res. Procedia **27**, 632–639 (2017). https://doi.org/10.1016/j.trpro.2017.12.131
9. Bertsimas, D., Jaillet, P., Martin, S.: Online vehicle routing: the edge of optimization in large-scale applications. Oper. Res. **67**(1), 143–162 (2019). https://doi.org/10.1287/opre.2018.1763
10. New York City Taxi and Limousine Commission: Trip record data, 2021 (2021). https://www1.nyc.gov/site/tlc/about/tlc-trip-record-data.page
11. Riley, C., Van Hentenryck, P., Yuan, E.: Real-time dispatching of large-scale ride-sharing systems: integrating optimization, machine learning, and model predictive control. arXiv preprint arXiv:2003.10942 (2020)
12. Balas, E., Padberg, M.: On the set-covering problem: II. An algorithm for set partitioning. Oper. Res. **23**(1), 74–90 (1975)
13. Zaghrouti, A., Soumis, F., El Hallaoui, I.: Integral simplex using decomposition for the set partitioning problem. Oper. Res. **62**(2), 435–449 (2014). https://doi.org/10.1287/opre.2013.1247
14. Rosat, S., Elhallaoui, I., Soumis, F., Lodi, A.: Integral simplex using decomposition with primal cuts. In: International Symposium on Experimental Algorithms, vol. 166, pp. 22–33 (2017). https://doi.org/10.1007/s10107-017-1123-x
15. Zaghrouti, A., El Hallaoui, I., Soumis, F.: Improving set partitioning problem solutions by zooming around an improving direction. Ann. Oper. Res. **284**(2), 645–671 (2020). https://doi.org/10.1007/s10479-018-2868-1
16. Tahir, A., Desaulniers, G., El Hallaoui, I.: Integral column generation for the set partitioning problem. EURO J. Transp. Logist. **8**(5), 713–744 (2019). https://doi.org/10.1007/s13676-019-00145-6
17. Tahir, A., Desaulniers, G., El Hallaoui, I.: Integral column generation for set partitioning problems with side constraints. INFORMS J. Comput. **34**(4), 2313–2331 (2022). https://doi.org/10.1287/ijoc.2022.1174
18. Messaoudi, M., El Hallaoui, I., Rousseau, L.-M., Tahir, A.: Solving a real-world multi-attribute VRP using a primal-based approach. In: Baïou, M., Gendron, B., Günlük, O., Mahjoub, A.R. (eds.) ISCO 2020. LNCS, vol. 12176, pp. 286–296. Springer, Cham (2020). https://doi.org/10.1007/978-3-030-53262-8_24
19. Ghilas, V., Cordeau, J.-F., Demir, E., Van Woensel, T.: Branch-and-price for the pickup and delivery problem with time windows and scheduled lines. Transp. Sci. **52**(5), 1191–1210 (2018). https://doi.org/10.1287/trsc.2017.0798
20. Trubin, V.: On a method of solution of integer linear programming problems of a special kind. Sov. Math. Dokl. **10**, 1544–1546 (1969)
21. Rosat, S., Elhallaoui, I., Soumis, F., Chakour, D.: Influence of the normalization constraint on the integral simplex using decomposition. Discrete Appl. Math. **217**, 53–70 (2017). https://doi.org/10.1016/j.dam.2015.12.015

A Constraint Programming Model
for the Electric Bus Assignment Problem
with Parking Constraints

Mathis Azéma[1(✉)], Guy Desaulniers[2], Jorge E. Mendoza[3], and Gilles Pesant[2]

[1] École Polytechnique, Palaiseau, France
mathis.azema@polytechnique.org
[2] Polytechnique Montréal, Montreal, Canada
{guy.desaulniers,gilles.pesant}@polymtl.ca
[3] HEC Montréal, Montreal, Canada
jorge.mendoza@hec.ca

Abstract. Electric buses serve as a key leverage in mitigating the transportation sector's carbon footprint. However, they pose a challenge, requiring transit agencies to adapt to a new operational approach. In particular, the assignment of buses to trips is more complex because it must consider the planning of the recharging activities. Unlike diesel buses, electric buses have less autonomy and take longer to refuel. In this paper, we address the assignment of electric buses to trips and the scheduling of charging events, taking into account parking constraints at the depot (a novelty in the literature). These constraints are particularly relevant in countries such as Canada where the buses are parked indoors to shelter them from harsh winter conditions. This problem, called the electric Bus Assignment Problem with Parking Constraints (eBAP-PC), is a feasibility problem. We propose a Constraint Programming model to solve it and compare it to mixed-integer linear programming approaches. In particular, we show its benefits for solving this problem with a one-day horizon and minimum end-of-day charge level constraints.

1 Introduction

There is no doubt Climate Change is the century's biggest challenge for humanity. It has followed the phenomenon of globalization, which has brought far-flung regions into economic proximity. The transportation sector plays a crucial role in a globalized world but also bears the responsibility for an important share of greenhouse gas emissions: it accounted for 20.2% of the global CO_2 emissions in 2021 (www.statista.com).

One avenue to decrease CO_2 emissions is the electrification of public transportation. However, replacing diesel buses with electric buses (EBs) is complex because it requires deploying a charging infrastructure and scheduling charging events. In addition, the strategies to assign buses to trips must also be deeply reviewed, because of two fundamental differences between EBs and diesel-powered buses. First, diesel buses can be fully refueled in a few minutes whereas, EBs may take hours to charge. Second, diesel buses can operate a whole day without refueling, but EBs often need to be charged during the day.

© The Author(s), under exclusive license to Springer Nature Switzerland AG 2024
B. Dilkina (Ed.): CPAIOR 2024, LNCS 14742, pp. 17–33, 2024.
https://doi.org/10.1007/978-3-031-60597-0_2

One strategy to plan operations with an electric fleet is to assign EBs to trips one day at a time, as before with diesel buses, solving an electric Vehicle Scheduling Problem (eVSP) that inserts a minimum number of charging events in the bus schedules. After this assignment, one has to schedule the night charging events to ensure each EB can perform all the trips of the following day. In such a restrictive context where charging is only allowed at night, the idea of multi-day planning was recently shown to bring potential benefits [25]. We consider here a more flexible strategy in which we allow to reoptimize the charging events occurring at the depot during the day. We study multi-day planning and one-day planning with minimum end-of-day charge level constraints.

To the best of our knowledge, parking constraints for EBs such as those considered in this paper have not been addressed yet in the literature. These constraints are particularly relevant in countries such as Canada where the buses are parked indoors to shelter them from harsh winter conditions. They can severely restrict solution feasibility. Indeed, a typical bus depot is composed of identical lanes that operate as queues because backing up is forbidden for security reasons. Consequently, the assignment must respect the *first-in, first-out policy* (FIFO), i.e., it ensures that the arrival order in a lane is the same as the departure order.

In this paper, we study the electric Bus Assignment Problem with Parking Constraints (eBAP-PC). Given a set of trips (more precisely, sequences of trips or STs) starting and ending at the depot, a set of identical EBs, a set of available chargers at the depot, and a set of identical parking lanes in the depot, the problem consists in i) assigning an EB to each trip and ii) a lane to each EB completing a trip, and iii) scheduling potential charging events at the depot such that the FIFO policy is satisfied, each EB respects its battery's limits, and the depot charging capacity is never exceeded.

As the goal is to find a feasible solution (there is no objective function), we consider a Constraint Programming (CP) approach. To assess our model, we use data provided by our industrial partner GIRO Inc., a world-leading developer of optimization solutions for public transit.

The rest of this paper is organized as follows. Section 2 reviews the related literature. Section 3 formally defines the eBAP-PC before presenting our CP model in Sect. 4. Section 5 describes the computational experiments and their results. Finally, concluding remarks are presented in Sect. 6.

2   Literature Review

The electrification of public transport has sprung numerous optimization problems in the last decade. Indeed, it is necessary to review the daily operations of public transport, including the planning of charging events, and to build new infrastructures, which questions the number and the location of charging stations. [8] and [22] surveyed the problems studied in the literature.

The eVSP has already been widely studied. The modeling of the charging function and the consideration of charging capacity constraints are the main issues. Early modeling considered constant-time charging events and full recharges [1,7,17,23]. For instance, [7] developed a genetic algorithm where batteries can be exchanged in 15 min at the depot. Then, several authors consider a linear charging time and the possibility of partially charging batteries. For instance, this is possible by discretizing time and considering the charging power on a time step as a decision variable for the problem. [24] used this approach with 15-minute discretization steps for a mixed fleet composed of diesel and electric buses. More recently, many articles studied the eVSP with a linear charging rate. For example, [11, 12, 26] proposed heuristics based on local search. However, the charging time of batteries is logarithmic with respect to the state-of-charge (SoC) variation, not linear. [19] demonstrated the significance of taking this aspect into account. eVSP models with non-linear charging functions have since multiplied [1, 13, 20].

There may be constraints on the charging capacity of the depot such as the number of parking slots equipped with a charger and the power available from the grid. To take these into account, [2] proposed a Benders decomposition to solve the eVSP with a linear charging time, a continuous time horizon, and these capacity constraints. The master problem considers the assignment of buses to trips and the subproblem checks whether the assignment is energetically feasible. Other articles considered time discretization [17, 24, 27] to deal with these capacity constraints.

A few authors studied parking constraints for diesel buses. Given a sequence of bus arrivals of known types and a sequence of morning departures with known requested types, the Bus Dispatching Problem (BDP) consists in assigning a parking slot to each bus such that the FIFO policy is respected and the type of the morning departures is matched. The BDP is often infeasible, so it is generally solved by minimizing the number of "mismatches" (i.e., wrong bus type assignments). [9, 10] showed that the BDP is NP-complete and proposed two mixed-integer linear programming (MILP) formulations for two problem variants that can be solved using a commercial MILP solver.

[18] studied a problem that, like in our paper, considers parking constraints for EBs, but in a very different way. On the one hand, the authors consider an arbitrary depot configuration and an objective function. They minimize first the number of EBs used and second the time-dependent charging costs. On the other hand, there are a few simplifications concerning the characteristics that make up the complexity of the eBAP-PC. Indeed, they first assume that an EB is assigned to a single parking slot during a stay at the depot, i.e., the EBs cannot be moved forward in a lane. Then, the time horizon is discretized in 15-minute periods compared with one minute in our paper. Finally, the authors used instances where the number of parking slots largely exceeds the number of EBs and the number of STs per EB used is around 1.3. During our computational experiments, the instances used consider a number of parking slots practically equal to the number of EBs and the number of STs per EB is around 3. All in all, [18] introduced a fast three-step MILP heuristic that is not applicable to our real-life context.

While for the most part mathematical programming-based approaches have been used to solve the eVSP, a few authors proposed CP models to solve problems close to the eVSP. [6] proposed two CP approaches to solve the electric vehicle routing problem with time windows. One uses alternative resources and the other uses an augmented horizon. [5] also proposed a model to solve a robot tasks allocation problem with the scheduling of charging activities. [4] studied a range-constrained variant of the multiple Unmanned Aerial Vehicles (UAVs) target search problem as well. In this problem, UAVs are used in tandem with ground-based mobile recharging vehicles that can travel, via the road network, to meet up with and recharge a UAV.

Even if this paper focuses on modeling parking constraints, our problem has other features: partial charging, linear charging time, and charging capacity constraints. As mentioned above, they have been studied in the literature but, to the best of our knowledge, considering all these features simultaneously in a problem is new. Note that in our problem there is no objective function, the fleet is homogeneous and the FIFO policy is respected throughout the time horizon, not only during the night.

3   Problem Statement

Consider a set of EBs $\mathcal{E} = \{b_1, b_2, \ldots\}$ with identical charging function and battery capacity. The SoC of an EB corresponds to the energy in the battery expressed as a percentage of the battery capacity. It increases with the charging operations and decreases as trips are performed. At the beginning of the time horizon, the SoC of EB b starts at SoC_b^{init} and must remain between SoC^{min} and SoC^{max} at all times. We assume a linear charging function with a rate $\lambda \in \mathbf{N}_+$.

Between two parking events at the depot, an EB performs a sequence of trips without any charging operation. These STs are predefined and are part of the input data. We denote by $\mathcal{S} = \{s_1, s_2, \ldots\}$ the set of STs. With a ST $s \in \mathcal{S}$ is associated a net energy consumption e_s, a departure time from the depot h_s^S, and an arrival time at the depot h_s^E. Each ST is performed by a unique EB from $\mathcal{E}$ and the EBs can only perform one ST at a time.

In between STs, an EB is parked at the depot. The latter is made up of L identical lanes, each with one entrance and one exit, that must respect the FIFO policy. We denote by $\mathcal{L} = \{1, \ldots, L\}$ the set of lanes and assume that they are independent, i.e., there are no precedence constraints between them. Each lane features V parking slots, denoted $\mathcal{V} = \{1, \ldots, V\}$. All the parking slots are identical so an EB can be parked anywhere in the depot. We assume that the depot is large enough to park all the EBs, i.e. $L \cdot V \geq |\mathcal{E}|$. At the beginning of the horizon, some EBs may be performing STs: we associate with each EB $b \in \mathcal{E}$ a time h_b when it enters the depot for the first time.

We consider a depot with an identical charger at each parking slot so that an EB may charge wherever it is parked. However all chargers cannot be used at the same time because the power consumed by the depot, which is proportional to the number of chargers in simultaneous use, has physical limitations. Transit companies also wish to avoid consumption peaks because in commercial electricity rate plans, there is a fixed cost, the so-called facility-related demand, that depends on the maximum power registered over the billing period [21]. We therefore introduce an upper limit $C \in \{1, 2, \ldots, L \cdot V\}$ on the number of chargers used simultaneously.

Figure 1 shows a depot with 4 lanes, 3 slots per lane, and $C = 3$. By examining the top and bottom numbers in each occupied slot, we see that the FIFO policy is respected in all lanes. Under each EB ID, there is the current SoC and the energy consumption of its next ST. Assuming that $SoC^{min} = 0\%$, EBs b_2, b_5, b_7, and b_8 do not have enough energy to perform their next ST and need to be charged. In this example the depot may operate at most 3 chargers at a time, so only EBs b_2, b_7, and b_8 are currently charging (See EBs in green in the figure).

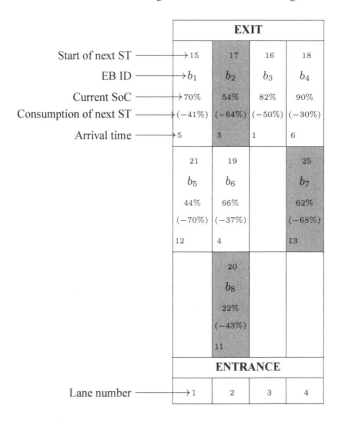

Fig. 1. Example of a depot at time $t = 14$. EBs in green slots are charging. (Color figure online)

We are now ready to define our problem formally. The eBAP-PC consists in assigning to each ST $\in \mathcal{S}$ an EB $\in \mathcal{E}$ and a lane $\in \mathcal{L}$ where this EB will be parked and possibly charged after completing the ST. When an EB is charging, it uses one charger. These assignments must be carried out in such a way that the EBs do not run out of energy, the number of chargers used simultaneously never exceeds C, and the FIFO policy is respected. EBs can be moved forward when parked in a lane and in this case charging is assumed to continue with a negligible interruption. Note that there is no objective function to optimize: our goal is only to find a feasible planning. The notation used is summarized in Table 1, which also includes notation used in the next sections.

4   Constraint Programming Formulation

In this section, we introduce a CP formulation for the eBAP-PC. CP mainly relies on logical inferences from each constraint, removing from the domain of variable values that are locally inconsistent with the satisfaction of some constraint. It is particularly appropriate to solve satisfaction problems and has had a lot of success on scheduling problems. So it is a natural candidate for the eBAP-PC. We describe next the main concepts [14, 15] we use in our model:

Table 1. Sets and parameters used in the model.

Sets	
$\mathcal{E}$	Set of EBs
$\mathcal{S}$	Set of STs
$\mathcal{L} = \{1, \ldots, L\}$	Set of parking lanes in the depot
$\mathcal{V} = \{1, \ldots, V\}$	Set of parking slots in a lane
Parameters	
h_s^S	Starting time of ST s
h_s^E	Ending time of ST s
h_b	Starting horizon of EB b
e_s	Net energy consumption along ST s
SoC_b^{init}	Initial SoC of EB b
SoC^{max}	Maximum battery SoC
SoC^{min}	Minimum battery SoC
C	Maximum number of simultaneous charging operations
λ	Charging rate
H	End of the planning horizon

- *Interval variables* v are defined by three fields: a boolean $\text{PRES}(v)$ indicating the presence of the variable v, and two integers $\text{START}(v)$ and $\text{LENGTH}(v)$ indicating the interval start and duration.
- *Sequence variables* are sets of interval variables on which we want to apply specific constraints. For example, we can enforce that there is no overlap or impose a certain order between variables with the constraints NOOVERLAP and PREVIOUS.
- *Cumulative functions* can be used to sum the contributions of interval variables to a more global variable. They are particularly useful for modeling resources consumed or produced by interval variables. For example if expression $\text{STEPATSTART}(v, d)$ is present in the definition of a cumulative function f, then f has an increment of d units at the start of v. Alternatively the expression $\text{PULSE}(v, d)$ indicates that f has an increment of d units at the start of v and a decrement of d units at the end of v.

Our CP model focuses on the three different activities which are modeled by interval variables: STs, charging, and parking. We will consider optional activities to determine the assignment to EBs and to lanes.

4.1 Variables

To define the constraints that determine the assignment of the activities to the EBs, we use the following variables:

- $\bar{x}_s$: interval variable which represents ST s, $\forall s \in \mathcal{S}$. It is always present, starts at h_s^S and ends at h_s^E.

- $x_{b,s}$: optional interval variable representing ST s if it is performed by EB b, $\forall s \in \mathcal{S}$, $\forall b \in \mathcal{E}$. It is present if and only if s is performed by b. It starts at h_s^S and ends at h_s^E.
- $c_{b,s}$: optional interval variable representing the charging activity on EB b after completing ST s, $\forall s \in \mathcal{S}$, $\forall b \in \mathcal{E}$. It is present if and only if b performs s.
- $y_{b,s}$: optional interval variable representing the parking activity for EB b after completing ST s, $\forall s \in \mathcal{S}$, $\forall b \in \mathcal{E}$. It is present if and only if b performs s. It starts at h_s^E.

We use the following variables to deal with the initialization of the planning, i.e., the first charging activity and the first parking activity:

- c_b: interval variable representing the first charging activity of EB b at the beginning of the horizon, $\forall b \in \mathcal{E}$. It is always present.
- y_b: interval variable representing the first parking activity at the beginning of the horizon of EB b, $\forall b \in \mathcal{E}$. It is always present and starts at h_b.

Next we use two sequence variables:

- π_b^{chg}: sequence variable of all the interval variables $x_{b,s}$, c_b, $c_{b,s}$ $\forall s \in \mathcal{S}$, $\forall b \in \mathcal{E}$. It represents the planning of STs and charging activities for each EB b.
- π_b^{park}: sequence variable of all the interval variables $x_{b,s}$, y_b, $y_{b,s}$ $\forall s \in \mathcal{S}$, $\forall b \in \mathcal{E}$. It represents the planning of STs and parking activities for each EB b.

Finally, to determine the assignment of the EBs to lanes after completing a ST, we use the following variables:

- $\underline{y}_s$: interval variable representing the parking activity after ST s, $\forall s \in \mathcal{S}$. It is always present and starts at h_s^E.
- $\bar{y}_{s,l}$: optional interval variable representing the parking activity after ST s in lane l, $\forall s \in \mathcal{S}$, $\forall l \in \mathcal{L}$. It is present if and only if the EB assigned to s is parked in lane l. It starts at h_s^E.
- $\bar{y}_{b,l}$: optional interval variable representing the parking activity at the beginning of the horizon of EB b in lane l, $\forall b \in \mathcal{E}$, $\forall l \in \mathcal{L}$. It is present if and only if b is parked in lane l at the beginning of the horizon. It starts at h_b.

All in all, there are $|\mathcal{E}|$ duplicates of interval variables representing each ST and its following charging activity, and there are $|\mathcal{E}| + L$ duplicates of the interval variables representing each parking activity.

4.2 Constraints

In this section, we break down the presentation of constraints in two parts: one on energy constraints and the other on parking constraints.

Energy Constraints

$$\text{ALTERNATIVE}(\bar{x}_s, \{x_{b,s}\}_{b\in\mathcal{E}}) \qquad\qquad \forall s \in \mathcal{S} \qquad (1)$$

$$\text{PRES}(x_{b,s}) = \text{PRES}(c_{b,s}) \qquad\qquad \forall b \in \mathcal{E}, \forall s \in \mathcal{S} \quad (2)$$

$$\text{NOOVERLAP}(\pi_b^{chg}) \qquad\qquad \forall b \in \mathcal{E} \qquad (3)$$

$$\text{PREVIOUS}(\pi_b^{chg}, x_{b,s}, c_{b,s}) \qquad\qquad \forall b \in \mathcal{E}, \forall s \in \mathcal{S} \quad (4)$$

$$\text{FIRST}(\pi_b^{chg}, c_b) \qquad\qquad \forall b \in \mathcal{E} \qquad (5)$$

$$\text{STARTOF}(c_b) \geq h_b \qquad\qquad \forall b \in \mathcal{E} \qquad (6)$$

$$
\begin{aligned}
Q_b = \;& \text{STEPATSTART}(c_b, SoC_b^{init} + \lambda\text{LENGTH}(c_b)) \\
& - \sum_{s\in\mathcal{S}} \text{STEPATSTART}(x_{b,s}, e_s) \\
& + \sum_{s\in\mathcal{S}} \text{STEPATSTART}(c_{b,s}, \lambda\text{LENGTH}(c_{b,s}))
\end{aligned}
\qquad \forall b \in \mathcal{E} \qquad (7)
$$

$$\text{ALWAYSIN}(Q_b, [0, H], [SOC_{min}, SOC_{max}]) \qquad \forall b \in \mathcal{E} \qquad (8)$$

$$\mathcal{C} = \sum_{b\in\mathcal{E}} \text{PULSE}(c_b, 1) + \sum_{s\in\mathcal{S},b\in\mathcal{E}} \text{PULSE}(c_{b,s}, 1) \qquad\qquad (9)$$

$$\text{ALWAYSIN}(\mathcal{C}, [0, H], [0, C]) \qquad\qquad (10)$$

Constraints (1) ensure that each ST s is assigned to a unique EB. Constraints (2) ensure that each charging activity is assigned to the EB which performed the associated ST. Constraints (3)–(6) ensure a correct order between trips and charging activities. Constraints (7)–(8) enforce the bounds of batteries' energy. Indeed, the cumulative function Q_b represents the SoC of the EB b throughout the horizon, whose end is denoted H. Constraints (9)–(10) ensure not to use more chargers than available.

Parking Constraints

$$\text{PRES}(x_{b,s}) = \text{PRES}(y_{b,s}) \qquad\qquad \forall b \in \mathcal{E}, \forall s \in \mathcal{S} \quad (11)$$

$$\text{FIRST}(\pi_b^{park}, y_b) \qquad\qquad \forall b \in \mathcal{E} \qquad (12)$$

$$\text{NOOVERLAP}(\pi_b^{park}) \qquad\qquad \forall b \in \mathcal{E} \qquad (13)$$

$$\text{STARTOFNEXT}(\pi_b^{park}, y_b) = \text{ENDOF}(y_b) \qquad \forall b \in \mathcal{E} \qquad (14)$$

$$\text{STARTOFNEXT}(\pi_b^{park}, y_{b,s}) = \text{ENDOF}(y_{b,s}) \qquad \forall b \in \mathcal{E}, \forall s \in \mathcal{S} \quad (15)$$

$$\text{ALTERNATIVE}(\underline{y}_s, \{y_{b,s}\}_{b\in\mathcal{E}}) \qquad\qquad \forall s \in \mathcal{S} \qquad (16)$$

$$\text{ALTERNATIVE}(\underline{y}_s, \{\bar{y}_{s,l}\}_{l\in\mathcal{L}}) \qquad\qquad \forall s \in \mathcal{S} \qquad (17)$$

$$\text{ALTERNATIVE}(y_b, \{\bar{y}_{b,l}\}_{l\in\mathcal{L}}) \qquad\qquad \forall b \in \mathcal{E} \qquad (18)$$

$$
\begin{aligned}
K_l = \;& \sum_{s\in\mathcal{S},l\in\mathcal{L}} \text{PULSE}(\bar{y}_{s,l}, 1) \\
& + \sum_{b\in\mathcal{E},l\in\mathcal{L}} \text{PULSE}(\bar{y}_{b,l}, 1)
\end{aligned}
\qquad \forall l \in \mathcal{L} \qquad (19)
$$

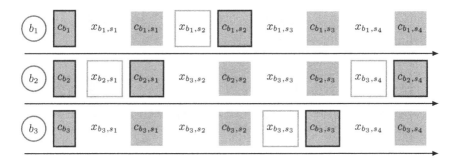

Fig. 2. Planning of STs and charging activities

$\text{ALWAYSIN}(K_l, [0, H], [0, V])$ $\qquad\qquad$ $\forall l \in \mathcal{L}$ $\qquad\qquad$ (20)

$\text{ENDOF}(\bar{y}_{s_i,l}) \leq \text{ENDOF}(\bar{y}_{s_j,l})$ $\qquad$ $\forall l \in \mathcal{L}, \forall (s_i, s_j) \in \mathcal{S}^2 \text{ s.t. } h^E_{s_i} \leq h^E_{s_j}$
$\qquad\qquad\qquad\qquad\qquad\qquad\qquad\qquad\qquad\qquad\qquad\qquad\qquad\qquad\qquad$ (21)

$\text{ENDOF}(\bar{y}_{b,l}) \leq \text{ENDOF}(\bar{y}_{s,l})$ $\qquad\qquad$ $\forall l \in \mathcal{L}, \forall b \in \mathcal{E}, \forall s \in \mathcal{S}$ $\qquad\qquad$ (22)

Constraints (11) ensure that each parking activity is assigned to the EB which performed the associated trip. Constraints (12)–(15) define the end of each parking activity. Constraints (16)–(18) assign each parking activity to only one parking lane. Constraints (19)–(20) ensure to respect the number of parking slots in each parking lane. Indeed, the cumulative function K_l indicates the number of EBs parked in lane l throughout the horizon. A parking activity in lane l produces one unit of the resource K_l whose maximum capacity is V. Constraints (21)–(22) enforce the FIFO policy for each lane.

We illustrate our model with three EBs, four STs, and a depot with $V = L = 2$. Figures 2 and 3 provide a simplified illustration of the assignment of STs to EBs and the scheduling of the charging and parking activities. In reality, all the STs and the charging and parking activities have different lengths and overlaps are possible. In these two figures, a frame means the interval variable is present. Then, in this example, EB b_1 is assigned to ST s_2, EB b_2 is assigned to STs s_1 and s_4, and EB b_3 is assigned to ST s_3. This implies that the subsequent charging and parking activities are also present. Finally, Fig. 4 represents the assignment of the parking activities to the parking lanes. For each lane, the parking activities are distributed according to start times, and above there is a representation of EBs parked in the lane throughout the horizon. For example, at the beginning of the horizon, lane 2 contains only EB b_2, then it is empty, then it receives EB b_1, and then it receives EB b_2. At the end, EB b_1 is represented above EB b_2 because it is the case in the depot. We note that this representation uses the assignment of STs to EBs made in Figs. 2 and 3. For instance, the ST s_2 is performed by the EB b_1 which will then be parked in lane 2. Thus, we can see in this example that the FIFO policy is well respected throughout the horizon. We can also see that the capacity of each lane ($V = 2$) is respected.

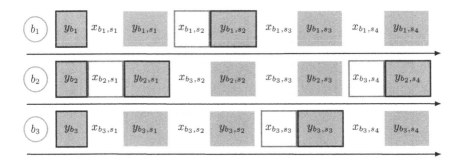

Fig. 3. Planning of STs and parking activities

5　Computational Experiments

Our implementation uses IBM ILOG CPLEX 22.1.1 to solve the instances with CP Optimizer. To assess the performance of our model, we conducted computational experiments on instances built from real-world data provided by our industrial partner, GIRO Inc., one of the world-leading optimization software providers for public transit companies. All computations were performed on a 64-bit computer equipped with an Intel® Core i7-12700 (2.10 GHz), 65 GB of RAM, running on CentOS Linux 9.2. In the remainder of this section, we first describe the instances and we then report and discuss our results.

5.1　Instance Generation

GIRO Inc. provided us with two real-world instances. The first is composed of 98 STs, 30 EBs, and a depot with 30 parking slots and 6 lanes. The second is composed of 125 STs, 42 EBs, and a depot with 42 parking slots and 6 lanes. The two instances are input for a 1-day eBAP-PC, and the STs are repeated every day if we want a multi-day eBAP-PC. The SoC of the EBs must remain between SoC^{min} = 1500 (for 15%) and SoC^{max} = 10000 (for 100%). The linear charging rate is $\lambda = 22/min$.

For a more comprehensive assessment of our model, we used these data to create other instances identified using the following convention $B/\tau/V/D/\gamma$, where B is the number of EBs, τ is the ratio between the number of STs per day and the number of EBs, V is the ceiling of the ratio between the number of EBs and the number of lanes in the depot, D is the number of days in the planning horizon, and γ is the ratio between the number of chargers and the number of EBs. To create an instance $B/\tau/V/D/\gamma$, we randomly draw $\lfloor B \cdot \tau \rfloor$ STs among the 223 provided by Giro Inc. and we make D copies of each of them separated by 24 h. Then, we set $C = \lfloor \gamma \cdot B \rfloor$, $L = \lceil B \cdot V \rceil$ and each lane contains V parking slots. The instances denoted by real_30/V/D/$gamma$ and real_42/V/D/$gamma$ represent the real ones with different values of V, D, and γ. Moreover, for all the following tests, we assume that all EBs have the same initial SoC, SoC^{init}, at the beginning of the horizon and are parked in the depot. Considering the depot as a table, we fill it from left to right, then from top (exit) to bottom (entrance) at the beginning of the horizon. For example, for 30 EBs and a depot with 7 lanes and 5

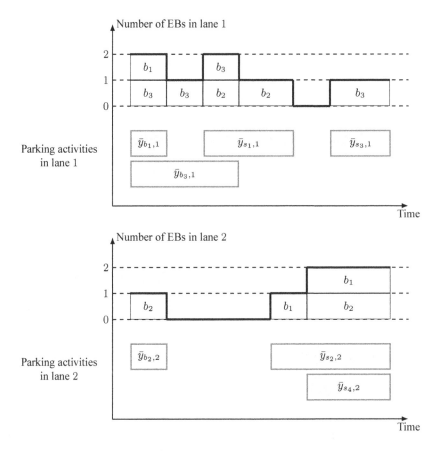

Fig. 4. Planning of parking activities in the depot

parking slots per lane, 2 lanes are full and the others contain 4 EBs. This initial filling is made by adding the following constraints to Model (1)–(22):

$$\text{PRES}(\bar{y}_{b_i,i\%L}) = \text{TRUE} \qquad \forall i \in \{1,\dots,|\mathcal{E}|\} \qquad (23)$$

$$\text{LENGTH}(y_{b_i}) \leq \text{LENGTH}(y_{b_{i+1}}) \qquad \forall i \in \{1,\dots,L-1\} \qquad (24)$$

$$\text{LENGTH}(y_{b_i}) \leq \text{LENGTH}(y_{b_{i+L}}) \qquad \forall i \in \{1,\dots,|\mathcal{E}|-L\} \qquad (25)$$

Constraints (23) impose in which parking lane the EB b is initially parked. Constraints (24) break symmetries between the parking lanes. Constraints (25) impose an initial order between the EBs in each parking lane.

5.2 Solution Methods

While developing our CP approach, we also designed a competing MILP model [3]. The model developed is based on a three-graph representation for assigning STs to EBs, and

Table 2. Computational times in seconds of the real instances for three-day planning.

Instance	CP	CP-searchPhase	MILP-decomp	MILP
real_30/5/3/0.5	643	396	13	X
real_30/6/3/0.5	287	X	41	X
real_30/7/3/0.5	X	2047	8	X
real_42/5/3/0.5	286	X	447	X
real_42/6/3/0.5	X	X	176	X
real_42/7/3/0.5	X	2972	212	X

their assigned EBs to lanes and to chargers. The last assignment to chargers ensures that the constraint on the maximum number of chargers used simultaneously is respected. In these three graphs, nodes are the STs and a path represents the STs assigned to either an EB or a lane or a charger. Thus, the model can be seen as a combination of three flow models (STs assignment, parking, charging) with additional constraints linking them. From this representation, the search space is reduced through a heuristic deletion of arcs in each graph. For example, it considers that an EB cannot stay in the depot for too long, and that it has to perform STs regularly. This reduces the set of next possible STs an EB can perform. Second, this MILP model adds an objective function encouraging the use of likely arcs in order to guide the solution process. More precisely, it assumes that estimators of the average parking time and the average time between two arrivals in a lane are known. The objective function minimizes the deviation from these estimators. Two variants from this MILP model are considered. *MILP* solves the model as a whole. *MILP-decomp* solves it in two steps: the first one assumes that there are as many chargers as EBs and determines the assignment of STs to EBs and lanes; the second one determines the assignment of STs to chargers.

In the spirit of that decomposition postponing the consideration of chargers, we consider two variants for our CP model, denoted *CP-searchPhase* and *CP*. *CP-searchPhase* adds a search phase with the variables $x_{b,s}$, $\bar{x}_s$, y_b, $\bar{y}_{b,l}$, $y_{b,s}$, $\bar{y}_{s,l}$, and $\underline{y}_s$, $\forall b \in \mathcal{E}, \forall s \in \mathcal{S}, \forall l \in \mathcal{L}$, meaning that these variables have priority over the charging variables. There is no search phase for *CP*.

5.3 Three-Day Horizon

The first series of instances considers 141 instances ($= 6 + 3 \times 3 \times 3 \times 5$) where $B \in \{30, 40, 50\}$, $\tau \in \{2.5, 3, 3.5\}$, $V \in \{5, 6, 7\}$, $D \in \{3\}$, and $\gamma \in \{0.5\}$. For each combination of B, τ and V values, we generated five instances. We assume that the EBs are fully charged at the beginning of the horizon, $SoC^{init} = 10000$. To avoid instabilities, we fixed the variables c_b such that $\forall b \in \mathcal{E}$, $\text{STARTOF}(c_b) = \text{ENDOF}(c_b) = h_b = 0$. Table 2 provides the computation times in seconds of the real instances by each solution method. Figure 5 shows for each solution method the number of instances solved in less time than the one displayed on the x-axis.

We observe that there is no significant advantage between *CP* and *CP-searchPhase* on this series. Approximately half of all instances have been solved by both variants.

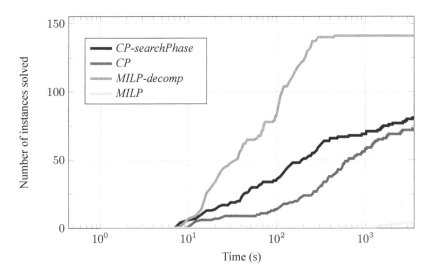

Fig. 5. Performance graph comparing the four solution methods for multi-day planning

In particular the most difficult ones, $B/\tau = 3.5/V/D/\gamma$, are not solved. Nevertheless, the easiest instances ($B/\tau = 2.5/V/D/\gamma$) are solved much faster by *CP-searchPhase* than *CP* (about three times faster). Moreover, we see a strong dependence between computational times and the parameter B for *CP-searchPhase*. Between the instances with $B = 30$ and $B = 50$, the computational times are multiplied by about ten. It is the same order between instances with $\tau = 2.5$ and $\tau = 3$, while there is no significant dependence for parameter V. Still this is much better than the *MILP* variant which also considers directly the whole problem, as it only solves 5 instances. However the two-step variant *MILP-decomp* manages to solve every instance with very steady computation times, with the exception perhaps of the second real-world instance. In particular, we observe that the dependence of the computational time on parameter τ is much less. Between the instances with $\tau = 2.5$ and $\tau = 3.5$, the computational times are only multiplied by about three.

5.4   One-Day Horizon with Minimum End-of-Day Charge Level Constraints

Because the same STs are repeated from one day to the next (typical on weekdays), another approach to solving the eBAP-PC is to consider a single day while ensuring that at the end of it EBs are sufficiently charged to repeat the solution on the following day or at the very least to successfully resolve it if this can be done quickly.

We investigate this approach by adding the following target energy constraints to Model (1)–(25):

$$\text{ALWAYSIN}(Q_b, [H, H], [\beta \cdot SoC_b^{init}, SoC^{max}]) \qquad\qquad \forall b \in \mathcal{E} \quad (26)$$

$$\text{ALWAYSIN}(\sum_{b \in \mathcal{E}} Q_b, [H, H], [\alpha \cdot \sum_{b \in \mathcal{E}} SoC_b^{init}, |\mathcal{E}| \cdot SoC^{max}]) \qquad\qquad (27)$$

Table 3. Computational times in seconds of the real instances for one-day planning.

Instance	α	CP	CP-searchPhase	MILP-decomp	MILP
real_30/6/1/0.5	0.8	3	1	4	2839
real_30/6/1/0.5	0.9	3	1	3	X
real_30/6/1/0.5	1	9	3	14	1257
real_42/6/1/0.5	0.8	8	3	41	X
real_42/6/1/0.5	0.9	8	3	62	X
real_42/6/1/0.5	1	7	3	51	X

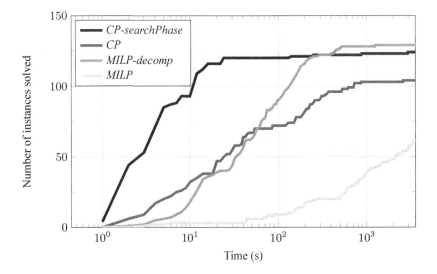

Fig. 6. Performance graph comparing the four solution methods for one-day planning

Constraints (26) impose a minimum SoC at the end of the planning for each EB b. Constraint (27) imposes a minimum for the average SoC of the fleet at the end of the planning.

The second series of instances considers 141 instances ($= 6 + 3 \times 3 \times 3 \times 5$) where $B \in \{30, 40, 50\}$, $\tau \in \{2.5, 3, 3.5\}$, $V \in \{6\}$, $D \in \{1\}$, $\gamma \in \{0.5\}$, $\alpha \in \{0.8, 0.9, 1\}$, and $\beta = 0.5$. For each combination of B, τ and α values, we generated five instances. We assume that the EBs are partially charged at the beginning of the horizon, $SoC^{init} = 8000$. The horizon starts at 240 and ends 24 h later at $H = 1680$. Table 3 provides the computation times in seconds of the real instances by each solution method. Figure 6 shows for each solution method the number of instances solved in less time than the one displayed on the x-axis.

Except for *MILP*, all the solution methods solved most of the instances: 104 for *CP*, 124 for *CP-searchPhase*, and 129 for *MILP-decomp*. If *CP* did not solve the hardest instances, the addition of the search phase allows *CP-searchPhase* to solve twenty more instances and to solve most instances in a very short time. In particular, we see that it solves the instances much faster than *MILP-decomp*. The real-world instances are solved in a few seconds by *CP-searchPhase* whereas *MILP-decomp* required up to a minute. This difference may seem insignificant for planning purposes, but in reality our instances represent the current situation of public transport with only part of the large fleet being replaced with electric buses. In a few years, there may be instances with 150 buses to consider and the time savings will become more essential. Moreover, except for instances $B/\tau = 3.5/\alpha = 1$, the computational times for *CP-searchPhase* do not show any dependence on the parameters of the instances, which is not the case for *MILP-decomp* and *CP*. This dependence corresponds to increasing computational times from around 10 s ($B = 30$, $\tau = 2.5$) to 300 s ($B = 50$, $\tau = 3.5$) for each $\alpha \in \{0.8, 0.9\}$ for *MILP-decomp*. In addition to computational times, this dependency affects the number of instances solved for *CP*. For instance, it did not solve any of the instances with $\tau = 3.5$. In the specific case of $B/\tau = 3.5/\alpha = 1$ instances, we hypothesize that with such a combination, the energy part of the problem becomes very hard: chargers must be used almost without interruption. The objective function of *MILP-decomp* in the first step drives EBs to stay parked long enough to recharge according to the number of chargers available in the second step. In contrast, the CP methods may waste time exploring insufficient parking durations.

6   Conclusion

We presented the eBAP-PC and introduced a CP model to solve it with CP Optimizer. We proposed two solution methods differentiated by the addition of a search phase focusing on parking constraints. We assessed the solution methods with multi-day planning instances and one-day planning instances with target energy constraints. For multi-day planning, we obtained good results for slightly easier instances than reality for both CP solution methods. However, the one-day planning tests showed a real benefit of the *CP-searchPhase* solution method, even compared to a sophisticated MILP approach. Indeed, most of the instances were solved in a few seconds.

One interesting avenue to explore for the multi-day problem is a hybrid approach based on Benders decomposition. We would get our inspiration from a recent previous work on electric vehicle routing where a branch-price-and-cut approach finds vehicle routes and ensures their feasibility with respect to energy consumption and charging station capacity by solving a subproblem modeled and solved with CP [16].

Possibilities for extending this work include the consideration of a heterogeneous EB fleet in terms of charging capacity, a non-linear charging rate, differentiated parking slots or chargers, and trip duration uncertainty. Even though in our context the number of buses (and drivers) is fixed and recharging costs are constant, this may not be the case in other settings and adding an objective function may then be advisable.

Acknowledgements. We are thankful to the personnel of GIRO Inc. for describing this problem to us and providing initial datasets. This work was funded by GIRO Inc. and the Natural Sciences and Engineering Research Council of Canada under the grant ALLRP 567169-21. This financial support was greatly appreciated.

References

1. Adler, J., Mirchandani, P.: The vehicle scheduling problem for fleets with alternative-fuel vehicles. Transp. Sci. **51**(2), 441–456 (2017)
2. Alvo, M., Angulo, G., Klapp, M.: An exact solution approach for an electric bus dispatch problem. Transp. Res. Part E Logist. Transp. Rev. **156**, 102528 (2021)
3. Azema, M., Desaulniers, G., Mendoza, J., Pesant, G.: Electric vehicle assignment problem with parking constraints. Working paper
4. Booth, K., Piacentini, C., Bernardini, S., Beck, J.: Target search on road networks with range-constrained UAVs and ground-based mobile recharging vehicles. IEEE Robot. Autom. Lett. **5**(4), 6702–6709 (2020)
5. Booth, K., Tran, T., Nejat, G., Beck, J.: Mixed-integer and constraint programming techniques for mobile robot task planning. IEEE Robot. Autom. Lett. **1**(1), 500–507 (2016)
6. Booth, K.E.C., Beck, J.C.: A constraint programming approach to electric vehicle routing with time windows. In: Rousseau, L.-M., Stergiou, K. (eds.) CPAIOR 2019. LNCS, vol. 11494, pp. 129–145. Springer, Cham (2019). https://doi.org/10.1007/978-3-030-19212-9_9
7. Chao, Z., Xiaohong, C.: Optimizing battery electric bus transit vehicle scheduling with battery exchanging: model and case study. Procedia. Soc. Behav. Sci. **96**, 2725–2736 (2013)
8. Dirks, N., Wagner, D., Schiffer, M., Walther, G.: A concise guide on the integration of battery electric buses into urban bus networks (2021)
9. Hamdouni, M., Desaulniers, G., Marcotte, O., Soumis, F., Putten, M.: Dispatching buses in a depot using block patterns. Transp. Sci. **40**(3), 364–377 (2006)
10. Hamdouni, M., Desaulniers, G., Marcotte, O., Soumis, F., Putten, M.: Parking buses in a depot with stochastic arrival times. Eur. J. Oper. Res. **183**(2), 502–515 (2006)
11. Jiang, M., Zhang, Y., Zhang, Y.: Multi-depot electric bus scheduling considering operational constraint and partial charging: a case study in Shenzhen, China. Sustainability **14**, 255 (2022)
12. Jovanovic, R., Bayram, I., Bayhan, S., Voß, S.: A GRASP approach for solving large-scale electric bus scheduling problems. Energies **14**, 6610 (2021)
13. van Kooten Niekerk, M., van den Akker, J., Hoogeveen, J.: Scheduling electric vehicles. Public Transp. **9**, 155–176 (2017)
14. Laborie, P., Rogerie, J., Shaw, P., Vilím, P.: IBM ILOG CP optimizer for scheduling. Constraints **23**(2), 210–250 (2018)
15. Laborie, P.: IBM ILOG CP optimizer for detailed scheduling illustrated on three problems. In: van Hoeve, W.-J., Hooker, J.N. (eds.) CPAIOR 2009. LNCS, vol. 5547, pp. 148–162. Springer, Heidelberg (2009). https://doi.org/10.1007/978-3-642-01929-6_12
16. Lam, E., Desaulniers, G., Stuckey, P.J.: Branch-and-cut-and-price for the electric vehicle routing problem with time windows, piecewise-linear recharging and capacitated recharging stations. Comput. Oper. Res. **145**, 105870 (2022)
17. Li, J.Q.: Transit bus scheduling with limited energy. Transp. Sci. Procedia Soc. Behav. Sci. **48**(4), 521–539 (2014)
18. Messaoudi, B., Oulamara, A.: Electric bus scheduling and optimal charging. In: Paternina-Arboleda, C., Voß, S. (eds.) ICCL 2019. LNCS, vol. 11756, pp. 233–247. Springer, Cham (2019). https://doi.org/10.1007/978-3-030-31140-7_15
19. Montoya, A., Guéret, C., Mendoza, J., Villegas, J.: The electric vehicle routing problem with nonlinear charging function. Transp. Res. Part B Methodol. **103**, 87–110 (2017)
20. Olsen, N., Kliewer, N.: Scheduling electric buses in public transport: modeling of the charging process and analysis of assumptions. Logist. Res. **13**, 4 (2020)
21. Pelletier, S., Jabali, O., Laporte, G.: Charge scheduling for electric freight vehicles. Transp. Res. Part B Methodol. **115**, 246–269 (2018)

22. Perumal, S., Lusby, R., Larsen, J.: Electric bus planning & scheduling: a review of related problems and methodologies. Eur. J. Oper. Res. **301**(6), 395–413 (2022)
23. Reuer, J., Kliewer, N., Wolbeck, L.: The electric vehicle scheduling problem: a study on time-space network based and heuristic solution. In: Proceedings of the Conference on Advanced Systems in Public Transport (CASPT), pp. 1–15 (2015)
24. Sassi, O., Oulamara, A.: Electric vehicle scheduling and optimal charging problem: complexity, exact and heuristic approaches. Int. J. Prod. Res. **55**(2), 519–535 (2017)
25. Vendé, P., Desaulniers, G., Kergosien, Y., Mendoza, J.: Matheuristics for a multi-day electric bus assignment and overnight recharge scheduling problem. Transp. Res. Part C **156**, 104360 (2023)
26. Zhang, A., Li, T., Zheng, Y., Li, X., Abdullah, M., Dong, C.: Mixed electric bus fleet scheduling problem with partial mixed-route and partial recharging. Int. J. Sustain. Transp. **16**(1), 73–83 (2022)
27. Zhang, L., Wang, S., Qu, X.: Optimal electric bus fleet scheduling considering battery degradation and non-linear charging profile. Transp. Res. Part E Logist. Transp. Rev. **154**, 102445 (2021)

Acquiring Constraints for a Non-linear Transmission Maintenance Scheduling Problem

Hugo Barral[1]([✉])([iD]), Mohamed Gaha[2]([iD]), Amira Dems[2]([iD]), Alain Côté[2]([iD]),
Franklin Nguewouo[3], and Quentin Cappart[1]([iD])

[1] Polytechnique Montréal, Montréal, Canada
{hugo.barral,quentin.cappart}@polymtl.ca
[2] Institut de recherche d'Hydro-Québec, Varennes, Canada
{gaha.mohamed,dems.amira,cote.alain7}@hydroquebec.com
[3] TransÉnergie Hydro-Québec, Montréal, Canada
nguewouo.franklin@hydroquebec.com

Abstract. Over time, power network equipment can face defects and must be maintained to ensure transmission network reliability. Once a piece of equipment is scheduled to be withdrawn from the network, it becomes unavailable and can lead to power outages when other adjacent equipment fails. This problem is commonly referred to as a *transmission maintenance scheduling* (TMS) problem and remains a challenge for power utilities. Numerous combinatorial constraints must be satisfied to ensure the stability and reliability of the transmission network. While most of these constraints can be naturally formalized in constraint programming (CP), there are some complex constraints like *transit-power limits* that are challenging to model because of their continuous and non-linear nature. This paper proposes a methodology based on active constraint acquisition to automatically approximate these constraints. The acquisition is carried out using a simulator developed by Hydro-Québec (HQ), a power utility to compute the power-flow of its transmission network. The acquired constraints are then integrated into a CP model to solve the HQ network's TMS problem. Our experimental results show the relevance of the methodology to approximate transit-power constraints in an automated way. It allows HQ to automatically schedule a maintenance plan for an instance that remained intractable until now. To our knowledge, it is the first time that active constraint acquisition has been used successfully for the TMS problem in an industrial setting.

Keywords: Transmission maintenance scheduling · Electric power network · Constraint programming · Constraint acquisition

1 Introduction

Hydro-Québec (HQ) is a power utility, responsible for the generation, transmission and distribution of electricity in the province of Québec, Canada. In order to

B. Dilkina (Ed.): CPAIOR 2024, LNCS 14742, pp. 34–50, 2024.
https://doi.org/10.1007/978-3-031-60597-0_3

deliver sustainable and reliable energy to its customers, HQ has to maintain its electrical assets (lines, power transformers, breakers, switches, etc.) and replace them when they reach the end of their service life. Recently, the *International Electrotechnical Commission* (IEC) highlighted the importance for power utilities of having a robust and specialized approach to asset management [18]. In response to this challenge, and following similar initiatives around the world [1], in 2019 HQ started a project aiming to develop an integrated decision support system including predictive modelling methods for identifying and prioritizing the replacement and maintenance of electrical equipment [12]. Given the complexity of this project, it is divided into sub-modules, such as cloud data warehouses, asset behavior, reliability simulator, transmission network simulator, risk and optimization. In the scope of this paper, we are interested in the reliability simulator module and more particularly in the scheduling aspect of maintenance operations. In fact, equipment needs to be periodically withdrawn for maintenance or replacement purposes and no outages must be experienced by customers. To achieve this, the withdrawal must be done without violating constraints that guarantee power network stability during maintenance. Such constraints are commonly referred to as *transit-power constraints*. Currently, this task is performed in most utilities by the network control center, which uses a contingency approach to ensure that energy is not interrupted [27]. In addition, traditional scheduling constraints (e.g., respecting a hard deadline) must be taken into account. In the related literature, the maintenance scheduling task refers to two well-known NP-hard problems: *generator maintenance scheduling* (GMS) and *transmission maintenance scheduling* (TMS). Efficiently solving both problems is crucial to automatically generate a coherent maintenance plan that ensure the stability of the transmission network. Although several approaches have been proposed for solving the GMS (see for instance the survey proposed by Froger et al. (2016) [11]), the TMS problem is less studied in the literature.

The objective of TMS is to create an annual maintenance plan for electric power transmission equipment while preserving the stability of the network and ensuring uninterrupted power supply for customers. It is important to note that the TMS is confined to transmission equipment and excludes distribution equipment. This decision is deliberate, as a majority of distribution assets are typically operated on a run-to-fail basis, necessitating a distinct resolution process. Pandžić et al. (2012) [19] propose a bi-objective mathematical model to solve the TMS problem. The main idea is to achieve a balance between power network transmission capacity and customer impact. This is achieved by reformulating the initially non-linear problem into a mixed integer linear program (MILP), and solving it by a branch-and-cut algorithm. In addition, Mei et al. (2021) [16] propose another MILP for maximizing the number of devices maintained while preserving stability constraints. A specific feature of this approach is the use of machine learning to speed up the processing time. However, a major limitation is that it is based on relatively simple IEEE 24-bus and IEEE 30-bus systems, and is thereby not representative of real networks involving complex electrical constraints. Rocha et al. (2021) [25] proposed a MILP to solve the

TMS problem on a more realistic network. The problem is solved with Benders' decomposition [24]. However, advanced constraints, related to the limitations of the energy flow in the network, are still not taken into account. To handle the power-flow constraints, Coffrin et al. propose to relax them, either linearly [10] or as a quadratic convex program [9]. Also, software libraries were introduced to facilitate the exploration of power-flow formulations [8]. In the field of constraint programming (CP), Popovic et al. (2022) [20] proposed a scheduling model to solve the TMS problem on a real transmission network. The transit-power constraints were not modeled but partially handled thanks to expert knowledge injected in the search phase as heuristics. This approach was able to come up with a compliant schedule inside the HQ power grid infrastructure for some *interface* (i.e., a strategic point of the power grid infrastructure) but failed for the most challenging interface.

Because of their continuous and non-linear nature, handling arbitrary transit-power constraints inside a scheduling problem is still a challenge today and is an open-question in the research community. This paper proposes to tackle this issue by introducing a new methodology based on active constraint acquisition. Briefly, the goal of constraint acquisition is to induce, from examples, constraints that adequately represent the target problem [7,22]. Algorithms are commonly referred to as *passive*, when we are only provided with a pool of examples [4,21], or *active*, where we can dynamically issue requests to the algorithm [2,17,29]. Our idea is to approximate the transit-power constraints thanks to an active interaction with QUACQ2 [5], a framework for active constraint acquisition, and an industrial power-flow simulator, able to detect if a proposed maintenance schedule is compliant with the transit-power constraints. By doing so, such constraints do not have to be manually encoded into the CP model but are dynamically *acquired* instead. A few successful applications of constraint acquisition are available [3,26] but none were used for solving the TMS problem. Our experiments are carried out on a real power grid infrastructure that involves more than 200 pieces of electrical equipment and 300 withdrawal requests. This is the same situation as Popovic et al. (2022) [20]. The computational results show that we obtained a maintenance schedule plan complying with all the constraints for each strategic point. Specifically, the most constrained strategic point was still unsolved by Popovic et al. (2022) and by experts from HQ. In practice, field experts from HQ had to accommodate this situation with a schedule that does not respect the constraints.

The paper is structured as follows. The next section introduces the TMS problem and propose a formalization of it. Our methodology is then presented in Sect. 3, which presents how constraint acquisition is used for our purpose, and in Sect. 4, which describes our global framework for solving the TMS. Experimental results and related analyses are presented in Sect. 5.

2   Problem Description

The transmission maintenance scheduling problem consists in designing a maintenance plan for all withdrawal requests that are scheduled during a given period (e.g., a year). As proposed by Popovic et al. (2022) [20], we formalize this problems as a constraint-based scheduling problem and use *time-interval variables*, *cumul*, and *state* functions [13,14]. In our context, a *cumul* function $f_1^c : \mathbb{N} \times \mathbb{N} \times E \to \mathbb{N}$ gives the number of piece of equipment withdrawn, from a set E, during a discretized period between a start and an end time. A *cumul* function $f_2^c : \mathbb{N} \times \mathbb{N} \times E \to \mathbb{N}$ gives the summed charge of equipment withdrawn during a discretized period between a start and an end time. On the other hand, a *state* function $f^s : \mathbb{N} \times \mathbb{N} \times E \to \mathsf{State}$ gives a state to each time step. The state depends on the equipment that is currently withdrawn. Let W be the set of planned withdrawal requests within the planning horizon, and let E be the set of electrical transmission equipment. Fulfilling a request $w \in W$ requires removing a subset E_w of equipment from E. For HQ, withdrawal requests are permitted for 246 consecutive days within a year, between d_0 and d_{245}. This corresponds to a period spanning roughly from March to October, and excludes the winter. To ensure the stability and availability of the power network, every year, HQ must solve every year the TMS formalized below. This formalization is inspired by the model of Popovic et al. (2022) [20], which we refer to for additional insights concerning this model. A running example is also proposed in Fig. 1.

satisfy

$$
\begin{aligned}
\text{s. t. } &\mathsf{alwaysIn}(f_1^c, d_a, d_b, 0, h) && \forall f_1^c(a, b, S), h \in F_1^c \quad (1)\\
&\mathsf{alwaysIn}(f_2^c, d_0, d_{245}, 0, \theta) && \forall f_2^c(0, 245, S), \theta \in F_2^c \quad (2)\\
&\mathsf{alwaysEqual}(f^s, S_i, i) && \forall S_i \in \Lambda, \forall f^s, \Lambda \in F^s \quad (3)\\
&\mathsf{transitPower}(\gamma, W_d, d) \geq \mathsf{minPower}(\gamma, d) && \forall d \in [d_0, d_{245}], \forall \gamma \in \Gamma \quad (4)
\end{aligned}
$$

Constraint 1. This constraint limits the simultaneous withdrawal of related equipment. Formally, it involves a set F_1^c of cumul functions (f_1^c) and their related maximum value (h). Each function monitors the number of piece of equipment from a set $S \subset E$ withdrawn during the period between d_a to d_b. Then, the constraint enforces that at most h pieces of equipment can be withdrawn at the same time during this period. The restriction is modelled using the $\mathsf{alwaysIn}(f, d_{min}, d_{max}, h_{min}, h_{max})$ constraint [14]. The constraint enforces the accumulated consumption of f to be between the value h_{min} and h_{max} for the period from d_{min} to d_{max}.

Constraint 2. This constraint states that the sum of the electric load of the withdrawn equipment from a set $S \subset E$ must always be below θ during the complete planning horizon (d_0 to d_{245}). Internally, the cumul function uses a function $ch : E \to \mathbb{N}$ giving the load associated to a piece of equipment. It is expressed by a set F_2^c of cumul functions (f_2^c) and their related maximum load (θ).

Constraint 3. Let $\Lambda = \{S_1, S_2, \ldots, S_K\}$ be a set containing K sets of equipment $S_k \subset E$. This constraint ensures that equipment associated with a set S_i and those with a set S_j, such that $i \neq j$, cannot be withdrawn together. We model this constraint by means of *state* functions and the $\mathsf{alwaysEqual}(f^s, a, v)$

constraint [14]. The constraint maps the value v to the function f^s when equipment a is withdrawn and enforces that f^s cannot take more than one value at a time. We simplify the notation with alwaysEqual(f^s, S, v) that maps the same value v to s whenever a piece of equipment from S is withdrawn. It is expressed by a set F^s of state functions (f^s) and their associated sets of equipment (Λ). We note that this formalization is a minor improvement from the model proposed by Popovic et al. (2022) [20], which used noOverlap constraints instead.

Constraint 4. This constraint, commonly referred to as a *transit-power constraint*, ensures that each interface impacted by the withdrawn equipment, has a sufficient transit-power generation during each day of the planned horizon. Handling this constraint is the main challenge of the TMS, as it involves continuous, differential, non-linear equations and cannot be easily integrated inside the model. Intuitively, each withdrawal causes a loss in the transit power and this loss is amplified when subsets of equipment are withdrawn at the same time. The challenge is that the relationship between withdrawals and related losses is not known. Let Γ be the set of interfaces of the power grid. Let transitPower : $\Gamma \times 2^E \times D \to \mathbb{R}$ be the generated transit power obtained at an interface when removing a set of equipment on a single day within an horizon D. Let minPower : $\Gamma \times D \to \mathbb{R}$ be the minimal requirement on transit power for each day of the horizon. Finally, let W_d be the set of withdrawn equipment at day d in a specific solution. The constraint ensures sufficient transit power for each interface during the complete maintenance horizon.

3   Handling the Transit-Power Constraint

As we do not have a closed-form for the transit-power constraint, we are still unable to solve this problem. However, the satisfaction of this constraint can be easily checked thanks to a simulator that mimics the impact of a maintenance schedule on the real power network. Popovic et al. (2022) [20] introduced a search heuristic to generate diverse solutions and used this simulator to check if the maintenance schedule obtained is compliant with the constraint. By doing so, a schedule complying with four interfaces from the power grid was obtained. However, the fifth interface, determined to be very challenging to comply with by field experts, was still unsolved. In this paper, we propose a new methodology based on *active constraint acquisition with partial queries* [5] to approximate the transit-power constraint in an automated way. The idea is to issue partial withdrawal requests to the simulator and to check if the transit-power is still satisfied. By leveraging the output of the simulator, we will be able to infer constraints mimicking the real transit-power dynamics. This section describes how active constraint acquisition is leveraged for this task.

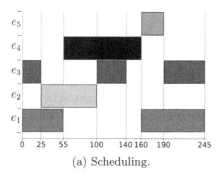
(a) Scheduling.

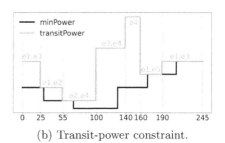
(b) Transit-power constraint.

Fig. 1. This figure shows a valid example of the TMS with five pieces of equipment (e_1 to e_5), one interface, and few constraints. The maintenance schedule obtained is represented on the left. First, Constraint 1 is enforced with $S = \{e_1, e_2, e_3\}$, $f_1^c(0, 245, S)$ and $h = 2$. This shows that e_1, e_2 and e_3 cannot be withdrawn together during the whole planning horizon. Second, Constraint 3 is enforced using $\Lambda = \{S_1, S_2\}$ with $S_1 = \{e_1, e_5\}$ and $S_2 = \{e_4\}$. It shows that e_1 and e_5 cannot be withdrawn if e_4 is withdraw. Finally, a possible transit-power constraint is shown on the right. The red curve represents the transit power while the blue curve shows the minimal requirement on transit power. Each withdrawal causes a drop of the transit power, but the exact dynamics are not known. (Color figure online)

3.1 Preliminaries on Active Constraint Acquisition

Active constraint acquisition features two distinct processes interacting and sharing common knowledge from the situation they have to learn [7]. Formally, let $\langle X, D \rangle$ be a *vocabulary*, where X is a set of variables and D is the set of corresponding domains. The first process, referred to as the *learner*, generates and sends candidate solutions, called *queries*, to the second process, referred to as the *user*. A candidate solution simply consists of an assignation of variables with values from their domain. The learner starts from a predetermined vocabulary and its goal is to acquire a set of constraints over this vocabulary. As there exist an infinite number of constraints, the acquisition is commonly restricted to a set of constraints, defined as the *language*. Only constraints belonging to the language can be acquired. When a language is instantiated to a specific vocabulary, we have a set $\mathcal{B}$ of eligible constraints (referred to as the *bias*) that can be obtained for our specific problem. Upon the reception of a query, the user replies with a positive or a negative answer. A positive answer indicates that no constraint is violated by the query and the learner can remove some constraints from bias $\mathcal{B}$. For instance, if the solution $\langle x = 1, y = 1 \rangle$ is accepted, the constraint $x \neq y$ is no longer eligible. On the other hand, a negative answer indicates that at least one constraint was violated by the query. On the learner side, this constitutes new information that can be used to acquire new constraints that are recorded in a set $\mathcal{L}$ (referred to as the *learned network*). This interaction is carried out iteratively until no more information can be added ($\mathcal{B} = \emptyset$).

The challenge is to efficiently identify which constraints from $\mathcal{B}$ are violated; there exist several framework for that, such as CONACQ.2 [6]. In this paper, we propose to leverage QUACQ2 [5], an acquisition framework based on *partial queries* [5]. This variant allows the learner to issue queries involving partial assignments (i.e., all the variables do not have to be assigned). The motivation is that partial queries allow for easier identification of violated constraints for the learner. Specifically, QUACQ2 is able to find a constraint in a logarithmic number of queries after a negative example and converges in a polynomial number of queries.

3.2   Redefining the Transit-Power Constraint

Our first step is to redefine the transit-power constraint into a formulation more amenable to constraint acquisition. From a combinatorial point of view, satisfying this constraint turns in knowing *which set of equipment can be withdrawn together without generating a power loss exceeding the threshold*. Let x_e be a binary variable indicating that equipment $e \in E$ is withdrawn. Let $\mathcal{F}$ be the set of all forbidden simultaneous withdrawals, i.e., a set of equipment that, when withdrawn together, violates the transit-power constraint. A configuration is given by solving the following subproblem. Constraint 5 states that when a set of n pieces of equipment is forbidden, we can only remove $n-1$ pieces of equipment from this set at the same time.

satisfy

$$\text{s. t.} \sum_{e \in F} x_e < |F| \qquad\qquad \forall F \in \mathcal{F} \ (5)$$

$$x_e \in \{0,1\} \qquad\qquad \forall e \in E \ (6)$$

Although the model is simple, the challenge is that we do not know what the forbidden sets in $\mathcal{F}$ are. Our idea is to acquire the constraints depicted in Eq. (5). When found, the resulting constraints are integrated into the main model defined in Sect. 2 instead of the transit-power constraint. It is important to note that there is currently no temporal aspect involved in this subproblem. Specifically, a forbidden set does not change over the planning horizon. This aspect will be handled later in our methodology.

4   Acquiring the Forbidden Sets

The benefit of the constraints in Eq. (5) is that we can control their arity by setting the maximum size of the forbidden sets in $\mathcal{F}$. This characteristic makes them amenable to be learned by efficient acquisition algorithms. Figure 2 gives a high-level overview of our framework to acquire such constraints. Each component is explained in the next subsections.

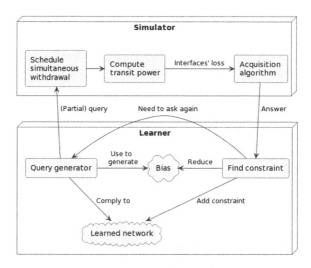

Fig. 2. Illustration of the interactions within the framework.

4.1   Learner Behavior

The vocabulary is defined as the set of all the variables of the subproblem (x_e with $e \in E$) with their binary domain. The bias $\mathcal{B}$ is the set of all the constraints in Eq. (5) involving at least l, and at most m pieces of equipment. This restriction is required to control the arity of the acquired constraints. Thanks to preliminary tests, we set l to be either 1 or 2, and m to 5. Finally, we enriched the learner with expert knowledge $\mathcal{K}$, i.e. sets of equipment that are known to be incompatible for a simultaneous withdrawal. This knowledge was obtained from HQ's field experts. Provided with this information, the learner carries out queries (i.e., a partial assignation of variables with values) to find new constraints or to eliminate candidates [5]. The bias is then reduced. This mechanism is referred to as FIND CONSTRAINT in Fig. 2. The query is determined thanks to heuristics $bdeg$ for variable ordering and max_v for value ordering [29] and must also comply with the known and acquired constraints at each generation. This mechanism is referred to as QUERY GENERATOR in Fig. 2. Once a constraint is acquired, it is removed from the bias, and the learner has converged when the bias is empty ($\mathcal{B} = \emptyset$).

4.2   Simulation of Withdrawals

Let $q = \{e_1, \ldots, e_n\}$ be the query sent by the learner, consisting of a set of n pieces of equipment to withdraw together. We use the same simulator as Popovic et al. (2022) [20] to analyze the impact of these withdrawals on the power grid. Briefly, the simulator is based on complex differential power-flow equations and was developed internally at HQ. It simulates power flow thanks to PSS/E software. The impact is measured in terms of the quantitative power loss on five geographical areas that divide the entire power grid infrastructure into strategic points (i.e., the *interfaces*). The output of the simulator is then a vector $\lambda : [\lambda_1, \lambda_2, \lambda_3, \lambda_4, \lambda_5]$ corresponding to the five losses. Note that this can be extended to an arbitrary number of interfaces, without any loss of generality.

4.3 Acquisition of Constraints

Provided with loss vector λ, the next step is to assess whether this stationary power is acceptable for maintaining the stability and availability of the power network, i.e., if it is above the threshold minPower for each interface in Constraint 4 during the whole planning horizon. From this information, we propose to acquire two types of constraints: (1) constraints that are relevant for the whole planning horizon, from day 0 to day 245, and (2) constraints that are relevant only for a specific period (e.g., a month). We refer to these constraints as *permanent constraints* and *temporary constraints*, respectively.

Permanent Constraints. To acquire permanent constraints, we first built a machine learning model dedicated to predicting whether loss vector λ is compliant for the five interfaces during the whole planning horizon. The model was trained through a set of historical data curated by field experts from HQ. In total we had a dataset of 2201 entries. Each entry i corresponded to a loss $\lambda^{(i)}$ generated when a set of equipment was withdrawn together, plus a label $y^{(i)} = \{0, 1\}$ indicating whether the loss violated the transit-power constraint. This was a relatively small dataset, and we therefore opted for a lightweight model based on a decision tree of depth 4 [23]. A proportion of 15% of the dataset was used for validation while the remaining constituted the training set. The tree accuracy on both sets was 86%. Once trained, the model achieved an accuracy of 85% for recovering known constraints in the background knowledge $\mathcal{K}$. This level of performances were considered sufficient for our goal. Note that building the best model was not the goal pursued in this paper. The constraint acquisition was then carried out by the internal algorithm of QUACQ2 from the answer given by the model. We only allowed the acquisition of constraints involving at least two pieces of equipment ($l = 2$) as a constraint with a single piece of equipment would forbid this withdrawal for the whole planning horizon, which is not desirable.

Temporary Constraints. These constraints were acquired directly using the minPower(γ, d) threshold. We recall that this function depends on the interface γ and on the day d. Let $d_{[a,b]}$ be a sequence of days, referred to as an *interval*, where the threshold does not change between day a and b on any interface. This information is available from internal data of HQ, and nine intervals matching this condition have been determined. Temporary constraints were defined and acquired for each interval separately. Let $\mathcal{I}$ be the set of intervals. Let $[a_i, b_i]$ be the bound for interval $i \in \mathcal{I}$. Let $\mu_i^\lambda \in \{0, 1\}$ be the answer if the loss λ is acceptable for interval i. Let Ψ_γ be the stationary power regime on interface

$\gamma \in \Gamma$ when no equipment is withdrawn. The answer is formalized in Eq. 7. For a given interval, it consists of verifying for each interface that the power flow does not drop below the threshold.

$$\forall i \in \mathcal{I} : \mu_i^\lambda = \begin{cases} 1 & \text{if } \Psi_\gamma - \lambda_\gamma \geq \mathsf{minPower}(\gamma, d_k) \quad \forall \gamma \in \Gamma, \forall k \in \{a_i, \dots, b_i\} \\ 0 & \text{otherwise} \end{cases}$$

(7)

The QUACQ2 algorithm is then executed nine times to obtain the temporary constraints associated with each interval.

4.4   Accelerating Convergence by Leveraging Dominances

While the approach described can already be used for acquiring the transit-power constraint through the forbidden sets, we integrated an additional improvement to speed up the convergence of the acquisition algorithm. If two equipment e_1 and e_2 cannot be removed together, it follows that we will not be able to remove larger sets that include both e_1 and e_2. In such cases, we have a *dominance* relation, which is defined as follows:

Definition 1 (Dominance). *Let c be a constraint from Eq. 5 with an arity of k_c, and let E_c be the set of equipment involved by this constraint. A constraint c' is dominated by c if and only if $k_{c'} > k_c$ and if $E_{c'} \supseteq E_c$.*

In other words, any superset of an invalid configuration of withdrawals is dominated by it. As a dominated constraint does not bring new information to the model, we modified the acquisition algorithm to prevent the acquisition of constraints which will be dominated by existing ones.

5   Application on a Real Power Grid

This section applies our methodology to schedule the maintenance of a real power network in Québec, Canada. The application is the same as the one analyzed by Popovic et al. (2022) [20]. In total, 359 withdrawal requests are considered and 271 pieces of electrical equipment are involved. This data corresponds to the year 2020. Each withdrawal request involved at most eight pieces of equipment, yielding a maximum of 2872 activities in the model, and there are five interfaces in the grid. Each interface has its own transit-power constraints. Concerning the grid topology, the production infrastructures are all located in the northern Québec while most of the load centers are in the South (e.g., Montréal). Four interfaces were successfully solved by Popovic et al. (2022) [20], but the last one, which was highly challenging, was still unsolved, even by field experts. The goal of the experiments was to assess the relevance of our methodology to learn relevant transit-power constraints, and to provide a schedule that is compliant with this last interface, which we did successfully. All experiments are executed on an Intel i7-11850H processor (2.5 GHz) with 64 GB of RAM. The CP model was implemented in C++ using CP Optimizer 20.1. Constraint acquisition was carried out with QUACQ2 [5].

5.1 Characteristics of Interfaces

Each interface has its own subproblem formalized in Sect. 3.2. In other words, there is specific transit-power constraint (i.e., a set of permanent and temporary constraints) to be acquired for each interface. An interface γ uses a set of equipment $E_\gamma \subset E$, but it is possible that the same equipment is used by several interfaces. This explains why each interface cannot be considered separately in the main problem of Sect. 2. Table 1 shows relevant characteristics for the five interfaces. Recall that the background knowledge $\mathcal{K}$ corresponds to constraints that are known before the acquisition, thanks to expert knowledge, and that l is the minimal arity of the constraints that can be acquired. The maximal arity is set to $m = 5$. The size of the bias is computed as the sum of combinations involving $\{l, \dots, m\}$ equipment among E_γ. We limit ourselves to specific pieces of equipment that are known by field experts for their significant impact on transit-power loss once withdrawn to decrease the size of bias $\mathcal{B}$. Interestingly, we observe that the proportion of background knowledge in relation to the number of pieces of equipment involved is smaller for the fifth interface (i.e., the challenging one). This observation stands even if all the equipment for all interfaces are involved.

Table 1. Characteristics of the five interfaces, with $m = 5$.

Interface (Γ)	1	2	3	4	5
No. of pieces of equipment (E_γ)	18	28	18	24	12
Background knowledge ($\mathcal{K}$)	83	184	70	110	21
Bias ($\mathcal{B}$) with $l = 2$	3008	46705	3021	25859	510
Bias ($\mathcal{B}$) with $l = 1$	3026	46733	3039	25883	522

5.2 Result: Magnitude of the Acquisition

Table 2 presents statistics about the acquisition of permanent and temporary constraints. Specifically, for each interface, we record the number of constraints acquired ($\mathcal{L}$), the number of constraints acquired that are not redundant with background knowledge ($\mathcal{L} \backslash \mathcal{K}$), the number of QUACQ2 queries required to converge, and the execution time in seconds for permanent constraints and in minutes for temporary constraints. First, we observe that we are always able to infer new constraints that are not redundant with the background knowledge. Second, for most interfaces, we acquire more temporary constraints than permanent ones. This is expected as the acquisition is executed on nine intervals, providing a higher opportunity for acquisition. Finally, the most interesting result is that we have significantly more constraints that are learned for the fifth interface compared to their respective bias size. This shows that non trivial knowledge could be acquired for this interface.

Table 2. Statistics of constraint acquisition for the five interfaces.

Interface (Γ)	1	2	3	4	5
Permanent constraints					
# constraints acquired ($|\mathcal{L}|$)	134	318	130	147	27
# new constraints ($|\mathcal{L}\backslash\mathcal{K}|$)	51	134	60	37	6
Number of queries	125	340	146	231	56
Runtime (seconds)	18	518	17	210	4
Temporary constraints					
# constraints acquired ($|\mathcal{L}|$)	89	4565	86	176	106
# new constraints ($|\mathcal{L}\backslash\mathcal{K}|$)	6	4381	16	66	84
Number of queries	458	12227	778	6317	412
Runtime (minutes)	6.45	859	9.73	255.18	0.83

5.3 Result: Solving the TMS with Acquired Constraints

To evaluate our acquired constraints, we generated 100 different and diverse maintenance schedules with data related to the year 2020 using the CP model from Sect. 2 and the acquired constraints. The acquired constraints were added as new tuples of cumul functions and maximal values to the set $\mathcal{F}_1^c$ of Constraint 1 from Sect. 2. The 100 schedules were obtained with a multi-point restart strategy mechanism and 4 parallel workers [14]. As a baseline, we generated schedules using the same objective functions as Popovic et al. (2022) [20]: balancing the withdrawal requests inside the planning horizon and maximizing the number of overlapping withdrawals of the same equipment. We refer to Popovic et al. (2022) for an extended explanation about these heuristics. We set a total timeout of 1500 s for each restart. We monitored the number of days in the schedule (245 days in total) where the transit-power went below the allowed threshold for each interface. Specifically, we highlight that if multiple interfaces went below the threshold during the same day, we accumulate these has multiple days (e.g. if two interface went bellow their threshold the same day, we counted them has two days). This value is referred to as *exceeded days* (ED).

Table 3 shows the results for the 100 schedules generated, for the baseline model of Popovic et al. (2022) and ours with the acquired constraints (permanent, temporary, and both). Results reported are the number of times a feasible schedule was found, statistics on the number of exceeded days among the 100 schedules generated, and the execution time to build the 100 schedules. First, we observe that our approach has always been able to find at least one feasible schedule among the 100 executions, which was not the case for the baseline proposed by Popovic et al. (2022) [20]. The solution found by the baseline still has two exceeded days at best, and 30.30 days on average. In our case, having both the permanent and temporary constraints provides the best results: 40 schedules compliant with all the constraints, 7.40 exceeded days on average, and the lowest variance. The only downside is the execution time which increases from 8 h for the baselines to 15 h for our best model. However, this increase has no practical impact, as the schedule needs only to be built once per year.

Table 3. Results of 100 generated schedules on withdrawal requests for year 2020.

Approaches	Baseline [20]	Our approach with acquired constraints		
		Permanent	Temporary	Both
# feasible schedules (/100)	0	4	12	**40**
Minimum ED	2	**0**	**0**	**0**
Total ED	3030	2059	1205	**740**
Average ED	30.30	20.59	12.05	**7.40**
Standard deviation ED	13.57	14.11	10.59	**10.20**
Runtime (hours)	**8**	9	17	15

5.4   Analysis: Visualizing the Transit-Power on the Fifth Interface

Figure 3 presents the maintenance schedule with the transit-power curves for the fifth interface. The left plot shows the schedule obtained by Popovic et al. (2022) [20] and the right one illustrates the schedule we obtained with both permanent and temporary constraints. We can see that there are few conflicting days with the Popovic et al. (2022) and none with ours. Note that this interface was still unsolved by Hydro-Québec and by Popovic et al. (2022). This paper provides a solution to it for the first time.

5.5   Analysis: Distribution of the Exceeded Days

Figure 4 shows the distribution of the exceeded days depending on the approach considered. The distribution is computed using the sum of the exceeded days of the 100 schedules. Interestingly, the more constraints are added, the more the distribution shifts to the left, meaning that there are fewer exceeded days.

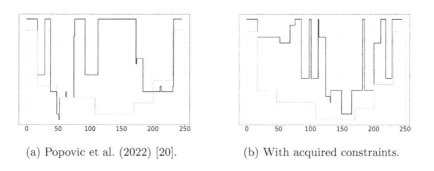

(a) Popovic et al. (2022) [20].   (b) With acquired constraints.

Fig. 3. Comparison of two schedules obtained for the fifth interface. The red curve is the transit-power computed by the simulator as a function of the proposed maintenance schedule and the current day. The blue curve represents the transit-power lower bound. The x-axis is the planning horizon, and the y-axis reports the related transit-power. The exact values are hidden for confidentiality purposes. (Color figure online)

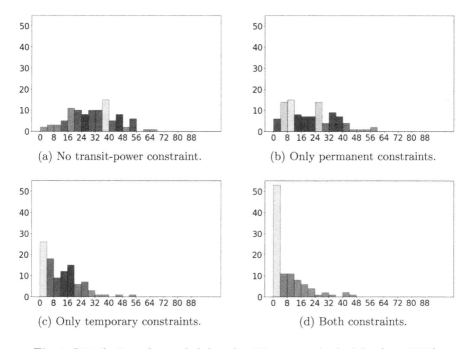

(a) No transit-power constraint.   (b) Only permanent constraints.

(c) Only temporary constraints.   (d) Both constraints.

Fig. 4. Distribution of exceeded days for 100 generated schedules (year 2020).

5.6   Analysis: Discussion of Limitations

The main limitation is the prior knowledge required to set the maximum arity of constraints desired for the acquisition (value m, which we set to 5 in our case). The size of the bias can grow significantly depending on this value, meaning that acquisition algorithms, such as QUACQ2 will take more time to remove unnecessary constraints. On the other hand, a value that is too low will prevent the acquisition of constraints involving more equipment, which will impact the quality of the solutions obtained. Note that background knowledge has been used to speed up acquisition. A second limitation is that our constraints are valid for a specific power-grid and cannot be easily transferred to other situations, e.g., another power-grid or a reshaping of the grid that changes the power loss of the withdrawn equipment. Although our methodology can be reused, new types of constraints considered for acquisition may need to be identified.

6   Conclusion

The transmission maintenance scheduling problem is a combinatorial problem that consists of creating an annual maintenance plan for electric power transmission equipment while preserving the stability of the network and ensuring uninterrupted power supply for customers. A challenging aspect of this problem is accommodating transit-power constraints. This paper proposes a methodology based on active constraint acquisition to automatically formalize these constraints in a combinatorial fashion. The acquired constraints are then added to a standard CP model to solve the problem. With this approach, we obtained a maintenance schedule complying with all the constraints for each strategic point of Hydro-Québec's power grid. Prior to our work, the most constrained strategic point on Hydro-Québec's grid was unsolved both by Popovic et al. (2022) and by experts from Hydro-Québec. Our methodology therefore provides a successful application of constraint acquisition. Further improvements could be made to increase the number of pieces of equipment considered during acquisition [28] and to include other types of constraints in the model (e.g., thermic constraints) through other learning paradigms, such as *empirical decision model learning* [15].

Acknowledgements. This work was supported by Mitacs IT33118 and the authors gratefully acknowledge the discussions with and input from Hydro-Québec researchers.

References

1. Alvarez, D.L., et al.: Optimal decision making in electrical systems using an asset risk management framework. Energies **14**(16), 4987 (2021). https://doi.org/10.3390/en14164987
2. Belaid, M.B., Belmecheri, N., Gotlieb, A., Lazaar, N., Spieker, H.: GEQCA: generic qualitative constraint acquisition. In: Proceedings of the AAAI Conference on Artificial Intelligence, vol. 36, no. 4, pp. 3690–3697 (2022). https://doi.org/10.1609/aaai.v36i4.20282
3. Beldiceanu, N., Ifrim, G., Lenoir, A., Simonis, H.: Describing and generating solutions for the EDF unit commitment problem with the ModelSeeker. In: Schulte, C. (ed.) CP 2013. LNCS, vol. 8124, pp. 733–748. Springer, Heidelberg (2013). https://doi.org/10.1007/978-3-642-40627-0_54
4. Beldiceanu, N., Simonis, H.: A model seeker: extracting global constraint models from positive examples. In: Milano, M. (ed.) CP 2012. LNCS, vol. 7514, pp. 141–157. Springer, Heidelberg (2012). https://doi.org/10.1007/978-3-642-33558-7_13

5. Bessiere, C., et al.: Learning constraints through partial queries. Artif. Intell. **319**, 103896 (2023). https://doi.org/10.1016/j.artint.2023.103896
6. Bessiere, C., Coletta, R., O'Sullivan, B., Paulin, M.: Query-driven constraint acquisition. In: IJCAI 2007: International Joint Conference on Artificial Intelligence, pp. 44–49 (2007)
7. Bessiere, C., Koriche, F., Lazaar, N., O'Sullivan, B.: Constraint acquisition. Artif. Intell. **244**, 315–342 (2017). https://doi.org/10.1016/j.artint.2015.08.001
8. Coffrin, C., Bent, R., Sundar, K., Ng, Y., Lubin, M.: Powermodels.jl: an open-source framework for exploring power flow formulations. In: 2018 Power Systems Computation Conference (PSCC), pp. 1–8. IEEE (2018)
9. Coffrin, C., Hijazi, H.L., Van Hentenryck, P.: The QC relaxation: a theoretical and computational study on optimal power flow. IEEE Trans. Power Syst. **31**(4), 3008–3018 (2015)
10. Coffrin, C., Van Hentenryck, P.: A linear-programming approximation of AC power flows. INFORMS J. Comput. **26**(4), 718–734 (2014). https://doi.org/10.1287/ijoc.2014.0594
11. Froger, A., Gendreau, M., Mendoza, J.E., Pinson, É., Rousseau, L.M.: Maintenance scheduling in the electricity industry: a literature review. Eur. J. Oper. Res. **251**(3), 695–706 (2016). https://doi.org/10.1016/j.ejor.2015.08.045
12. Gaha, M., et al.: Global methodology for electrical utilities maintenance assessment based on risk-informed decision making. Sustainability **13**(16), 9091 (2021)
13. Laborie, P., Rogerie, J.: Reasoning with conditional time-intervals. In: FLAIRS Conference, pp. 555–560 (2008)
14. Laborie, P., Rogerie, J., Shaw, P., Vilím, P.: IBM ILOG CP optimizer for scheduling. Constraints **23**(2), 210–250 (2018). https://doi.org/10.1007/s10601-018-9281-x
15. Lombardi, M., Milano, M., Bartolini, A.: Empirical decision model learning. Artif. Intell. **244**, 343–367 (2017). https://doi.org/10.1016/j.artint.2016.01.005
16. Mei, J., Zhang, G., Qi, D., Zhang, J.: Accelerated solution of the transmission maintenance schedule problem: a Bayesian optimization approach. Glob. Energy Interconnection **4**(5), 493–500 (2021). https://doi.org/10.1016/j.gloei.2021.11.001
17. Menguy, G., Bardin, S., Lazaar, N., Gotlieb, A.: Active disjunctive constraint acquisition. In: Proceedings of the 20th International Conference on Principles of Knowledge Representation and Reasoning, pp. 512–520 (2023). https://doi.org/10.24963/kr.2023/50
18. Mitsuhiro, N., Shigeyuki, T., Masao, N., Kouji, M.: Approach to asset management of substation equipment in Japan. In: 2020 International Symposium on Electrical Insulating Materials (ISEIM), pp. 68–71 (2020)
19. Pandzic, H., Conejo, A., Kuzle, I., Caro, E.: Yearly maintenance scheduling of transmission lines within a market environment. IEEE Trans. Power Syst. **27**, 407–415 (2012). https://doi.org/10.1109/TPWRS.2011.2159743
20. Popovic, L., Côté, A., Gaha, M., Nguewouo, F., Cappart, Q.: Scheduling the equipment maintenance of an electric power transmission network using constraint programming. In: Solnon, C. (ed.) 28th International Conference on Principles and Practice of Constraint Programming, CP 2022. LIPIcs, Haifa, Israel, 31 July–8 August 2022, vol. 235, pp. 34:1–34:15. Schloss Dagstuhl - Leibniz-Zentrum für Informatik (2022). https://doi.org/10.4230/LIPICS.CP.2022.34
21. Prestwich, S.: Robust constraint acquisition by sequential analysis. In: ECAI 2020, pp. 355–362. IOS Press (2020)
22. Prestwich, S.: Unsupervised constraint acquisition. In: 2021 IEEE 33rd International Conference on Tools with Artificial Intelligence (ICTAI), pp. 256–262 (2021). https://doi.org/10.1109/ICTAI52525.2021.00042

23. Quinlan, J.R.: C4.5: Programs for Machine Learning. Morgan Kaufmann Publishers Inc., San Francisco (1993)
24. Rahmaniani, R., Crainic, T.G., Gendreau, M., Rei, W.: The benders decomposition algorithm: a literature review. Eur. J. Oper. Res. **259**(3), 801–817 (2017). https://doi.org/10.1016/j.ejor.2016.12.005
25. Rocha, M., Anjos, M., Gendreau, M.: Optimal planning of preventive maintenance tasks on power transmission systems. Les Cahiers du GERAD (2022). ISSN: 0711-2440
26. Tanaka, M., Ishida, T.: Predicting and learning executability of composite web services. In: Bouguettaya, A., Krueger, I., Margaria, T. (eds.) ICSOC 2008. LNCS, vol. 5364, pp. 572–578. Springer, Heidelberg (2008). https://doi.org/10.1007/978-3-540-89652-4_48
27. Trudel, G., Gingras, J., Pierre, J.: Designing a reliable power system: hydro-Quebec's integrated approach. Proc. IEEE **93**, 907–917 (2005). https://doi.org/10.1109/JPROC.2005.846332
28. Tsouros, D.C., Berden, S., Guns, T.: Guided bottom-up interactive constraint acquisition. In: Yap, R.H.C. (ed.) 29th International Conference on Principles and Practice of Constraint Programming (CP 2023). Leibniz International Proceedings in Informatics (LIPIcs), vol. 280, pp. 36:1–36:20. Schloss Dagstuhl – Leibniz-Zentrum für Informatik, Dagstuhl (2023). https://doi.org/10.4230/LIPIcs.CP.2023.36
29. Tsouros, D.C., Stergiou, K.: Efficient multiple constraint acquisition. Constraints **25**(3–4), 180–225 (2020). https://doi.org/10.1007/s10601-020-09311-4

Efficiently Mining Closed Interval Patterns with Constraint Programming

Djawad Bekkoucha[✉], Abdelkader Ouali, Patrice Boizumault,
and Bruno Crémilleux

Normandie Univ, UNICAEN, ENSICAEN, CNRS, GREYC, Caen, France
{djawad.bekkoucha,abdelkader.ouali,patrice.boizumault,
bruno.cremilleux}@unicaen.fr

Abstract. Constraint programming (CP) has become increasingly prevalent in recent years for performing pattern mining tasks, particularly on binary datasets. While numerous CP models have been designed for mining on binary data, there does not exist any model designed for mining on numerical datasets. Therefore these kinds of datasets need to be pre-processed to fit the existing methods. Afterward a post-processing is also required to recover the patterns into a numerical format. This paper presents two CP approaches for mining closed interval patterns directly from numerical data. Our proposed models seamlessly execute pattern mining tasks without any loss of information or the need for pre- or post-processing steps. Experiments conducted on different numerical datasets demonstrate the effectiveness of our proposed CP models compared to other methods.

Keywords: Constraint Programming · Pattern Mining · Numerical Data

1 Introduction

Pattern mining is a crucial task in data mining. The constraint-based pattern mining task [15] embraces a wide range of queries and methods aimed at extracting characterized patterns from data. These patterns can be interpreted by domain experts or serve as inputs for downstream tasks such as classification or clustering [5].

Datasets commonly encountered in many application domains often comprise a variety of value types, including binary, discrete, numerical or mixed values. There are a plethora of pattern mining methods on binary data, sequences or graphs [15] but few methods address numerical [8,18] or mixed values [4]. A simple approach to cope with numerical data is to reuse existing methods by first converting data into a binary representation [6]. However, it is well-known that the binarization process often leads to a loss of information. Considering the example of interval patterns defined by numerical values, MININTCHANGE [8] is a method aiming to mine such patterns directly from numerical data. However it is devoted to closed interval patterns and cannot deal with other data mining tasks.

B. Dilkina (Ed.): CPAIOR 2024, LNCS 14742, pp. 51–67, 2024.
https://doi.org/10.1007/978-3-031-60597-0_4

The aim of our work is to propose a comprehensive declarative framework to efficiently tackle numerical data. Our work is guided by two principles. First, we keep the whole original information expressed by the numerical data. Second, we choose to set our approach in a declarative paradigm, the Constraint Programming (CP) framework. The declarative paradigm enables to specify pattern mining tasks in an easy way by using general constraint primitives. In the last few years, many declarative constraint-based methods were proposed for mining tasks on binary data or sequences [5,9,16], to the best of our knowledge, so far, there is no declarative method tackling numerical datasets. The core of our contribution is a declarative method to mine the complete set of closed interval patterns from numerical data without any discretization process. A closed pattern is essential in data mining tasks as it captures the maximum amount of similarity within a dataset. These patterns have crucial properties and are widely used in data mining applications. Moreover the whole set of interval patterns can be regenerated following the principle of the pattern condensed representations [2]. Our CP models go beyond interval patterns by combining primitives to write queries addressing complex mining tasks on numerical data. As an illustration, we combine the no overlapping between patterns and the cover of the set of the mined patterns to perform the conceptual clustering task.

Our contributions are the following. We define two CP models to mine the complete set of closed interval patterns. The first model, called CP4CIP, is based on reified constraints and offers a general encoding of the closure relation to mine closed patterns. In CP, global constraints can capture hidden relations between a set of variables to improve the efficiency. We design a global constraint, called GC4CIP, using new specific filtering rules, and we give a second CP model enhancing the mining efficiency. We provide an extensive empirical evaluation comparing the efficiency of our CP models with respect to other declarative methods and the ad hoc method MinIntChange. Finally, we illustrate the interest of the declarativity by using GC4CIP in a scenario of data processing chain to find conceptual clustering from numerical data.

Section 2 introduces notations and basic concepts. Section 3 is devoted to related work. Section 4 presents our first CP model based on reified constraints. Section 5 and Sect. 6 depicts our second CP model based on global constraints. Section 7 provides an extensive empirical evaluation on benchmark datasets and Sect. 8 concludes.

2   Preliminaries

2.1   Interval Pattern Mining

Numerical Dataset. A *numerical dataset* $\mathcal{N}$ is defined by a set of objects $\mathcal{G}$ where each object is described by a set of attributes $\mathcal{M}$. Each attribute $m \in \mathcal{M}$ has a range $\mathcal{N}_m$ which is a finite set containing all the possible values of m in $\mathcal{N}$. An object $g \in \mathcal{G}$ is defined by a vector of numerical values $< v_{g,m} >_{m \in \{1,\ldots,|\mathcal{M}|\}}$. A dataset where the values of all attributes are binary $\mathcal{N}_m = \{0,1\}, \forall m \in \mathcal{M}$, is a special case of a numerical dataset and referred as a *binary dataset*.

Table 1. A running example of a numerical dataset $\mathcal{N}$

	m_1	m_2	m_3
g_1	2	8	130
g_2	4	12	102
g_3	3	7	91
g_4	2	9	101
g_5	6	12	110

Example 1. Table 1 shows a running example of a numerical dataset containing 5 objects $\mathcal{G} = \{g_1, g_2, g_3, g_4, g_5\}$, each object is described by 3 attributes $\mathcal{M} = \{m_1, m_2, m_3\}$.

Closed Interval Pattern. Patterns in numerical datasets can be represented in many ways, we use the notion of Interval Pattern [8] which is defined as a vector of intervals $\mathcal{V} = \langle [\underline{w_m}, \overline{w_m}] \rangle_{\forall m \in \mathcal{M}}$, where $\underline{w_m}, \overline{w_m} \in \mathcal{N}_m$. Each dimension of the vector $\mathcal{V}$ corresponds to an attribute following a canonical order on the set of attributes $\mathcal{M}$. We denote $\mathcal{B}[g] = \langle [v_{g,m}, v_{g,m}] \rangle_{m \in \{1,...,|\mathcal{M}|\}}$ as the vector of intervals corresponding to an object identified by g. An object g is an occurrence of the interval pattern $\mathcal{V}$ if each interval in the vector $\mathcal{B}[g]$ is included in the interval of $\mathcal{V}$, i.e. $\mathcal{B}[g] \sqsubseteq \mathcal{V} \iff [v_{g,m}, v_{g,m}] \subseteq [\underline{w_m}, \overline{w_m}], \forall m \in \{1, ..., |\mathcal{M}|\}$. The cover of $\mathcal{V}$ in $\mathcal{N}$ is the set of objects $g \in \mathcal{G}$ occurring in $\mathcal{V}$, i.e. $cover(\mathcal{V}) = \{g \in \mathcal{G} \mid \mathcal{B}[g] \sqsubseteq \mathcal{V}\}$.

Example 2. From Table 1, $\mathcal{V} = \langle [3,4], [7,12], [91,130] \rangle$ is an interval pattern covering the objects $\{g_2, g_3\}$. $\mathcal{B}[g_2] = \langle [4,4], [12,12], [102,102] \rangle$ is the vector of intervals identified by the object g_2 and an occurrence of $\mathcal{V}$.

The frequency of $\mathcal{V}$ is the cardinal of its cover, i.e. $freq(\mathcal{V}) = |cover(\mathcal{V})|$. Given a minimum frequency threshold θ, the interval pattern $\mathcal{V}$ is frequent if and only if $freq(\mathcal{V}) \geq \theta$. A description of a subset of objects $G \subseteq \mathcal{G}$ is an interval pattern $\mathcal{V}$ where for each $g \in G$, g is an occurrence of $\mathcal{V}$, i.e. $desc(G) = < [a_m, b_m] >_{m \in \{1,...,|\mathcal{M}|\}}$ such that $a_m = min(\{v_{g,m} \mid g \in G\})$ and $b_m = max(\{v_{g,m} \mid g \in G\})$.

Exact pattern condensed representations (such as the *closed patterns* [19]) enable to reduce the large number of patterns extracted from the datasets without loss of information [2]. The key idea of pattern condensed representations is to take advantage of the redundancy of a collection of patterns to construct a concise representation of the patterns instead of mining all patterns. A closed interval pattern (CIP) is defined by the closure of an interval pattern that is the vector of intervals of its cover (i.e. $close(\mathcal{V}) \iff desc(cover(\mathcal{V})) = \mathcal{V}$).

Example 3. From Table 1, the set of objects $\{g_2, g_3\}$ is described by the interval pattern $\mathcal{V} = desc(\{g_2, g_3\}) = \langle [3,4], [7,12], [91,102] \rangle$. The interval pattern $\mathcal{V} = \langle [3,4], [7,12], [91,102] \rangle$ is closed since $desc(\{g_2, g_3\}) = \langle [3,4], [7,12], [91,102] \rangle$ and the $cover(\langle [3,4], [7,12], [91,102] \rangle) = \{g_2, g_3\}$.

2.2   Problem Statement

Our goal is to discover all frequent closed interval patterns (FCIP). More formally, given a numerical dataset $\mathcal{N}$ and a minimum frequency threshold θ, the closed frequent interval pattern mining problem is the problem of finding *all* interval patterns $\mathcal{V}$ such that $freq(\mathcal{V}) \geq \theta$ and $close(\mathcal{V})$.

There are a few things one should note about this statement. First, the search space contains $\sum_{k=1}^{|\mathcal{G}|} \binom{|\mathcal{G}|}{k}$ candidates. This size is gigantic and a naive search that consists of enumerating and testing the frequency of all interval pattern candidates is not practical.

Second, mining FCIP can be solved by the following process: transform the numerical data using Interordinal Scaling (IS) technique to get binary data, mine closed itemsets from the binary data and post-processing closed itemsets to obtain closed interval patterns [8]. Results are equivalent because IS preserve all the information of the original data by creating pairs of binary attributes of the following form: $m \leq v_{g,m}$ and $m \geq v_{g,m}, \forall m \in \mathcal{M}, g \in \mathcal{G}$. Each element of these pairs is then used as a binary attribute on the set of objects. The value of the attribute on each object is set to 1 if the condition holds, otherwise the value is set to 0. However, this approach produces a large dataset having $\sum_{m \in \mathcal{M}} 2|\mathcal{N}_m|$ items. Moreover, the post-processing step is highly expensive. For each itemset found, it needs to determine the interval of each attribute by calculating the minimum and the maximum values that are present in the itemset. The time complexity of the post-processing is in the worst case $\mathcal{O}(\mathcal{C} \cdot |\mathcal{M}| \cdot \mathcal{U})$ where $\mathcal{C}$ is the total number of mined closed itemsets and $\mathcal{U}$ is the number of distinct values for each attribute.

Finally, using the declarative paradigm easily enables us to combine CIP with other constraints such as the overlapping or the cover of the set of returned patterns [3,7,9]. We illustrate the use of these constraints with our CP models in Sect. 7.3.

2.3   Constraint Programming

A CSP consists of a set of variables $X = \{x_1, \ldots, x_n\}$, a set of domains $\mathcal{D}$ mapping each variable $x_i \in X$ to a finite set of possible values $\mathcal{D}(x_i)$, and a set of constraints $\mathcal{C}$ on X. A constraint $c \in \mathcal{C}$ is a relation specifying the allowed combinations of values for its variables $X(c)$. An assignment on a set $Y \subseteq X$ of variables is a mapping from variables in Y to values in their domains. A solution is an assignment on X satisfying all the constraints. CP solvers typically use backtracking tree search to explore the search space of partial assignments and attempts to extend them to consistent ones with the objective of finding solutions. The main technique used to speed-up the search is the constraint propagation by a filtering algorithm. Each constraint filtering should remove as many variable domain values as possible by enforcing local consistency properties like *domain* or *bound consistency*. Global constraints are constraints capturing a relation between a fixed number of variables. These constraints provide the solver with a better view of the structure of the problem. Dedicated filtering algorithms are designed to achieve better time complexity.

3   Related Work

Mining patterns in numerical data started with quantitative association rule mining [18]. A lot of work is discussed in [17]. Many of them are based on a natural approach where each numerical attribute is discretized according to some interest functions, e.g. support, class values. Then patterns are mined from the discretized data. This family of approaches leads to a loss of information. More recently, in the field of subgroup discovery, a very common kind of patterns in modern pattern mining, Nguyen et al. [14] creates a discretization process with the goal to maximize the average quality of the patterns. [12] provides a thorough comparison of existing methods to deal with numerical attributes in subgroups. By considering the notion of closed interval patterns, OSMIND [13] finds optimal subgroups according to an interestingness measure in purely numerical data. Our work takes advantage of the principle of the closed interval patterns and it is not limited to subgroups.

Approaches based on Minimum Description Length are used for discovering useful patterns and returning a set of non-redundant overlapping patterns with well-defined boundaries [11,20]. In order to design relevant intervals on the fly based on numerical data, MinIntChange [8] introduces a framework that enumerates all closed interval patterns starting with the largest one, then explores the search space through minimal changes on the interval bounds. The principle has been reused to search for patterns corresponding to convex polygons [1] but the technique is limited to two dimensions. All these methods are dedicated to specific patterns and require to rewrite algorithms when the problem at hands changes.

There are many declarative methods for constraint-based mining tasks under declarative frameworks for binary data or sequences [5,9,16]. CP-based approaches have been proposed to mine closed itemsets in a binary context as CP4IM [16] by using reified constraints or the global constraint closedPattern [10]. However, to the best of our knowledge, there is no declarative method to discover patterns directly from numerical data without requiring pre and post processing steps.

4   First Model Using Reified Constraints

This section starts by presenting the variables used to model interval patterns and their associated domains. Then, we describe our first model named CP4CIP.

4.1   Variables and Domains

The bounds of an interval pattern are modelled by introducing two variables $\underline{x}_m$, $\overline{x}_m$ for each attribute $m \in \mathcal{M}$. These variables represent respectively the lower bound and the upper bound. The domains of these variables is the set of values $\mathcal{N}_m$ in the dataset, i.e. $\mathcal{D}(\underline{x}_m) = \mathcal{D}(\overline{x}_m) = \mathcal{N}_m$.

Additionally, we introduce another set of variables, denoted by Y, to capture the coverage of an object by an interval pattern. The variable $y_g \in Y$ is associated to the object $g \in \mathcal{G}$ and has a binary domain, i.e. $\mathcal{D}(y_g) = \{0, 1\}$. The variable y_g takes the value 1 if and only if the object g is covered by the candidate interval pattern. A FCIP is found by setting the variables $\underline{x}_m$ and $\overline{x}_m$ to a value in $\mathcal{N}_m$ and y_g to 0 or 1.

Example 4. From Table 1, the interval pattern $\langle [3, 4], [7, 12], [91, 102] \rangle$ is modelled by the following assignment $\{\underline{x}_1 = 3, \overline{x}_1 = 4, \underline{x}_2 = 7, \overline{x}_2 = 12, \underline{x}_3 = 91, \overline{x}_3 = 102, y_1 = 0, y_2 = 1, y_3 = 1, y_4 = 1, y_5 = 0\}$.

4.2   Reified Constraints

Coverage Constraints. An object is covered by a FCIP iff all values of its attributes are found in the intervals. To formulate the cover of a FCIP, we introduce in our model Boolean variables $B_{g,m}$, for each value m and each object g in the database such that:

$$B_{g,m} = 1 \iff \underline{x}_m \leq v_{g,m} \leq \overline{x}_m, \ \forall m \in \mathcal{M}, \forall g \in \mathcal{G} \tag{1}$$

The following constraint exploits the variables $B_{g,m}$ to enforce the cover on each object. An object $g \in \mathcal{G}$ is covered iff the value of each attribute m for the object g is bounded by the interval $[\underline{x}_m, \overline{x}_m]$. We can thus process the frequency of the interval pattern by summing the y_g variables such that the sum must be greater or equal than a minimum support θ (i.e. $\sum_{g \in \mathcal{G}} yg \geq \theta$).

$$y_g = 1 \iff \sum_{m \in \mathcal{M}} B_{g,m} = |\mathcal{M}|, \ \forall g \in \mathcal{G} \tag{2}$$

Proof. Let $\mathcal{V}$ be a candidate interval pattern. The variable y_g models the cover of object $g \in \mathcal{G}$ by $\mathcal{V}$. The cover of $\mathcal{V}$ is given by the following: $y_g = 1 \iff \underline{x}_m \leq v_{g,m} \leq \overline{x}_m, \forall \, m \in \mathcal{M}$. Since, $B_{g,m} = 1 \iff \underline{x}_m \leq v_{g,m} \leq \overline{x}_m$. So, $y_g = 1 \iff B_{g,m} = 1, \forall m \in \mathcal{M}$. It follows that $y_g = 1 \iff \sum_{m \in \mathcal{M}} B_{g,m} = |\mathcal{M}|$.

Closure Constraints. The closure requires that each interval associated to each attribute should contain all the values of the covered objects while each value of an uncovered object should be outside of the interval. Let $\mathcal{N}_m^\uparrow$ (resp. $\mathcal{N}_m^\downarrow$) be the maximum (resp. the minimum) value over the objects on the attribute m, the closure relation can be expressed in our model by introducing the new variables $\underline{H}_{g,m}$ and $\overline{H}_{g,m}$, where $\mathcal{D}(\underline{H}_{g,m}) = \{v_{g,m}\} \cup \{\mathcal{N}_m^\uparrow + 1\}$, and $\mathcal{D}(\overline{H}_{g,m}) = \{v_{g,m}\} \cup \{\mathcal{N}_m^\downarrow - 1\}$. The upper value $\{\mathcal{N}_m^\uparrow + 1\}$ (resp. lower value $\{\mathcal{N}_m^\downarrow - 1\}$) is added in the domain of $\underline{H}_{g,m}$ (resp. $\overline{H}_{g,m}$) in order to avoid selecting the minimum (resp. maximum) on the uncovered objects.

Lower Bound Closure. To capture the minimum value of covered objects, we use for each attribute $m \in \mathcal{M}$ a minimum constraint on the set variables $\{\underline{H}_{g,m}, \forall g \in \mathcal{G}\}$. The variables $\underline{H}_{g,m}$ of uncovered objects have value greater than all the values in the data. Thus, the minimum cannot be selected on uncovered objects.

$$\forall g \in \mathcal{G}, m \in \mathcal{M}, y_g = 1 \implies \underline{H}_{g,m} = v_{g,m} \quad (3)$$

$$\forall g \in \mathcal{G}, m \in \mathcal{M}, y_g = 1 \implies \overline{H}_{g,m} = v_{g,m} \quad (6)$$

$$\forall g \in \mathcal{G}, m \in \mathcal{M}, y_g = 0 \implies \underline{H}_{g,m} = \mathcal{N}_m^\uparrow + 1 \quad (4)$$

$$\forall g \in \mathcal{G}, m \in \mathcal{M}, y_g = 0 \implies \overline{H}_{g,m} = \mathcal{N}_m^\downarrow - 1 \quad (7)$$

$$\forall m \in \mathcal{M}, \underline{x}_m = min(\underline{H}_{1,m}, \underline{H}_{2,m}, ..., \underline{H}_{|\mathcal{G}|,m}) \quad (5)$$

$$\forall m \in \mathcal{M}, \overline{x}_m = max(\overline{H}_{1,m}, \overline{H}_{2,m}, ..., \overline{H}_{|\mathcal{G}|,m}) \quad (8)$$

Upper Bound Closure. Similarly, to find the maximum value on the covered objects, we use for each attribute m a maximum constraint on the set of variables $\{\overline{H}_{g,m}, \forall g \in \mathcal{G}\}$. The variables $\overline{H}_{g,m}$ of uncovered objects have value smaller than all the values in the data. Thus, the maximum cannot be selected on uncovered objects.

Example 5. Table 2 shows the values taken by the closure variables $\underline{H}_{g,m}$ (left) and $\overline{H}_{g,m}$ (right) to find the closed interval pattern $\langle [3,4], [7,12], [91,102] \rangle$.

Proof. Consider the subset G of objects covered by the interval pattern $\langle [a_m, b_m] \rangle_{\forall m \in \mathcal{M}}$. According to the closure definition, the lower bound can be expressed as $\forall m \in \mathcal{M}, a_m = min(\{v_{g,m} \mid g \in G\})$. This is equivalent to $\forall m \in \mathcal{M}, a_m = min(\{v_{g,m} \mid \forall g \in \mathcal{G}, y_g = 1\})$. Consequently, we deduce that $\forall m \in \mathcal{M}, a_m = min(\{v \mid \forall g \in \mathcal{G}, (y_g = 1 \implies v = v_{g,m}) \wedge (y_g = 0 \implies v = \mathcal{N}_m^\uparrow + 1)\})$. The proof of the upper bound is similar to lower bound.

Table 2. Values of closure variables $\underline{H}_{g,m}$ and $\overline{H}_{g,m}$ for the running example.

	y_g	$\underline{H}_{g,1}$	$\underline{H}_{g,2}$	$\underline{H}_{g,3}$		y_g	$\overline{H}_{g,1}$	$\overline{H}_{g,2}$	$\overline{H}_{g,3}$
g_1	0	7	13	131	g_1	0	1	6	90
g_2	1	4	12	102	g_2	1	4	12	102
g_3	1	3	7	91	g_3	1	3	7	91
g_4	0	7	13	131	g_4	0	1	6	90
g_5	0	7	13	131	g_5	0	1	6	90
min:		3	7	91	max:		4	12	102

CP4CIP Model. FCIP mining can be modeled by the conjunction of the coverage constraints (cf. Eq. 1 and Eq. 2) and the closure constraints (Eqs. 3 to 8). This model uses $2.|\mathcal{M}|$ variables for interval representation, $|\mathcal{G}|$ variables for objects coverage, and $3.|\mathcal{G}|.|\mathcal{M}|$ variables. It involves $|\mathcal{G}|.|\mathcal{M}|$ inclusion constraints, $|\mathcal{G}|$ coverage constraints, and $4.|\mathcal{G}|.|\mathcal{M}| + 2.|\mathcal{M}|$ closure constraints. The main limitation of such a model lies in its number of variables and constraints leading to scaling challenges on large datasets.

5   Second Model Using a Global Constraint

Similarly to Sect. 4.1, this second model uses the sets of variables $\underline{X}$ and $\overline{X}$ to represent the lower bounds and the upper bounds of an interval pattern. The

$GC4CIP_{\mathcal{N},\theta}(\underline{X}, \overline{X})$ global constraint holds if and only if $\mathcal{V}$ is closed, i.e. $close(\mathcal{V})$ and $\mathcal{V}$ is frequent, i.e. $freq(\mathcal{V})) \geq \theta$. In the following, we describe the new specific filtering rules associated to closure and frequency of an interval pattern.

5.1   Closure Filtering Rules

Let $\mathcal{V}^* = \langle [\min(\mathcal{D}(\underline{x}_1)), \max(\mathcal{D}(\overline{x}_1))], \ldots, [\min(\mathcal{D}(\underline{x}_{|\mathcal{M}|})), \max(\mathcal{D}(\overline{x}_{|\mathcal{M}|}))] \rangle$ be the largest interval pattern from the domains. Proposition 1 states that values occurring only in objects non covered by $\mathcal{V}^*$ must be removed.

Proposition 1. *Let* $\mathcal{V}^* = \langle [\min(\mathcal{D}(\underline{x}_1)), \max(\mathcal{D}(\overline{x}_1))], \ldots, [\min(\mathcal{D}(\underline{x}_{|\mathcal{M}|})), \max(\mathcal{D}(\overline{x}_{|\mathcal{M}|}))] \rangle$. *Let* $m \in \mathcal{M}, g \in \mathcal{G},$

$$\begin{cases} v_{g,m} \notin \mathcal{D}(\underline{x}_m), \\ v_{g,m} \notin \mathcal{D}(\overline{x}_m) \end{cases} if : \begin{cases} \exists\, m' \in \mathcal{M}, m \neq m', v_{g,m'} < \min(\mathcal{D}(\underline{x}_{m'})) \lor v_{g,m'} > \max(\mathcal{D}(\overline{x}_{m'})) \\ \land \\ \forall g' \in \mathcal{G},\ g \neq g'\ such\ that\ g'\ is\ covered\ by\ \mathcal{V}^*,\ v_{g,m} \neq v_{g',m} \end{cases} \tag{9}$$

Proof. Let $m, m' \in \mathcal{M}$ and $m \neq m'$. Suppose that $v_{g,m} \in \mathcal{D}(\underline{x}_m)$ and $v_{g,m'} < \min(\mathcal{D}(\underline{x}_{m'}))$ or $v_{g,m'} > \max(\mathcal{D}(\overline{x}_{m'}))$. If $v_{g,m'} < \min(\mathcal{D}(\underline{x}_{m'})) \lor v_{g,m'} > \max(\mathcal{D}(\overline{x}_{m'}))$, this means that g is not covered by $\mathcal{V}^*$, i.e. $g \notin cover(\mathcal{V}^*)$. Thus following the closed interval definition in Sect. 2.1, $\underline{x}_m = min(\{v_{g',m} \mid g' \in cover(\mathcal{V}^*)\})$. If there does not exist $g' \in \mathcal{G}$ where g' is covered and $v_{g',m} = v_{g,m}$, then $v_{g,m}$ will never be a bound in $\mathcal{V}^*$, therefore $v_{g,m} \notin \mathcal{D}(\underline{x}_m)$ which contradicts the assumption. The proof for the upper bound is similar.

Example 6. Consider the dataset in Table 1, and the following domains: $\mathcal{D}(\underline{x}_1) = \{4, 6\}$, $\mathcal{D}(\overline{x}_1) = \{4, 6\}$, $\mathcal{D}(\underline{x}_2) = \{7, 8, 9, 12\}$, $\mathcal{D}(\overline{x}_2) = \{7, 8, 9, 12\}$, $\mathcal{D}(\underline{x}_3) = \{91, 101, 102, 130\}$, $\mathcal{D}(\overline{x}_3) = \{91, 101, 102, 130\}$. Since the values 2 and 3 for the attribute m1 are not in $\mathcal{D}(\underline{x}_1)$ and $\mathcal{D}(\overline{x}_1)$, the objects g1, g3 and g4 are not covered. Therefore, the values 7, 8, 9 will be removed from $\mathcal{D}(\underline{x}_2)$ and $\mathcal{D}(\overline{x}_2)$ because 7 appears in g_3, 8 appears in g_1 and 9 in g_4, and these values do not occur in any covered object.

For defining the next filtering rule, we first consider the joint between the domains of two attributes m and m'. For each value $v_{g,m}$ in $D(x_m)$, we consider the objects g having such a value for attribute m, i.e. $I_m = \{g \mid g \in \mathcal{G}, v_{g,m} \in D(x_m)\}$. Then, for each object g in I_m, we determine its value for the attribute m' So, $join(x_{m'}, x_m) = \{v_{g,m'} \mid v_{g,m'} \in D(x_{m'}), g \in I_m\}$. Then, every value outside the bounds of this set has to be removed.

Proposition 2. *Let* $m, m' \in \mathcal{M}, m \neq m'$. *For simplicity, we denote by* x_m *as either a lower bound* $\underline{x}_m$ *or an upper bound* $\overline{x}_m$.

$$\begin{cases} v_{g,m} \notin \mathcal{D}(\underline{x}_m)\ if: & v_{g,m} > max(join(x_{m'}, \underline{x}_m)) \\ v_{g,m} \notin \mathcal{D}(\overline{x}_m)\ if: & v_{g,m} < min(join(x_{m'}, \overline{x}_m)) \end{cases} \tag{10}$$

Proof. Let $m, m' \in \mathcal{M}$ and $m \neq m'$. Suppose that $v_{g,m} \in \mathcal{D}(\underline{x}_m)$ and $v_{g,m} > \max(join(\underline{x}_{m'}, \overline{x}_m))$. $\mathcal{D}(\underline{x}_{m'})$ contains all possible lower bounds for the attribute m', therefore $join(\underline{x}_{m'}, \overline{x}_m)$ returns all the values which are potential lower bounds for the attribute m. If $v_{g,m} > \max(join(\underline{x}_{m'}, \overline{x}_m))$, there exists an object g which is either uncovered or is strictly inside the closed intervals. In both cases, all the values occurring in object g are not located in the closure's lower border. Consequently, the $v_{g,m} \notin \mathcal{D}(\underline{x}_m)$ which contradicts the assumption. The proof for the upper bound is similar.

Example 7. Consider the database in Table 1 and the following domains: $\mathcal{D}(\underline{x}_1) = \{2, 3, 4, 6\}$, $\mathcal{D}(\overline{x}_1) = \{4, 6\}$, $\mathcal{D}(\underline{x}_2) = \{7, 8, 9, 12\}$, $\mathcal{D}(\overline{x}_2) = \{7, 8, 9, 12\}$, $\mathcal{D}(\underline{x}_3) = \{91, 101, 102, 110, 130\}$, $\mathcal{D}(\overline{x}_3) = \{91, 101, 102, 110\}$. The objects associated to $\mathcal{D}(\underline{x}_1)$ are g_4 and g_6. The values of these objects for attribute m_3 are 102 and 110. So $join(\overline{x}_1, \underline{x}_3) = \{102, 110\}$. In the same way, $join(\overline{x}_1, \overline{x}_3) = \{102, 110\}$. Following the filtering rules in 10, value 130 will be removed from $\mathcal{D}(\underline{x}_3)$, because $130 > max(join(\overline{x}_1, \underline{x}_3))$ and values 91, 101 will be removed from $\mathcal{D}(\overline{x}_3)$ since 91 and 101 are smaller than $min(join(\overline{x}_1, \overline{x}_3))$.

5.2   Frequency Filtering Rules

Values not occurring in any frequent interval pattern have to be removed.

Proposition 3. *Let $m \in \mathcal{M}$, and $\mathcal{V}^p = \langle [\min(\mathcal{D}(\underline{x}_i)), \max(\mathcal{D}(\overline{x}_i))] \rangle_{1 \leq i \neq m \leq |\mathcal{M}|}$ be a partial interval pattern.*

$$\begin{cases} a_m \notin \mathcal{D}(\underline{x}_m) \ if: \ freq(\mathcal{V}^p \ +\!\!+ \ [a_m, \max(\mathcal{D}(\overline{x}_m))]) < \theta \\ b_m \notin \mathcal{D}(\overline{x}_m) \ if: \ freq(\mathcal{V}^p \ +\!\!+ \ [\min(\mathcal{D}(\underline{x}_m)), b_m]) < \theta \end{cases} \tag{11}$$

Proof. We prove that $a_m \notin \mathcal{D}(\underline{x}_m)$ if the candidate interval on the attribute m using the maximum value as an upper bound will always lead to a frequency less than θ. Suppose that $a_m \in \mathcal{D}(\underline{x}_m)$. We know that $cover(\mathcal{V}^p) \cap cover([a_m, b_m]) = cover(\mathcal{V}^p +\!\!+ [a_m, b_m])$. Since $\mathcal{V}^p$ is a consistent partial solution, $freq(\mathcal{V}^p) \geq \theta$. This means that the frequency of the candidate interval pattern $(\mathcal{V}^p +\!\!+ [a_m, b_m])$ is determined only by $freq([a_m, b_m])$. Consequently, if $freq([a_m, \max(\mathcal{D}(\underline{x}_m))]) < \theta$, then the value a_m as a lower bound will always lead to infrequent candidate interval pattern with current domains. Thus $a_m \notin \mathcal{D}(\underline{x}_m)$ which contradicts the assumption. The proof establishing the inconsistency of the value b_m in the upper bound domain is similar.

Example 8. Consider the dataset in Table 1, a minimum frequency threshold $\theta = 2$, and the following variable domains: $\mathcal{D}(\underline{x}_1) = \{2\}$, $\mathcal{D}(\overline{x}_1) = \{3\}$, $\mathcal{D}(\underline{x}_2) = \{8, 9\}$, $\mathcal{D}(\overline{x}_2) = \{8, 9, 12\}$, $\mathcal{D}(\underline{x}_3) = \{91, 101, 102, 130\}$, $\mathcal{D}(\overline{x}_3) = \{91, 101, 102, 130\}$. Following the frequency filtering rule, and considering the partial assignement $\mathcal{V}^p = \langle [2, 3], [91, 130] \rangle$, the values 8 and 9 will be removed from $\mathcal{D}(\underline{x}_2)$ and $\mathcal{D}(\overline{x}_2)$ respectively, since $freq(\mathcal{V}^p +\!\!+ [9, max(\mathcal{D}(\overline{x}_2))]) < 2$ and $freq(\mathcal{V}^p +\!\!+ [min(\mathcal{D}(\underline{x}_2)), 8]) < 2$ (Eq. 11)

GC4CIP Model. Our global constraint $\text{GC4CIP}_{\mathcal{N}, \theta}(\underline{X}, \overline{X})$ can be used with additional constraints to handle more complex mining tasks (see Sect. 7.3).

Algorithm 1: GC4CIP coverage update

---

1 **Function** $PushDown(T: tree, depth, n: current node, a: value in \mathcal{N}_m, b: value in \mathcal{N}_m, Cov: coverage array)$

2   **if** $depth = maxdepth(T) \;\wedge\; (inf(n) < a \;\vee\; sup(n) > b)$ **then**

3       $inf(n) \longleftarrow +\infty; \quad sup(n) \longleftarrow -\infty;$

4   Cov[ObjectIndexFromLeaf(n)] $\longleftarrow 0;$

5   **else**

6   **foreach** $c \in child(n)$ **do**

7   **if** $inf(c) < a \;\wedge\; sup(c) > b$ **then**

8   PushDown(T, c, a, b, depth+1)

9   **else if** $inf(c) < a$ **then**

10   PushDown(T, c, a, $sup(c)$, depth+1)

11   **else if** $sup(c) > b$ **then**

12   PushDown(T, c, $inf(c)$, b, depth+1)

---

6 GC4CIP Filtering Algorithm

In this section, we introduce our algorithm designed to implement the filtering rules outlined in Sect. 5, ensuring the domain consistency of the GC4CIP constraint. Our filtering algorithm leverages an internal data structure to certify candidate solutions and optimize the filtering process.

A Specific Data Structure. Consider a tree represented as $T = (N, E, r, inf, sup)$ where N is the set of nodes in T. Each node $n \in N$ contains an interval $[a_n, b_n]$ with a_n and b_n being values returned by the operators $inf(n)$ and $sup(n)$ respectively. The root node of T is r, and E is the set of edges in the tree. The collection of such trees is called forest and denoted as F. In our algorithm, this forest is represented as a list of trees, where each tree is associated with an attribute $m \in \mathcal{M}$ from the dataset. Additionally, we introduce a Boolean array, denoted as Cov, designed to compute the current coverage of the set of objects (i.e. leaf nodes). For $g \in \mathcal{G}$, $Cov[g]$ is set to 1 if g can be covered, otherwise it is set to 0. The synchronization of trees within the forest is maintained through the coverage array.

Initially, for each attribute m and object g in the dataset, a leaf node $n \in N$ is created, such that $n = [v_{g,m}, v_{g,m}]$ where $v_{g,m} \in \mathcal{N}_m$. A parent node p is created over a subset of leaf nodes, where $inf(p)$ is set to the minimum a_n of its children and $sup(p)$ is set to the maximum b_n of its children. The interval $[a_r, b_r]$ associated with the root node of each tree establishes coherent bounds for the variables corresponding to the attribute $m \in \mathcal{M}$ (i.e. $\underline{x}_m$ and $\overline{x}_m$).

To maintain the trees and the coverage array with respect to changes in the variables domains, we introduce a function $PushDown()$, see Algorithm 1. The $PushDown$ function allows to update leaves nodes having their intervals (i.e. $[v_{g,m}, v_{g,m}]$) not included in the interval of the root node, lines 3 and 12. If this condition holds, the interval of the leaf node is set to $[+\infty, -\infty]$, making the corresponding object uncovered.

Algorithm 2: GC4CIP closure filtering

1 **Input:** $\underline{X}_{\mathcal{M}} = \{\underline{x}_1, \underline{x}_2, ..., \underline{x}_{|\mathcal{M}|}\}$, $\overline{X}_{\mathcal{M}} = \{\overline{x}_1, \overline{x}_2, ..., \overline{x}_{|\mathcal{M}|}\}$, θ :Frequency
   threshold, $x_{m'}$: modified variable, $F = [T_1, ..., T_{|M|}]$: Forest corresponding to
   databases' attributes, $Cov[|\mathcal{G}|]$:Coverage vector

2 **Output:** Consistent $\underline{X}_{\mathcal{M}} = \{\underline{x}_1, \underline{x}_2, ..., \underline{x}_{|\mathcal{M}|}\}$ and $\overline{X}_{\mathcal{M}} = \{\overline{x}_1, \overline{x}_2, ..., \overline{x}_{|\mathcal{M}|}\}$

3 **begin:**

4 PushDown(F[m'],depth=0, $r \in$ F[m], $min(\underline{x}_{m'})$, $max(\overline{x}_{m'})$)

5 **foreach** $m \in \mathcal{M}$ **do**

    // synchronize the tree leaves

6   **foreach** *leaf* $n \in F[m]$ **do**

7   **if** $Cov[n] = 0$ **then**

8             $inf(n) \longleftarrow +\infty; \quad sup(n) \longleftarrow -\infty$

    // update nodes intervals of the tree

9     $level \longleftarrow Maxdepth(F[m]) - 1$

10   **while** $level \geq 0$ **do**

11   **foreach** $n \in F[m]$ *where* $depth(n) = level$ **do**

12             $inf(n) \longleftarrow min(\{inf(c) \mid c \in child(n)\})$
            $sup(n) \longleftarrow max(\{sup(c) \mid c \in child(n)\})$

13         $level \longleftarrow level - 1$

    // Maintaining interval coherence

14     $\mathcal{D}(x_m) \longleftarrow \mathcal{D}(x_m) \setminus \{v \mid v \in \mathcal{D}(x_m) \wedge v > sup(r)\}$

15     $\mathcal{D}(x_m) \longleftarrow \mathcal{D}(x_m) \setminus \{v \mid v \in \mathcal{D}(x_m) \wedge v < inf(r)\}$

    // Closure filtering

16     $\mathcal{D}(\overline{x}_m) \longleftarrow \mathcal{D}(\overline{x}_m) \setminus \{v \mid v \in \mathcal{D}(\overline{x}_m) \wedge \forall g \in \mathcal{G}, v \neq v_{g,m} \text{ where } Cov[g] = 1\}$

17     $\mathcal{D}(\underline{x}_m) \longleftarrow \mathcal{D}(\underline{x}_m) \setminus \{v \mid v \in \mathcal{D}(\underline{x}_m) \wedge \forall g \in \mathcal{G}, v \neq v_{g,m} \text{ where } Cov[g] = 1\}$

18   **if** $m \neq m'$ **then**

19         $\mathcal{D}(\overline{x}_m) \longleftarrow \mathcal{D}(\overline{x}_m) \setminus \{v \mid v \in \mathcal{D}(\overline{x}_m) \wedge v < min(join(x_{m'}, \overline{x}_m))\}$

20         $\mathcal{D}(\underline{x}_m) \longleftarrow \mathcal{D}(\underline{x}_m) \setminus \{v \mid v \in \mathcal{D}(\underline{x}_m) \wedge v > max(join(x_{m'}, \underline{x}_m))\}$

    // Frequency filtering

21 $\mathcal{V}^p \longleftarrow \langle [min(\mathcal{D}(\underline{x}_i)), max(\mathcal{D}(\overline{x}_i))] \rangle_{1 \leq i \neq m' \leq |\mathcal{M}|}$

22 $\mathcal{D}(\underline{x}_{m'}) \longleftarrow \mathcal{D}(\underline{x}_{m'}) \setminus \{v \mid v \in \mathcal{D}(\underline{x}_{m'}) \wedge freq(\mathcal{V}^p ++ [v, max(\overline{x}_{m'})]) < \theta\}$

23 $\mathcal{D}(\overline{x}_{m'}) \longleftarrow \mathcal{D}(\overline{x}_{m'}) \setminus \{v \mid v \in \mathcal{D}(\overline{x}_{m'}) \wedge freq(\mathcal{V}^p ++ [min(\underline{x}_{m'}), v]) < \theta\}$

Implementing the Filtering Rules. Algorithm 2 applies the filtering rules described in Sect. 5. When the domain of a bound variable of an interval x_m (i.e. $\underline{x}_m$ or $\overline{x}_m$) changes, we first update the leaves through the $PushDown()$ function described in Algorithm 1. The leaves of all the trees within the forest are then synchronized through the cover array such that each leaf corresponding to a newly uncovered object is set to $[+\infty, -\infty]$, lines 6 to 8 in Algorithm 2. The new interval borders are updated using a bottom-up approach which sets for each parent node a new lower and upper bound consisting respectively, of the smallest value from its children's lower bound and the highest value among its children's upper bound, lines 10 to 13. The intervals coherence is maintained in lines 14 and 15 by filtering values that are greater than $sup(r)$ and smaller than $inf(r)$ from both $\mathcal{D}(\underline{x}_{m'})$ and $\mathcal{D}(\overline{x}_{m'})$. Lines 16 and 17 implements Proposition 1 using the cover array by filtering values that appears only in uncovered objects.

Afterward lines 19 and 20 are a straightforward application of Proposition 2 as they filter all the values greater than $max(join(x_m, \underline{x}_{m'}))$ and smaller than $min(join(x_m, \overline{x}_{m'}))$. Finally, Lines 22 and 23 implements Proposition 3 since it filters values that only leads to infrequent closed interval patterns.

Time Complexity Analysis. Algorithm 1 has a worst-case time complexity of $O(|\mathcal{G}|)$, which correspond to the number of leaves in the tree. This is simplified from $O(S^{log_S|\mathcal{G}|})$, where S is the maximal number of children of a parent node. The time complexity of the Algorithm 2 in the worst case is given by $O(|\mathcal{G}| + (|\mathcal{M}| \cdot |\mathcal{G}|^3 \cdot log_S|\mathcal{G}|) + |\mathcal{G}|)$. Consequently the complexity of GC4CIP is bounded by $O(|\mathcal{M}| \cdot |\mathcal{G}|^3 \cdot log_S|\mathcal{G}|)$.

7 Experiments and Results

The experimental evaluation is designed to address the following questions:

1. What are the results in terms of CPU time for our two models compared to:
   (a) CP approaches such as CP4IM and CLOSEDPATTERN outlined in Sect. 3 which require a pre and post processing step ?
   (b) the ad hoc method MININTCHANGE described in Sect. 3?
2. Can our CP model be seamlessly extended to address other data mining tasks on numerical data (e.g. clustering) while being competitive with ad hoc approaches (e.g. K-MEANS)?

7.1 Benchmark Datasets

We selected several difficult numerical datasets for declarative approaches which were used in [8]. We also selected two other datasets (Cancer and Yacht) from the UC Irvine archive[1]. The names of datasets are given by standard abbreviations used in the database of Bilkent University. All the selected datasets come in different sizes and types, most of them containing real values and one of them negative values. Table 3 provides a comprehensive summary of dataset characteristics, including the following key metrics: the number of objects denoted by $|\mathcal{G}|$, the number of items represented by $|\mathcal{M}|$, the sum of distinct values which is calculated by the formula $\sum_{m=1}^{|\mathcal{M}|} |\mathcal{N}_m|$, the count of binary attributes resulting from the process of interordinal scaling's binarization denoted as |Binary attributes| and the density of the binarized datasets. The dataset's density, after applying IS method, is always greater than 50%. This is a direct result of the IS method which assigns for each numerical value at least one binary attribute to 1.

The implementation of our approach was carried out using the OR-tools solver v9.0[2]. All experiments were conducted on an AMD Opteron 6174 with

---

[1] https://archive.ics.uci.edu/datasets.
[2] https://github.com/google/or-tools/.

Table 3. Datasets characteristics

	NT	AP	BK	Cancer	CH	Yacht	LW
$|\mathcal{M}|$	3	5	5	9	8	7	10
$|\mathcal{G}|$	130	135	96	116	209	308	189
distinct values	67	674	313	900	396	322	253
Interordinal scaled datasets							
|Binary attributes|	134	1348	626	1800	792	644	506
density (%)	52.23	50.37	50.79	50.50	51.01	51.08	51.97

2,2 GHz of CPU and 256 GB of RAM. with a timeout of 12 h. For each dataset, we decreased the (relative) frequency threshold until it is impossible to extract all closed interval patterns within the allocated time/memory. The source code and datasets are available at https://github.com/djawed-bkh/CPAIOR2024.

7.2 Mining CIP

Comparing with Other CP Approaches. Table 4 presents the computation time required to discover all closed interval patterns under different minimum frequency thresholds across various datasets. In these experiments, for both the CP4IM and CLOSEDPATTERN approaches, we provide the cpu-time of pre-processing and post-processing steps for using itemset mining. The pre-processing time is negligible compared to the post-processing time due to the high number of closed itemsets found.

If we consider the two reified models, CP4CIP and CP4IM, we can observe that CP4CIP consistently outperforms CP4IM across most of the selected datasets. In terms of CPU times, an average speed-up of 8.36, 10.11, 14.12, and 32.64 is observed for the BK, CH, Cancer, and AP datasets, respectively. For the LW, and Yacht datasets, we observe a speedup of 1.26 and 2.17 respectively. Finally in the NT dataset the results are more balanced as in the high frequencies CP4IM slightly overcome CP4CIP whereas in low frequencies CP4CIP is better. This can be explained with the low number of distinct values, resulting in fewer binary attributes generated with the IS method which enables CP4IM to enhance performance.

If we consider the use of global constraints GC4CIP and CLOSEDPATTERN, GC4CIP consistently outperforms CLOSEDPATTERN across all datasets. Regarding CPU times, an average speed-up of 1.20, 1.67, 3.98, and 4.83 is noted for the NT, LW, Cancer, and BK datasets, respectively. For the Yacht, CH, and AP datasets, a substantial speed-up of 10.70, 13.15, and 18.31 is observed, showcasing the efficiency of GC4CIP which emerges as the most efficient approach among all the compared declarative methods. Its main competitor, CLOSEDPATTERN, is consistently outperformed in the majority of instances, even when comparing only resolution times (without including pre-processing and post-processing times).

Table 4. Closed Interval Pattern Mining Methods vs Itemset Mining Methods with Interordinal Scaling

$\mathcal{N}$	θ	# Sol	Time (s)						
	(%)	($\approx$)	(1)	(2)	(3)	(1 + 3)	(2 + 3)	(4)	(5)
BK	80	10^6	1840.21	148.91	176.65	2016.86	325.56	271.10	89.63
	70	10^7	15132.87	1457.99	1326.58	16459.45	2784.57	1770.22	655.63
	60	10^7	TO	8643.34	6713.25	TO	15356.59	7311.24	2879.54
	50	10^8	TO	28302.62	19307.70	TO	47610.32	18471.23	7780.65
	20	10^8	TO	TO	TO	TO	TO	TO	34598.10
Cancer	95	10^4	170.14	6.19	13.69	183.83	19.88	18.42	5.80
	94	10^5	568.00	18.21	38.88	606.88	57.09	45.43	15.66
	92	10^5	6944.07	294.14	542.82	7486.89	836.96	486.87	190.84
	90	10^6	29787.19	1190.42	2348.45	32135.64	3538.87	1806.19	786.25
AP	80	10^5	783.92	175.02	55.21	839.13	230.23	28.55	19.18
	70	10^6	5909.86	189.30	415.76	6325.62	605.06	194.64	128.83
	60	10^6	18479.87	7995.84	1275.85	19755.72	9271.69	548.12	373.01
	50	10^7	TO	23252.89	2964.71	TO	26217.60	1223.79	770.83
	20	10^7	TO	43199.73	3052.93	TO	46252.66	5129.20	2891.55
	0	10^7	TO	TO	TO	TO	TO	5867.37	2343.98
CH	95	10^6	25.59	1.16	29.93	55.52	31.09	5.98	1.60
	90	10^5	608.94	36.58	224.70	833.64	261.28	89.81	38.42
	85	10^6	4753.35	331.08	835.24	5588.59	1166.32	671.49	256.86
	80	10^6	19154.96	1444.64	18009.40	37164.36	19454.04	2739.85	890.82
	50	TO	TO	TO	TO	TO	TO	TO	TO
LW	80	10^6	1612.68	96.91	174.46	1787.14	271.37	1638.03	181.81
	70	10^6	12904.12	757.02	1279.34	14183.34	2036.36	9886.90	1269.50
	60	10^7	TO	3436.91	5236.91	TO	8673.82	33 148.24	4,965.20
	50	10^8	TO	11060.23	15588.10	TO	26648.33	TO	14298.64
	20	TO	TO	TO	TO	TO	TO	TO	TO
NT	80	10^3	0.87	0.06	0.07	0.97	0.13	1.80	0.13
	50	10^4	7.08	0.41	0.50	7.58	0.91	11.01	0.91
	20	10^4	28.13	1.53	1.83	29.96	3.36	28.77	2.89
	10	10^5	41.75	2.51	2.61	44.36	5.12	32.50	4.02
	0	10^5	62.48	2.88	3.13	65.61	6.01	33.72	3.81
Yacht	80	10^4	40.12	2.03	83.20	123.32	85.23	90.92	2.45
	50	10^6	7277.85	336.03	268.28	7546.13	604.31	4090.63	181.63
	40	10^6	30519.66	1282.32	727.09	31246.75	2009.41	9380.16	501.52
	30	10^7	TO	4265.71	1695.63	TO	5961.34	20464.22	1179.13
	20	10^7	TO	12898.20	2874.08	TO	15772.28	33294.36	2487.68
	0	10^7	TO	TO	TO	TO	TO	TO	4116.60

(**1**): CP4IM   (**2**): CLOSEDPATTERN
(**3**): Preprocessing + Postprocessing   (**4**): CP4CIP   (**5**): GC4CIP

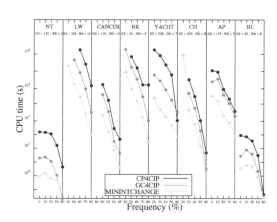

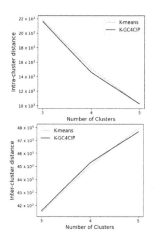

Fig. 1. MinIntChange compared to declarative approaches

Fig. 2. Cluster quality of GC4CIP and K-MEANS

Comparing with the Ad Hoc Approach MININTCHANGE. Figure 1 compares CPU times of our two proposals with the adhoc method MININTCHANGE. As expected, MININTCHANGE outperforms the declarative approaches across all databases, with speedup between GC4CIP and MININTCHANGE ranging from 2 for the NT database to 7 in Yacht. This outcome is due to the dedicated character of MININTCHANGE, tailored specifically for the FCIP mining task. However, in contrast to our generic declarative approaches, MININTCHANGE struggles to adapt to other data mining tasks due to its need of rewriting the solving algorithm. In the following section, we illustrate the genericity of our declarative approaches by applying them to another data mining task.

7.3 Modelling a K-Clustering Problem Using GC4CIP

We demonstrate the genericity of our approach by considering a clustering task as a use case. Our objective is to find a clustering which forms a partition over the objects in the dataset as akin to K-MEANS method. Closed interval patterns form the set of clusters, making this task known as conceptual clustering, where each cluster is defined by a unique concept (i.e. closure property). This task of finding k interval closed patterns $\{\mathcal{V}^1, ..., \mathcal{V}^k\}$ is formally described by the following:

$$
\begin{cases}
\textbf{Closure} & close(\mathcal{V}^i), & \forall 1 \leq i \leq k \\
\textbf{No overlapping} & cover(\mathcal{V}^i) \cap cover(\mathcal{V}^j) = \emptyset, \forall 1 \leq i < j \leq k \\
\textbf{Total coverage} & \bigcup_{1 \leq i \leq k} cover(\mathcal{V}^i) = \mathcal{G}
\end{cases}
$$

To carry this task, we extend our global constraint GC4CIP to handle the conceptual clustering directly from the numerical data. This involves adding a new set of variables denoted Y as parameter in our global constraint i.e. $GC4CIP_{\mathcal{N},\theta}(\underline{X}, \overline{X}, Y)$. The set of binary variables $Y = \{y_1, ..., y_{|\mathcal{G}|}\}$ indicates whether an object is covered by $\mathcal{V}$ or not. The filtering rules over the variables Y in the global constraint are described in supplementary material https://github.com/djawed-bkh/CPAIOR2024. Our CP based k-Clustering model requires three sets of variables. $\underline{X}^i$ and $\overline{X}^i$ to represent each interval pattern associated to a cluster and Y^i to indicate whether an object is included in a cluster. In term of constraints, our model requires the conjunction of three distinct types of constraints. The first one is a closure constraint, represented by GC4CIP$_{\mathcal{N},\theta}(\underline{X}^i, \overline{X}^i, Y^i)$ which is applied k times to generate k closed interval patterns each one representing a cluster. Then, to ensure the non-overlapping coverage of our interval patterns, we introduce a partitioning constraint which ensures that each object $g \in \mathcal{G}$ can be covered with at most one interval pattern, thus guaranteeing that an object belongs to a single clustering at most (i.e. $\sum_{1 \le i \le k} y_g^i = 1$). Finally, we introduce a constraint enforcing the k-clusters to cover all the objects in the database (i.e. $\bigcup_{1 \le i \le k} cover(\mathcal{V}^i) = \mathcal{G}$).

In Fig. 2, we showcase the results of clustering on the NT database with 3, 4, and 5 clusters for both the K-MEANS and K-GC4CIP approaches, with a 12 h time-out. To measure the results quality we use the intra-cluster and inter-cluster Euclidean distances. Notably, for cluster numbers 3 and 4, the K-GC4CIP method outperforms K-MEANS in both intra-cluster distance and inter-cluster distance as the former is smaller and the latter is greater than the heuristic approach. However, for a number of clusters of 5, we note that K-MEANS performed slightly better than K-GC4CIP in both measures, which can be explained by the timeout of K-GC4CIP, as it was unable achieve an exhaustive search within the allocated time.

8   Conclusion

In this paper, we introduced two CP models (CP4CIP and GC4CIP) for mining FCIP directly from numerical data without requiring any pre- or post-processing step. We demonstrated the efficiency of GC4CIP compared to existing declarative methods and an ad hoc one. Finally, we illustrated the genericity of our approach by combining GC4CIP with other constraints to tackle the conceptual k-clustering problem. The declarative nature of our contributions enables straightforward combination with other declarative approaches. For example, combining GC4CIP with CP4IM allows effortless mining of closed patterns from heterogeneous data, where the former handles numerical data and the later focuses on binary data.

Acknowledgement. The first author is supported by the French National Research Agency (ANR) and Region Normandie under grant HAISCoDe.

References

1. Belfodil, A., Kuznetsov, S.O., Robardet, C., Kaytoue, M.: Mining convex polygon patterns with formal concept analysis. In: Sierra, C. (ed.) IJCAI, pp. 1425–1432 (2017)
2. Calders, T., Rigotti, C., Boulicaut, J.F.: A survey on condensed representations for frequent sets. In: Boulicaut, J.F., De Raedt, L., Mannila, H. (eds.) Constraint-Based Mining and Inductive Databases. LNCS, vol. 3848, pp. 64–80. Springer, Heidelberg (2005). https://doi.org/10.1007/11615576_4

3. Chabert, M., Solnon, C.: A global constraint for the exact cover problem: application to conceptual clustering. J. Artif. Intell. Res. **67**, 509–547 (2020)
4. Codocedo, V., Napoli, A.: A proposition for combining pattern structures and relational concept analysis. In: ICFCA, pp. 96 – 111 (2014)
5. Dao, T., Vrain, C., Duong, K., Davidson, I.: A framework for actionable clustering using constraint programming. In: ECAI, pp. 453–461. Frontiers in Artificial Intelligence and Applications (2016)
6. Dougherty, J., Kohavi, R., Sahami, M.: Supervised and unsupervised discretization of continuous features. In: Machine Learning: Proceeding of the Twelfth International Conference, pp. 194–202. Morgan Kaufmann (1995)
7. Guns, T., Nijssen, S., De Raedt, L.: k-pattern set mining under constraints. IEEE Trans. Knowl. Data Eng. **25**(2), 402–418 (2013)
8. Kaytoue, M., Kuznetsov, S., Napoli, A.: Revisiting numerical pattern mining with formal concept analysis. IJCAI (2011)
9. Khiari, M., Boizumault, P., Crémilleux, B.: Constraint programming for mining n-ary patterns. In: Cohen, D. (ed.) CP 2010. LNCS, vol. 6308, pp. 552–567. Springer, Heidelberg (2010). https://doi.org/10.1007/978-3-642-15396-9_44
10. Lazaar, N., et al.: A global constraint for closed frequent pattern mining. In: Rueher, M. (ed.) CP 2016. LNCS, vol. 9892, pp. 333–349. Springer, Heidelberg (2016). https://doi.org/10.1007/978-3-319-44953-1_22
11. Makhalova, T., Kuznetsov, S.O., Napoli, A.: Mint: MDL-based approach for mining interesting numerical pattern sets. Data Min. Knowl. Discov. **36**, 108–145 (2022)
12. Meeng, M., Knobbe, A.J.: For real: a thorough look at numeric attributes in subgroup discovery. Data Min. Knowl. Discov. **35**(1), 158–212 (2021)
13. Millot, A., Cazabet, R., Boulicaut, J.: Optimal subgroup discovery in purely numerical data. In: Lauw, H., Wong, R.W., Ntoulas, A., Lim, E.P., Ng, S.K., Pan, S. (eds.) PAKDD 2020. LNCS, vol. 12085, pp. 112–124. Springer, Heidelberg (2020). https://doi.org/10.1007/978-3-030-47436-2_9
14. Nguyen, H.V., Vreeken, J.: Flexibly mining better subgroups. In: Venkatasubramanian, S.C., Jr., W.M. (eds.) Proceedings of the SIAM International Conference on Data Mining, USA, pp. 585–593. SIAM (2016). https://doi.org/10.1137/1.9781611974348.66
15. Nijssen, S., Zimmermann, A.: Constraint-based pattern mining. In: Aggarwal, C., Han, J. (eds.) Frequent Pattern Mining, pp. 147–163. Springer, Heidelberg (2014). https://doi.org/10.1007/978-3-319-07821-2_7
16. Raedt, L.D., Guns, T., Nijssen, S.: Constraint programming for data mining and machine learning. In: AAAI (2010)
17. Salleb-Aouissi, A., Vrain, C., Nortet, C.: Quantminer: a genetic algorithm for mining quantitative association rules. In: Veloso, M.M. (ed.) IJCAI, pp. 1035–1040 (2007)
18. Song, C., Ge, T.: Discovering and managing quantitative association rules. In: CIKM 2013, pp. 2429–2434 (2013)
19. Uno, T., Asai, T., Uchida, Y., Arimura, H.: LCM: an efficient algorithm for enumerating frequent closed item sets. In: Proceedings of the ICDM Workshop on Frequent Itemset Mining Implementations (2003)
20. Witteveen, J., Duivesteijn, W., Knobbe, A.J., Grünwald, P.: Realkrimp - finding hyperintervals that compress with MDL for real-valued data. In: IDA, pp. 368–379 (2014)

Local Alterations of the Lagrange Multipliers for Enhancing the Filtering of the AtMostNValue Constraint

Frédéric Berthiaume and Claude-Guy Quimper$^{(\boxtimes)}$ (iD)

Université Laval, Québec, Canada
`frederic.berthiaume.1@ulaval.ca, claude-guy.quimper@ift.ulaval.ca`

Abstract. The reduced cost filtering is a technique that consists in filtering a constraint using the reduced cost of a linear program that encodes this constraint. Sellmann [16] shows that while doing a Lagrangian relaxation of a constraint, suboptimal Lagrange multipliers can provide more filtering than optimal ones. Boudreault and Quimper [5] make an algorithm that locally altered the Lagrange multipliers for the WEIGHTEDCIRCUIT constraint to enhance filtering and achieve a speedup of 30%. We seek to design an algorithm like Boudreault and Quimper, but for the AtMostNValue constraint. Based on the work done by Cambazard and Fages [7] on this constraint, we use a subgradient algorithm which takes into consideration the reduced cost to boost the Lagrange multipliers in the optimal filtering direction. We test our methods on the dominating queens and the p-median problem. On the first, we record a speedup of 71% on average. On the second, there are three classes of instances. On the first two, we have an average speedup of 33% and 8%. On the hardest class, we find up to 13 better solutions than the previous algorithm on the 30 instances in the class.

Keywords: Global constraint · Lagrangian relaxation · N values

1 Introduction

Global constraints contributed to the success of constraint programming. These constraints involve a non-fixed number of variables [2]. A filtering algorithm prunes the values from the variable domains. To aid the filtering process, Foccaci et al. [9] introduced a technique called *reduced cost filtering*. It studies the variation of the objective function after a consistent variable change of values. There is also Lagrangian relaxation, which is a technique to rewrite the problem without some constraints while penalizing when they are violated. During the process of Lagrangian relaxation, Sellmann [16] showed that the suboptimal Lagrange multipliers can provide better filtering than the optimal ones. Boudreault and Quimper [5] proposed an algorithm for the WEIGHTEDCIRCUIT constraint where they locally alter the Lagrange multipliers to enhance the filtering.

B. Dilkina (Ed.): CPAIOR 2024, LNCS 14742, pp. 68–83, 2024.
https://doi.org/10.1007/978-3-031-60597-0_5

Cambazard and Fages [7] treated the Lagrangian relaxation of the sub constraint AtMostNValue. We propose to augment their algorithm based on a Lagrangian relaxation by locally altering the Lagrangian multipliers to enhance the filtering. We use the dominating queens problem and the facility location problem in order to test our algorithm.

Section 2 introduces the notation and defines the NValue constraint leading to the starting point of this paper, the filtering algorithm of the AtMostNValue constraint based on a Lagrangian relaxation. Section 2.5 proposes an addition to the previous algorithm. Section 3 presents this addition along with the theoretical justification. Section 4 presents the experiments.

2 Background

2.1 Notation

We note column and row vectors with the symbols v and $v^\top$. The vectors $\mathbf{0}$ and $\mathbf{1}$ are the vectors with only zeros and ones. The columns and rows of a matrix A are noted a_{cj} and $a_{ri}^\top$ respectively. So, a_{ri} is the column vector that, when transposed, gives the i-th row of A. The components of A are written a_{ij}.

We note the gradient of a function $f : \mathbb{R}^n \to \mathbb{R}$ with ∇f. For a *piece-wise* continuous linear function $f : \mathbb{R}^n \to \mathbb{R}$, we note one of its *subgradient* with $\tilde{\nabla} f$. Briefly, a *subdifferential* is the generalization of a *derivative*. For example, the function $f(x) = |x|$ is differentiable everywhere except at $x = 0$. Although, the function $f(x) = |x|$ does not admit a derivative at $x = 0$, it admits a *subdifferential* at $x = 0$ labeled $\partial f(0)$. $\partial f(x_0)$ is the set of all v that satisfies $f(x) - f(x_0) \geq v(x - x_0)$; for $f(x) = |x|$, $\partial f(0) = [-1, 1]$. One may view an ordinary derivative as a subdifferential with a unique value. The subdifferential of $f(x)$ at x_0, $\partial f(x_0)$, is the set of all vectors v, the subgradients, that satisfies $f(x) - f(x_0) \geq v^\top (x - x_0)$. The symbol $\tilde{\nabla} f(x)$ represents an element of this set. If there is ambiguity on which variables the gradient (or subgradient) is taken, we specify them with ∇_x (or $\tilde{\nabla}_x$).

We use two types of variables that are fundamentally different while related. There are constraint satisfaction problem (*CSP*) variables, written in uppercase (X), and each has a set of *values*, called a domain $(\mathrm{dom}(X))$. There are linear program (LP) variables that we write as vectors (x).

2.2 The NValue Constraint

Definition 1. *The NValue constraint bounds the number of distinct values taken by a set of variables. It is written* $\mathrm{NValue}([X_1, \ldots, X_n], N)$*, where N is the cardinality variable. The constraint is satisfied when* $N = |\{X_1, \ldots, X_n\}|$*.*

Enforcing domain consistency on the NValue constraint is *NP-hard* [4]. The constraint can be decomposed into two other constraints: AtLeastNValue and AtMostNValue. Enforcing domain consistency on the former can be done in polynomial time [4,14]. Enforcing domain consistency on the AtMostNValue is *NP-hard* [4]. We will only consider this constraint. The AtMostNValue constraint is satisfied when $|\{X_1, \ldots, X_n\}| \leq N$.

2.3   The LP of the ATMOSTNVALUE Constraint

Bessiere et al. [4] present three approaches to propagate the NVALUE constraint. The first one, an algorithm proposed by Beldiceanu [1], enforces bounds consistency and is based on interval graphs. The second approach is based on the independence number of a graph. The third approach is a LP formulation of the ATMOSTNVALUE constraint. Bessiere et al. show that the LP formulation offers a better estimation of the lower bound of N.

In the CSP, each value $j \in \bigcup_{i=1}^n \text{dom}(X_i) = \{1, ..., m\}$ is paired with a Boolean variable Y_j. The CSP variable Y_j encodes whether the value j belongs to $\{X_1, ..., X_n\}$. In the LP that encodes the ATMOSTNVALUE constraint, each of the CSP variable Y_j has an analog LP variable y_j. If Y_j is instantiated, then y_j takes the same value as Y_j. Otherwise, the LP variables $\boldsymbol{y}$ are free and are assigned while solving the LP of the ATMOSTNVALUE constraint:

$$\min_{\boldsymbol{y}} \quad h(\boldsymbol{y}) = \mathbf{1}^\top \boldsymbol{y}, \quad \text{s.t.} \quad A\boldsymbol{y} \geq \mathbf{1}, \quad \boldsymbol{y} \leq \mathbf{1}, \quad \boldsymbol{y} \geq \mathbf{0}, \tag{1}$$

where $a_{ij} = 1$ if $j \in \text{dom}(X_i)$ and $a_{ij} = 0$ otherwise. The i-th row $\boldsymbol{a_{ri}}$ encodes in a vector the set $\text{dom}(X_i)$. These constraints ensure that for every variable X_i, there is at least one value $j \in \text{dom}(X_i)$ for which $y_j = 1$.

2.4   The Propagator of the ATMOSTNVALUE Constraint

Cambazard and Fages [7] designed a propagator for the ATMOSTNVALUE constraint. The algorithm takes as input the domains of the following CSP variables: the integer variables $\{X_1, ..., X_n\}$, the Boolean variables $\{Y_1, ..., Y_m\}$ and the cardinality variable N. This algorithm is based on a **Lagrangian relaxation** [8,11,13] of the LP (1) and estimates a lower bound on N.

A **Lagrangian relaxation** is a twofold process. First, there is a transformation from an initial constrained optimization problem $\mathcal{P}$ into another $\mathcal{P}'$. The *relaxed* problem $\mathcal{P}'$ is defined on a subset of constraints $C' \subset C$. Each constraint $c_i \in C \setminus C'$ is paired with a *Lagrange multiplier* in the new objective function, a *Lagrangian* function, $f'(\boldsymbol{x}, \boldsymbol{\lambda})$. The Lagrangian function depends on the original objective function $f(\boldsymbol{x})$ and each relaxed constraint weighted by its Lagrange multipliers. The transformation of (1), done by Cambazard and Fages, is

$$\min_{\boldsymbol{y}} \quad h'(\boldsymbol{y}, \boldsymbol{\lambda}) = \mathbf{1}^\top \boldsymbol{y} + \boldsymbol{\lambda}^\top (\mathbf{1} - A\boldsymbol{y}), \quad \text{s.t.} \quad \boldsymbol{y} \leq \mathbf{1}, \quad \boldsymbol{y} \geq \mathbf{0}, \quad \boldsymbol{\lambda} \geq \mathbf{0}. \tag{2}$$

The constraints that are relaxed in (2) are those that ensure that at least one value from every domain $\text{dom}(X_i)$ is selected.

The Lagrange multiplier λ_i associated to the i-th constraint must be positive to penalize the function when the constraint $\boldsymbol{a_{ri}}^\top \boldsymbol{y} \geq 1$ is violated. Constraints violation result in an increase of the Lagrangian function, which is opposite to the original objective of minimization.

The second phase of a Lagrangian relaxation is the optimization of the Lagrangian function, where one tries to solve $\max_{\boldsymbol{\lambda}}(\min_{\boldsymbol{y}} h'(\boldsymbol{y}, \boldsymbol{\lambda}))$. This translates into trying to find the greatest lower bound of $h'(\boldsymbol{y}, \boldsymbol{\lambda})$. Before addressing this phase, we must clarify what happens with $\min_{\boldsymbol{y}} h'(\boldsymbol{y}, \boldsymbol{\lambda})$.

The LP (2) is *solved* when the LP variables y_j are fixed. We first need the coefficients q_j coupled with the LP variables y_j. After rewriting the function $h'(\boldsymbol{y}, \boldsymbol{\lambda})$ as $h'(\boldsymbol{y}, \boldsymbol{\lambda}) = (\boldsymbol{1} - A^\top \boldsymbol{\lambda})^\top \boldsymbol{y} + \boldsymbol{\lambda}^\top \boldsymbol{1}$, the coefficient vector is:

$$q(\boldsymbol{\lambda}) = \boldsymbol{1} - A^\top \boldsymbol{\lambda}. \tag{3}$$

The coefficient $q_j(\boldsymbol{\lambda})$ is the reduced cost of y_j, the next section explains the meaning of the reduced cost. The LP variables y_j can be fixed either from their CSP variables counterpart Y_j or the value of their coefficient in $h'(\boldsymbol{y}, \boldsymbol{\lambda})$:

$$\text{if} \quad |\mathrm{dom}(Y_j)| = 1 \quad \text{then} \quad y_j = Y_j \quad \text{else,} \quad y_j(\boldsymbol{\lambda}) = \begin{cases} 1 \text{ if } q_j(\boldsymbol{\lambda}) < 0 \\ 0 \text{ otherwise} \end{cases} \tag{4}$$

The LP variables $\boldsymbol{y}$ associated with unfixed CSP variables Y_j can be viewed as functions of $\boldsymbol{\lambda}$, because they depend on the sign of $q_j(\boldsymbol{\lambda})$. Afterwards the problem $\min_{\boldsymbol{y}} h'(\boldsymbol{y}, \boldsymbol{\lambda})$ is considered solved for the current $\boldsymbol{\lambda}$. To enhance that the LP variables $\boldsymbol{y}$ are fixed, we write the current solution as $h''(\boldsymbol{\lambda}) = \min_{\boldsymbol{y}} h'(\boldsymbol{y}, \boldsymbol{\lambda})$.

The initial Lagrange multipliers $\boldsymbol{\lambda}$ generally need improvement to give the greater relaxed lower bound on N and to respect the relaxed constraints. The $\boldsymbol{\lambda}$ must be updated with care because $\nabla h''(\boldsymbol{\lambda})$ is not continuous.

Definition 2. $L_c(f) = \{\boldsymbol{x} \mid f(\boldsymbol{x}) = c\}$ *is the level set* c *of function* f.

Lemma 1. $h''(\boldsymbol{\lambda})$ *is continuous* $\forall \boldsymbol{\lambda}$, *but not* $\nabla h''(\boldsymbol{\lambda})$.

Proof. $h''(\boldsymbol{\lambda}) = q(\boldsymbol{\lambda})^\top \boldsymbol{y} + \boldsymbol{1}^\top \boldsymbol{\lambda}$. The LP variables $\boldsymbol{y}$ depend on $\boldsymbol{\lambda}$ (4) and are discontinuous functions of $\boldsymbol{\lambda}$. The term $q_j(\boldsymbol{\lambda})y_j$ is similar to a *RELU function* ($\max\{0, x\}$), which has no derivative at $x = 0$ (from the left the derivative is 0 and 1 from the right). The term $q_j(\boldsymbol{\lambda})y_j$ has no derivative on $L_0(q_j)$ (Definition 2). So each term in $q(\boldsymbol{\lambda})^\top \boldsymbol{y}$ has no derivative on its $L_0(q_j)$. Thus, $h''(\boldsymbol{\lambda})$ is not derivable, but continuous, on $\bigcup_{j=1}^m L_0(q_j)$.   □

Hence the Lagrange multipliers are updated according to the *subgradient* of $h''(\boldsymbol{\lambda})$ instead of the traditional gradient. For a given $\boldsymbol{\lambda}$, $\tilde{\nabla} h''(\boldsymbol{\lambda}) = \boldsymbol{1} - A\boldsymbol{y}$ is a subgradient of $h''(\boldsymbol{\lambda})$ [10]. Then step-sizes α_k are carefully chosen to update the $\boldsymbol{\lambda}$ with the following equation

$$\lambda_i^{(k+1)} = \max(0, \lambda_i^{(k)} + \alpha_k [\tilde{\nabla} h''(\boldsymbol{\lambda})]_i), \tag{5}$$

Studies of step-sizes rules are listed here [3,6,10,15]. The *max* operator in (5) ensures that $\boldsymbol{\lambda} \geq \boldsymbol{0}$. Here $[\tilde{\nabla} h''(\boldsymbol{\lambda})]_i$, which is equal to $1 - \boldsymbol{a_{ri}}^\top \boldsymbol{y}$, represents the i-th constraint satisfaction. If the i-th component is 1, it means that the CSP variable X_i has no value in its domain. This procedure is repeated until the optimal solution is found or some other convergence criterion is satisfied.

Reduced cost filtering (introduced by Focacci et al. [9]) is a technique to filter values from CSP variable domains based on their contribution in the current state of the solution. In the LP (2), the contribution of the LP variable y_j to the solution is its coefficient $q_j(\boldsymbol{\lambda})$. When

$$h''(\boldsymbol{\lambda}) + |q_j(\boldsymbol{\lambda})| = h''(\boldsymbol{\lambda}) + |1 - \boldsymbol{a_{cj}}^\top \boldsymbol{\lambda}| > \max(\mathrm{dom}(N)) \tag{6}$$

Table 1. The three step-size rules and their parameters used in [7]. The denominator in the *Newton* column is an analog of the squared norm of $\tilde{\nabla}_\lambda h(\boldsymbol{y}, \boldsymbol{\lambda})$.

Harmonic	Geometric	Newton
$\alpha_k = 1/k$	$\alpha_k = 10^3 (0.95)^k$	$\alpha_k = \dfrac{5}{2^{\lfloor k/10 \rfloor}} \dfrac{\max(\mathrm{dom}(N)) - h''(\boldsymbol{\lambda})_k}{\sum_{i=1}^{n} \gamma_i (\tilde{\nabla} h''(\boldsymbol{\lambda})_i)^2}$ $\gamma_i = \begin{cases} 1 & \text{if } \lambda_i > 0 \vee \tilde{\nabla} h''(\boldsymbol{\lambda})_i = 1 \\ 0 & \text{otherwise} \end{cases}$

the complementary value of the LP variable y_j, $1 - y_j$, can be filtered from the domain of the CSP variable Y_j. This is because the LHS of (6) is the relaxed lower bound on N, if the LP variable y_j was fixed to the value $1 - y_j$. If, (6) is satisfied, it means that $1 - y_j$ leads to $\min(\mathrm{dom}(N)) > \max(\mathrm{dom}(N))$, which is impossible. The $|q_j(\boldsymbol{\lambda})|$ is the reduced cost of the LP variables y_j.

Algorithm 1, designed by Cambazard and Fages [7], filters the constraint AtMostNValue. To perform the transformation from (1) to (2), the LP variables $\boldsymbol{y}$ (not the CSP variables Y_j) and $\boldsymbol{\lambda}$ are initialized at line 1 of Algorithm 1. On lines 2–14, the problem $\min_{\boldsymbol{y}} h'(\boldsymbol{y}, \boldsymbol{\lambda})$ is solved by fixing the LP variables y_j according to (4). The relaxed lower bound on N, $h''(\boldsymbol{\lambda})$, is computed at line 8. On lines 9–10 the reduced cost filtering phase is performed for each non-instantiated LP variable y_j. Line 10 flags the value for future filtering. As Sellman [16] explains in Section *A disturbing example*, filtering values during a Lagrangian relaxation lead to inconsistencies because the current problem, $\min_{\boldsymbol{y}} h'(\boldsymbol{y}, \boldsymbol{\lambda})$, might change too much between consecutive iterations $k \to k + 1$. Thus, they are filtered at the end of the Algorithm (line 16). Finally, the Lagrangian optimization is ended if the maximum number of iterations is reached or if the change of $\boldsymbol{\lambda}$ between two consecutive iterations is too small. This last condition represents a situation where the algorithm converges to an optimal $\boldsymbol{\lambda}$, for the current state of the CSP variables at the start of Algorithm 1. Cambazard and Fages compared three types of step-size (Table 1) to update the Lagrange multipliers in (5).

2.5 Which Lagrange Multipliers are the Most Effective?

Sellmann [16] presents a dichotomy between *getting the best lower bound on LP like* (2) and *filtering many values from the variables' domain*. The main result is the *Filtering Penalties theorem* which justifies checking between iteration steps of (5) for inconsistent values (Algorithm 1 lines 9–10).

From this result, Boudreault and Quimper [5] presented a new algorithm for the WeightedCircuit constraint, where they model the constraint as a LP, relax it using a Lagrangian relaxation, and locally modify the Lagrange multipliers to augment the filtering. Their algorithm filters more values and is faster than the state-of-the-art algorithm for this constraint.

Algorithm 1: AtMostNValue LR-based propagator ($\{X_1, \ldots, X_n\}$, $\{Y_1, \ldots, Y_m\}$, N, IterMax)

1 $\boldsymbol{y} \leftarrow \underbrace{[0, \ldots, 0]}_{m}$, $\boldsymbol{\lambda} \leftarrow \underbrace{[0.0, \ldots, 0.0]}_{n}$, $\boldsymbol{q} \leftarrow \underbrace{[0.0, \ldots, 0.0]}_{m}$, $k \leftarrow 0$

2 **repeat**
3      $\boldsymbol{q} \leftarrow \mathbf{1} - A^\top \boldsymbol{\lambda}$
4   **for** $j \leftarrow 1$ **to** m **do**
5   **if** $(\min(\mathrm{dom}(Y_j)) = 1)$ **or** $((\max(\mathrm{dom}(Y_j)) = 1)$ **and** $(q_j < 0))$ **then**
6              $y_j \leftarrow 1$
7   **else** $y_j \leftarrow 0$
8      $h''(\boldsymbol{\lambda}) \leftarrow \boldsymbol{q}^\top \boldsymbol{y} + \mathbf{1}^\top \boldsymbol{\lambda}$
9   **for** $j \leftarrow 1$ **to** m **do**
10   **if** $(h''(\boldsymbol{\lambda}) + |q_j| > \max(\mathrm{dom}(N)))$ **then** Flag value $1 - y_j$ in $\mathrm{dom}(Y_j)$
11     $\tilde{\boldsymbol{\nabla}} h''(\boldsymbol{\lambda}) \leftarrow \mathbf{1} - A\boldsymbol{y}$
12   Update α_k // With Table 1
13     $\boldsymbol{\lambda}' \leftarrow \boldsymbol{\lambda}$, $\boldsymbol{\lambda} \leftarrow \max(\mathbf{0}, \boldsymbol{\lambda} + \alpha_k \tilde{\boldsymbol{\nabla}} h''(\boldsymbol{\lambda}))$, $k \leftarrow k + 1$
14 **until** $k = \text{IterMax}$ **or** $\max_{1 \leq i \leq n} |\lambda_i - \lambda_i'| \leq 0.0001$
15 // The local alterations algorithm (Algorithm 2) is called here
16 Filter out all the flag values

---

3 Improving the Filtering of AtMostNValue

We present how Lagrange multipliers can be altered to increase the level of filtering. First, we write the quantity $h''(\boldsymbol{\lambda}) + |q_j(\boldsymbol{\lambda})|$ as a function.

Definition 3. *The* ***forced lower bound function***, *associated with the LP variable* y_j, *is given in* (7) *and its (sub)gradient with respect to* $\boldsymbol{\lambda}$ *in* (8)

$$\rho_j(\boldsymbol{\lambda}) = h''(\boldsymbol{\lambda}) + |q_j(\boldsymbol{\lambda})| = h''(\boldsymbol{\lambda}) + |1 - \boldsymbol{a_{cj}}^\top \boldsymbol{\lambda}|, \tag{7}$$

$$\tilde{\boldsymbol{\nabla}} \rho_j(\boldsymbol{\lambda}) = \tilde{\boldsymbol{\nabla}} h''(\boldsymbol{\lambda}) + \tilde{\boldsymbol{\nabla}} |q_j(\boldsymbol{\lambda})| = 1 - A\boldsymbol{y} - \mathrm{sgn}(q_j(\boldsymbol{\lambda}))\boldsymbol{a_{cj}}. \tag{8}$$

We introduce Algorithm 2 that locally alters the Lagrange multipliers in a way to enhance the filtering of Algorithm 1 for the AtMostNValue constraint. It performs several short Lagrangian optimization processes. Each Lagrangian optimization process has a different forced lower bound function $\rho_j(\boldsymbol{\lambda})$ as the objective function. Only a subset of forced lower bound functions are chosen for this second optimization process. The choice is based on how close $\rho_j(\boldsymbol{\lambda})$ is from $\max(\mathrm{dom}(N))$. If the difference is under a threshold τ, the algorithm is triggered. We optimize the *local* Lagrange multipliers $\bar{\boldsymbol{\lambda}}$ with these steps

$$\bar{\boldsymbol{\lambda}}^{(k+1)} = \max(\mathbf{0}, \quad \bar{\boldsymbol{\lambda}}^{(k)} + \beta_k \tilde{\boldsymbol{\nabla}} \rho_j(\bar{\boldsymbol{\lambda}}^{(k)})), \tag{9}$$

$$\beta_k = \omega_k \frac{(\max(\mathrm{dom}(N)) + \eta) - \rho_j(\bar{\boldsymbol{\lambda}}^{(k)})}{||\tilde{\boldsymbol{\nabla}}_{\bar{\lambda}} \rho_j(\bar{\boldsymbol{\lambda}}^{(k)})||^2}, \quad \omega_k \in (0, 2), \quad \eta > 0. \tag{10}$$

We chose a *Newton* type for the step size β_k with a *constant* coefficient ω_k as opposed to the variable coefficient of Table 1 where $\omega_k = \frac{5}{2^{\lfloor k/10 \rfloor}}$. The second parameter η prevents a null step size β_k at the start of the optimization.

Algorithm 2: Algorithm to enhance filtering

---

1 **for** $j = 1..m$ **if** $|\operatorname{dom}(Y_j)| \neq 1$ **and** $\max(\operatorname{dom}(N)) - \rho_j(\lambda) < \tau$ **do**

2 $\quad$ $y' \leftarrow y, \bar{\lambda} \leftarrow \lambda, q' \leftarrow q, h''(\lambda)' \leftarrow h''(\lambda), k \leftarrow 0$

3 $\quad$ $\tilde{\nabla}\rho_j(\bar{\lambda}) \leftarrow 1 - Ay - \operatorname{sgn}(q_j(\lambda))a_{cj}$

4 $\quad$ **repeat**

5 $\quad\quad$ Update β_k // With equation (10)

6 $\quad\quad$ $\bar{\lambda}' \leftarrow \bar{\lambda}, \bar{\lambda} \leftarrow \max(0, \bar{\lambda} - \beta_k \tilde{\nabla}\rho_j(\bar{\lambda}))$

7 $\quad\quad$ **if** $(\max|\bar{\lambda}_i - \bar{\lambda}'_i| > 0.0001)$ **then**

8 $\quad\quad\quad$ $q' \leftarrow 1 - A^\top \bar{\lambda}$

9 $\quad\quad\quad$ **for** $j' \leftarrow 1$ **to** m **do**

10 $\quad\quad\quad\quad$ **if** $(\operatorname{dom}(Y_{j'}) = \{1\})$ **or** $((|\operatorname{dom}(Y_{j'})| \neq 1)$ **and** $(q_{j'} < 0))$ **then**

11 $\quad\quad\quad\quad\quad$ $y'_{j'} \leftarrow 1$

12 $\quad\quad\quad\quad$ **else** $y'_{j'} \leftarrow 0$

13 $\quad\quad\quad$ $h''(\bar{\lambda}) \leftarrow q^\top y' + 1^\top \bar{\lambda}$

14 $\quad\quad\quad$ **if** $(\rho_j(\bar{\lambda}) > \max(\operatorname{dom}(N)))$ **then** Flag the value $1 - y_j$ in $\operatorname{dom}(Y_j)$

15 $\quad\quad\quad$ $\tilde{\nabla}\rho_j(\bar{\lambda}) \leftarrow 1 - Ay'$

16 $\quad\quad\quad$ **if** $\rho_j(\bar{\lambda}) < \max(\operatorname{dom}(N))$ **then** $\tilde{\nabla}\rho_j(\bar{\lambda}) \leftarrow \tilde{\nabla}\rho_j(\bar{\lambda}) - \operatorname{sgn}(q_j(\lambda))a_{cj}$

17 $\quad\quad$ $k \leftarrow k + 1$

18 $\quad$ **until** $k = \text{numberOfSteps}$ **or** $\max_{1 \leq i \leq n}|\bar{\lambda}_i - \bar{\lambda}'_i| \leq 0.0001$ **or**
$\quad\quad$ $\rho_j(\bar{\lambda})) > \max(\operatorname{dom}(N))$

It happens whenever $\rho_j(\bar{\lambda}^{(k)}) = \max(\operatorname{dom}(N))$. A certain number of steps (numberOfSteps) is allowed to optimize the local Lagrange multipliers. Otherwise, Algorithm 2 is very similar to Algorithm 1. The order of some operations is different because we want to start by updating the $\bar{\lambda}$. The other difference is the line 16 of Algorithm 2. Here there is a choice of subgradient used to guide the search. The details are explained in Sect. 3.1.

3.1   The Theory

This section justifies the process used in Algorithm 2. The algorithm overcomes many difficulties and particularities of the problem. The forced lower bound functions (7) are neither convex nor concave. The first term, $h''(\lambda)$, is concave and the second term, $|q_j(\lambda)|$, is convex. This is a problem because subgradient methods converge to a local optimum. In Algorithm 1, the concavity of $h''(\lambda)$ saved the situation; here the situation requires some creativity because $\nabla\rho_j(\lambda)$ might not lead to a global optimal solution. Algorithm 2 overcomes this major issue by choosing between two directions for the optimization of $\bar{\lambda}$ (Algorithm 2 lines 15-16). To understand why this change overcomes the concavity issue, we start by analyzing some propriety of $\rho_j(\lambda)$ and $|q_j(\lambda)|$.

Theorem 1. $\rho_j(\lambda)$ *is concave on both side of* $L_0(q_j)$.

Proof. Given two concave functions $f_1(x)$ and $f_2(x)$, the function $f(x) = f_1(x) + f_2(x)$ is concave. The function $|q_j(\lambda)|$ is a convex function, but on each side of

$L_0(q_j)$ it is a linear function, which is both convex and concave. Since the sum of two concave function is concave, $\rho_j(\boldsymbol{\lambda})$ is concave on each side of $L_0(q_j)$. □

Theorem 1 alone is not enough to justify the optimization of $\rho_j(\boldsymbol{\lambda})$ with a subgradient. From the perspective of $\mathbb{R}^n$, there still exist two regions that provide local maxima of $\rho_j(\boldsymbol{\lambda})$, one on each side of $L_0(q_j)$. But if a value can be filtered from $\mathrm{dom}(Y_j)$, then one of the two local optimal regions is a global maximum of $\rho_j(\boldsymbol{\lambda})$. The rest of this section is dedicated to explaining how the properties of $\rho_j(\boldsymbol{\lambda})$ and Theorem 1 can be used to find a global maximum of $\rho_j(\boldsymbol{\lambda})$ and justify a subgradient method on $\rho_j(\boldsymbol{\lambda})$. In the following, we assume that it is possible to filter one value from $\mathrm{dom}(Y_j)$.

Before studying how to find a global maximum of $\rho_j(\boldsymbol{\lambda})$, we introduce some concepts. Since we consider both sides of $L_0(q_j)$ and we assume that we can filter one of the values from $\mathrm{dom}(Y_j)$, we need to define a side of $L_0(q_j)$ where y_j has the correct value and a side where it does not. We say that a vector $\boldsymbol{\lambda}$ *generates* $\boldsymbol{y}$ if $\boldsymbol{y}$ is obtained from $\boldsymbol{\lambda}$ using (4).

Definition 4. *$\boldsymbol{y}^*$ is an optimal solution of the LP (1), given the current state of the CSP variables at the start of Algorithm 1, but without insurance that $\boldsymbol{y}^*$ can be generated by some $\boldsymbol{\lambda}$ in the LP (2).*

In Definition 4, the optimal solution is *optimal* with regard to the original function $h(\boldsymbol{y})$ and the current state of the CSP variables Y_k. Meaning $\boldsymbol{y}^*$ minimizes the number of values used to cover the integer variables $X_1, ..., X_n$. The following lemmas show why there is no certainty, yet, that $\boldsymbol{y}^*$ can be generated.

Lemma 2. *Given $\boldsymbol{y}^*$, if there is $y_j^* = 1$ and there exists another $j' \neq j$ with $\boldsymbol{a_{cj'}} = \boldsymbol{a_{cj}}$, then $y_{j'}^* = 0$.*

Proof. If $\boldsymbol{a_{cj'}} = \boldsymbol{a_{cj}}$, with $j \neq j'$, it means that for every CSP variable X_i with j in its domain, X_i also has j'. Because $y_j^* = 1$, the value j is used to cover a subset of $\{X_1, ..., X_n\}$, and every variable with j and j' in their domain will either be fixed to j or another value in their domain, but never j'. Choosing j' and j only increases the number of values used in the solution. Thus $y_{j'}^* = 0$. □

Lemma 3. *If in $\boldsymbol{y}^*$, there is $y_j^* = 1$ and there exists another $j' \neq j$ with $\boldsymbol{a_{cj'}} = \boldsymbol{a_{cj}}$, then there exists no $\boldsymbol{\lambda}$ that generates this solution.*

Proof. From Lemma 2, $y_{j'}^* = 0$. Suppose there exist a $\boldsymbol{\lambda}$ that generates $\boldsymbol{y}^*$. Then for q_j and $q_{j'}$ the equations that generate $y_j^* = 1$ and $y_{j'}^* = 0$ are $1 - \boldsymbol{a_{cj}}^\top \boldsymbol{\lambda} < 0$ and $1 - \boldsymbol{a_{cj'}}^\top \boldsymbol{\lambda} \geq 0$. Since $\boldsymbol{a_{cj'}} = \boldsymbol{a_{cj}}$, the system has no solution. □

From Lemma 3, if we remove duplicated columns, we ensure that an optimal solution $\boldsymbol{y}^*$ can be generated by some $\boldsymbol{\lambda}$. In practice, we did not observe that phenomena often. From now on, we assume that the columns are unique and that the optimal solution $\boldsymbol{y}^*$ can be generated by some Lagrange multipliers $\boldsymbol{\lambda}^*$. We can compute the gradient of $h''(\boldsymbol{\lambda})$ in the region that produces $\boldsymbol{y}^*$. We use this gradient to explain why we can change the search direction in Algorithm 2 (line 16) and eventually get to a global maximum of $\rho_j(\boldsymbol{\lambda})$.

Definition 5. *The vector $\nabla h''(\boldsymbol{\lambda}^*) = 1 - A\boldsymbol{y}^*$ is the gradient of h'' for $\boldsymbol{y}^*$.*

Normally the gradient of a function at an extremum is zero, but because of how we assign values to the LP variable y_k (4), it is not always the case.

Lemma 4. $\nabla h''(\boldsymbol{\lambda}^*)$ *is not always* $\boldsymbol{0}$.

Proof. Whenever a variable X_i has in its domain more than one value in common with the values of the optimal solution $|\operatorname{dom}(X_i) \cap \{j \mid y_j^* = 1\}| = \boldsymbol{a_{ri}}^\top \boldsymbol{y}^* > 1$, then the gradient at that index is non-zero $[\nabla h''(\boldsymbol{\lambda}^*)]_i = 1 - \boldsymbol{a_{ri}}^\top \boldsymbol{y}^* < 0$.   □

Once the CSP variables $X_1, \ldots, X_n$ are all fixed to a value, then $\nabla h''(\boldsymbol{\lambda}^*) = 0$, but until that moment, it is not. We have all the information from the contribution of $h''(\boldsymbol{\lambda})$ to $\rho_j(\boldsymbol{\lambda})$. We analyze the last part of $\rho_j(\boldsymbol{\lambda})$, which is $|q_j(\boldsymbol{\lambda})|$.

Lemma 5. $|q_j(\boldsymbol{\lambda})|$ *has no derivative on* $L_0(q_j)$ *because for* $\boldsymbol{\lambda}$ *and* $\boldsymbol{\lambda}'$ *on two different sides of* $L_0(q_j)$ *we have* $\nabla|q_j(\boldsymbol{\lambda})| = -\nabla|q_j(\boldsymbol{\lambda}')|$.

Proof. First, we note that $\forall \boldsymbol{\lambda}'' \notin L_0(q_j)$, $\nabla|q_j(\boldsymbol{\lambda}'')| = \operatorname{sgn}(q_j(\boldsymbol{\lambda}''))\nabla(1 - \boldsymbol{a_{cj}}^\top\boldsymbol{\lambda}'') = -\operatorname{sgn}(q_j(\boldsymbol{\lambda}''))\boldsymbol{a_{cj}}$. Let $q_j(\boldsymbol{\lambda}) < 0$ and $q_j(\boldsymbol{\lambda}') > 0$. Then, we have $\nabla|q_j(\boldsymbol{\lambda})| = +\boldsymbol{a_{cj}}$ and $\nabla|q_j(\boldsymbol{\lambda}')| = -\boldsymbol{a_{cj}}$, which completes the proof.   □

From Lemmas 1 and 5, $\rho_j(\boldsymbol{\lambda})$ inherits the same set of points as $h''(\boldsymbol{\lambda})$ where $\nabla\rho_j(\boldsymbol{\lambda})$ does not exist, namely $\bigcup_{j'=1}^m L_0(q_{j'})$. Now we have all the tools to justify the use of a subgradient method for $\rho_j(\boldsymbol{\lambda})$.

Theorem 1 ensures that each gradient of $\rho_j(\boldsymbol{\lambda})$ on either of the two sides of $L_0(q_j)$ points toward the local maximum of $\rho_j(\boldsymbol{\lambda})$ of that side. On the side of $L_0(q_j)$ where $y_j = y_j^*$, the optimal region of $h''(\boldsymbol{\lambda})$ is present. From the definition of $\rho_j(\boldsymbol{\lambda})$, for any $\boldsymbol{\lambda}$ we have the following relation $\rho_j(\boldsymbol{\lambda}) \geq h''(\boldsymbol{\lambda})$. Which means that for any $\boldsymbol{\lambda}^*$ in the optimal region of $h''(\boldsymbol{\lambda})$, we have $\rho_j(\boldsymbol{\lambda}^*) \geq h''(\boldsymbol{\lambda}^*)$. In fact, for any $\boldsymbol{\lambda}^* \notin L_0(q_j)$ we have $\rho_j(\boldsymbol{\lambda}^*) > h''(\boldsymbol{\lambda}^*)$ which means filtering whenever $h''(\boldsymbol{\lambda}^*) = \max(\operatorname{dom}(N))$. The gradient of $\rho_j(\boldsymbol{\lambda})$ on this side of $L_0(q_j)$ points toward a $\boldsymbol{\lambda}$ that maximizes $\rho_j(\boldsymbol{\lambda}) > h''(\boldsymbol{\lambda}^*)$.

For the case where $y_j \neq y_j^*$, we need two other theorems to explain how we can find a direction to leave this local maximum of $\rho_j(\boldsymbol{\lambda})$ and move toward a global maximum. We show that there exist a region on the side where $y_j \neq y_j^*$ of $L_0(q_j)$, for which the gradient of $\rho_j(\boldsymbol{\lambda})$ is equal to the gradient $\nabla h''(\boldsymbol{\lambda}^*)$. This region corresponds to the set of global maxima on this side of $L_0(q_j)$.

Theorem 2. *Given* $\boldsymbol{y}^*$, *let* $\boldsymbol{\lambda}^*$ *be one of the vectors that generates* $\boldsymbol{y}^*$, *let* $\boldsymbol{\lambda}$ *be one of the vectors that generates* $\boldsymbol{y}$ *where* $\forall k \neq j, y_k = y_k^*$ *and* $y_j \neq y_j^*$. *Then* $\nabla\rho_j(\boldsymbol{\lambda}) = \nabla h''(\boldsymbol{\lambda}^*)$.

Proof. We can write the vector $\boldsymbol{y}$ generated as $\boldsymbol{y} = \boldsymbol{y}^* + \operatorname{sgn}(q_j(\boldsymbol{\lambda}^*))\hat{\boldsymbol{u}}_j$, where $\hat{\boldsymbol{u}}_j$ is a unit vector with only the j-th entry equal to one and the other entries equal to zero. For $\boldsymbol{\lambda}$, the gradient of h'' is

$$\nabla h''(\boldsymbol{\lambda}) = 1 - A(\boldsymbol{y}^* + \operatorname{sgn}(q_j(\boldsymbol{\lambda}^*))\hat{\boldsymbol{u}}_j) = 1 - A\boldsymbol{y}^* - \operatorname{sgn}(q_j(\boldsymbol{\lambda}^*))\boldsymbol{a_{cj}},$$
$$= \nabla h''(\boldsymbol{\lambda}^*) + \nabla|q_j(\boldsymbol{\lambda}^*)| = \nabla h''(\boldsymbol{\lambda}^*) - \nabla|q_j(\boldsymbol{\lambda})|$$

where we change the last term of the second equation into a more intuitive object. We use Lemma 5 to change the last term to $-\boldsymbol{\nabla}|q_j(\boldsymbol{\lambda})|$. The gradient of ρ_j for $\boldsymbol{\lambda}$ is

$$\boldsymbol{\nabla}\rho_j(\boldsymbol{\lambda}) = (\boldsymbol{\nabla}h''(\boldsymbol{\lambda}^*) - \boldsymbol{\nabla}|q_j(\boldsymbol{\lambda})|) + \boldsymbol{\nabla}|q_j(\boldsymbol{\lambda})| = \boldsymbol{\nabla}h''(\boldsymbol{\lambda}^*).$$

$\square$

Theorem 2 justifies when should Algorithm 2 change the subgradient on line 16. From Theorem 1, $\rho_j(\boldsymbol{\lambda})$ is concave on either side of $L_0(q_j)$, therefore for each of these two regions, the subgradient method converges to an optimal in each region. If y_j^* can only be 0 or only be 1 in all possible optimal solutions, then the value $1 - y_j^*$ must be filtered. The region of Lagrange multipliers that generates $\boldsymbol{y}$ from Theorem 2 is the local optimal region of $\rho_j(\boldsymbol{\lambda})$ on the side of $L_0(q_j)$ where $y_j \neq y_j^*$. In the next theorem, we formally explain why Algorithm 2 uses two subgradients.

Theorem 3. *If value $1 - y_j^*$ needs to be filtered from $\mathrm{dom}(Y_j)$, then in the region where $\boldsymbol{\lambda}$ generates $y_j = 1 - y_j^*$ but every other $y_k = y_k^*$, following $\boldsymbol{\nabla}h''(\boldsymbol{\lambda})$ bring the next $\boldsymbol{\lambda}'$ in the optimization closer to $L_0(q_j)$ and the filtering zone.*

Proof. The direction that brings $y_j \neq y_j^*$ to a sate $y_j = y_j^*$ is $-\boldsymbol{\nabla}|q_j(\boldsymbol{\lambda})|$, Lemma 5. Since $h''(\boldsymbol{\lambda})$ is concave then every region that is not the optimal region points toward this zone. So $(-\boldsymbol{\nabla}|q_j(\boldsymbol{\lambda})|)^\top(\boldsymbol{\nabla}h''(\boldsymbol{\lambda})) \geq 0$. Hence the next $\boldsymbol{\lambda}$ in the optimization is closer to $L_0(q_j)$ and the optimal region. The optimal region is in the filtering zone because, $\forall\boldsymbol{\lambda}$ we have $\rho_j(\boldsymbol{\lambda}) \geq h''(\boldsymbol{\lambda})$, we can conclude that for any $\boldsymbol{\lambda}^*$ which generates $\boldsymbol{y}^*$ we have filtering. $\square$

Algorithm 2 can alternate between the two sides of $L_0(q_j)$. This situation is due to a step-size that is to big in (9). Eventually, when the step-size is small enough, the algorithm stops cycling. We conclude that Algorithm 2 always approaches the filtering zone.

4   Experimentations and Results

We implemented Algorithm 1, denoted LR^0, based on the code that Cambazard and Fages [7] shared to us. We implemented our approach which we denote LR^+. It consists of the implementation of Algorithms 1 and 2. The code is accessible on GitHub[1]. We fixed the threshold to 0.4 and the number of steps to 40. The experiments ran on a MacBook Pro with M2 chip and with 8 Gb of RAM.

We test the algorithms on two problems: the dominating queens problem and the p-median problem. The instances for the second problem come from the discrete locations problem library [12] "*instances with a large duality gap*". There are three classes of instances (easier to harder): *A*, *B* and *C*. There are thirty instances within each class.

For all the experiences, we have chosen our constant step-size for our experiment: $\omega_k = 1.97$ and $\eta = 0.0001$ to determine the step size β_k in Eq. (10).

---

[1] https://github.com/frbert3/LocalAlterationsAlgorithmAtMostNValue.git.

Table 2. Dominating queens instances on $n \times n$ chessboards with v queens. If the instance is *feasible* we wrote Y and if not N. The time is in seconds.

	Harmonic				Geometric				Newton			
n/v F	Nodes	Fails	Iters	Time	Nodes	Fails	Iters	Time	Nodes	Fails	Iters	Time
7/4 Y LR^0	142	112	170k	0.165	166	135	84k	0.089	176	145	19k	0.026
LR^+	31	4	19k	0.024	31	4	6k	0.011	31	4	2k	0.005
8/5 Y LR^0	267	225	329k	0.301	362	320	186k	0.197	283	241	33k	0.045
LR^+	41	6	26k	0.040	76	41	24k	0.028	41	6	3k	0.007
8/4 N LR^0	2k	2k	2.4M	3.497	3k	3k	1.6M	2.461	2k	2k	279k	0.765
LR^+	1.2k	1.2k	1.5M	2.066	1k	1k	698k	1.177	1k	1k	146k	0.362
9/5 Y LR^0	1k	953	1.1M	2.188	1k	969	544k	1.315	956	909	113k	0.443
LR^+	490	443	576k	0.880	561	514	304k	0.556	441	394	57k	0.167
10/5 Y LR^0	113k	113k	133M	300	195k	195k	107M	300	944k	944k	91M	300
LR^+	9.4k	9.4k	11M	26.9	761	725	471k	3.711	844	808	121k	2.59
11/5 Y LR^0	32k	32k	38M	102.2	54k	54k	30M	83.1	83k	83k	7.34M	37.4
LR^+	4k	4k	5.6M	28.7	3.4k	3.4k	2.1M	11.6	4k	4k	461k	9.61

4.1   The Dominating Queens Problem

The dominating queens problem consists of placing the minimal number v of chess queens on a chessboard of dimensions $n \times m$ in order that every square is occupied or attacked by at least one queen. We focus on the case where $n = m$ which has more symmetries and is harder to solve.

We generate a chessboard with squares numbered from 1 to n^2 from the top-left corner to the bottom-right corner. We associate each square i with an integer variable X_i whose domain is the set of squares a queen would be able to attack from position i. We also associate a boolean variable Y_i to each square, representing whether there is a queen on the i-th square. The variable N has for domain $\{0, \ldots, v\}$. There are $2n^2 + 1$ variables in total. The objective is to minimize N. Therefore the model is

$$\textbf{Minimize}\quad N \tag{11}$$

$$\textsc{AtMostNValue}(\mathcal{X}, \mathcal{Y}, N) \tag{12}$$

$$(Y_j = 1 \Leftrightarrow \exists X_i \in \mathcal{X} \mid X_i = j)\quad \forall j \in \{1, \ldots, n^2\}, \tag{13}$$

(12) ensures that all squares are covered by the N queens. (13) ensures that if $Y_j = 1$ then there is at least one variable with $X_i = j$. Branching on the variables in $\mathcal{X}$ is performed lexicographically. A timeout of 300 s is fixed.

Table 2 contains the results of 6 instances of the *dominating queens problem*, five are feasible and one is not. The 4 metrics are *Nodes*, the numbers of explored nodes; *Fails*, the failures during the filtering, *Time*, the solving time in seconds; and *Iters*, the number of iterations done in the Lagrangian relaxation.

LR^0 does not find the solution for the 10/5 instance within the time limit. This instance is excluded from the next averages. LR^+ explores 71.8% fewer nodes, does 77.7% fewer failures, does 73.3% fewer iterations in the Lagrangian relaxation, and is 71.2% faster than LR^0. The local alterations algorithm significantly reduces the number of explored nodes. For the three-step size rules, the algorithm seems to reduce the nodes and fails metrics by closely the same amount for a given instance. We use a *Newton* type of step size, so it is not surprising that the additional filtering reduces those metrics in a similar way.

4.2   The p-Median Problem

The p-median problem is a variant of the facility location problem which consists of opening *at most* p facilities on a territory to deserve n clients at a minimal cost. In this variation, the problem focuses on disposing the p facilities in a way to minimize the total transportation cost from the facilities to the clients.

We use the same model as Cambazard and Fages [7]. There are m candidate facility locations. Each client X_i can be served by a subset of facilities, noted $P(i) \subset \{1, \ldots, m\}$. We associate a Boolean variable Y_j to each facility location j to represent if we open the facility j or not. The transportation cost of the client X_i for the facility j is contained in the matrix $C_{n \times m}$, with entries ∞ if $j \notin P(i)$. The variables T_{Ci} represent the transportation costs for the clients X_i. Their values are subject to an ELEMENT($Index, table, Value$) constraint, where $Index$ and $Value$ are variables and $table$ are constants. This constraint ensures that $Value = table[Index]$.

$$\text{Minimize} \quad \sum_{i=1}^{n} T_{Ci} \tag{14}$$

$$\text{ATMOSTNVALUE}(\mathcal{X}, \mathcal{Y}, N) \tag{15}$$

$$\text{ELEMENT}(X_i, [c_{i1}, \ldots, c_{im}], T_{Ci}) \quad \forall i \in \{1, \ldots, n\} \tag{16}$$

$$\sum_{j=1}^{m} Y_j = N \tag{17}$$

$$\text{if} \quad ((c_{ia} < c_{ib}) \vee ((c_{ia} = c_{ib}) \wedge (a < b)) \text{ then } Y_a = 1 \Rightarrow X_i \neq b \tag{18}$$

$$X_i \in P(i), \quad T_{Ci} \geq 0 \quad \forall i \in \{1, \ldots, n\}, \quad N \in [1, p] \tag{19}$$

The constraints (18) (there is one for each facility) help to choose which clients are served by which facility: the cheapest facility. We fix a timeout of 300 s.

We use two search heuristics: *standard* and *cost*. Both heuristics branch on the facilities Y_j first, because once these variables are fixed, the constraints (18) filter the variables X_i. The heuristic branch on the client variables X_i second applying the first fail principle, starting with the smallest value in the domains.

The *standard* heuristic follows the first fail principle for the facilities variables starting with the smallest domain size.

The *cost* based heuristic, introduced by Cambazard and Fages [7], is also based on the first fail principle. It branches on the facility variable Y_j which belongs to the client X_i who is served by the smallest number of facilities, the minimum domain size. It breaks ties with the cheapest facility j in average, $\frac{\sum_{i|j \in \text{dom}(X_i)} c_{ij}}{|\{i|j \in \text{dom}(X_i)\}|}$.

Table 3 shows the results for the p-median problem. The column *Optimal* reports the number of times the value of the optimal solution is found, without necessarily proving that the solution is optimal. The optimal value is given in the benchmark. The column *Proof optimal* is the number of times the solution is proven optimal. The column *# of better solutions* is the number of times that the algorithm finds a solution with a smaller objective value than other algorithm finds. The column *Av. time gain of LR^+ on LR^0* is the average percentage speedup by LR^+ over LR^0 for the instances proven optimal. The column *Av. node reduction from LR^0 to LR^+* is the average percentage of nodes not explored by LR^+ compared to LR^0 for the instances on which we prove the optimality.

We see in Table 3 that for the *easy* instances class A, only LR^+ achieves to find all the optimal solutions. It is also on this class of instances that we record the greatest speedup on instances for which both algorithms prove the optimality. For class A, the speedups of the standard and cost-based heuristics are 28.4% (in average 18 s) and 33.2% (in average 17 s).

Figure 1 presents a visual comparison of LR^0 and LR^+ on the time to solve the instances from the classes A and B. Figure 1a shows the instances that LR^0 and LR^+ achieve the proof of optimality. Figure 1b displays all instances.

Figure 1a shows that for class A, LR^+ is always faster than LR^0. We stress that LR^+ finds solutions faster and with fewer nodes than LR^0.

For instances of class B, both LR^0 and LR^+ find the same number of optimal solutions and achieve, on the same instances, the proof of optimality. With the standard heuristic, LR^+ finds two better solutions than LR^0 when neither can find the optimal solution. The average speedup for the standard and cost-based heuristics are 15.6% (in average 20 s) and 8.10% (in average 9 s).

Figure 1b shows that when using a more accurate search heuristic, as the cost based heuristic, LR^+ becomes less effective. In other words, LR^+ wastes time at filtering values for which the heuristic has no intention to branch on. Nevertheless, LR^+ is still faster than LR^0 and explores 72.6% fewer nodes with the standard heuristic (in average 94764 fewer nodes) and 71.3% fewer nodes with the cost based heuristic (which represents in average 85251 fewer nodes).

On the hardest instances in class C, no algorithms are able to prove the optimality within the time limit. Although, with both search heuristics, LR^+

Table 3. Results of the three classes of problem of the facility location problem. Each class has 30 instances and the solutions are known.

Problem	Heuristics	Optimal		Proof optimal		# of better solutions		Av. time gain of LR^+ on LR^0*	Av. node reduction from LR^0 to LR^+*
		LR^0	LR^+	LR^0	LR^+	LR^0	LR^+	[%]	[%]
A	standard	27	27	27	27	0	1	28.4	71.5
	cost	29	**30**	27	27	0	1	33.2	72.0
B	standard	28	28	18	18	0	**2**	15.6	72.6
	cost	28	28	18	18	0	0	8.10	71.3
C	standard	5	**6**	0	0	0	**8**	–	–
	cost	8	**12**	0	0	0	**13**	–	–

On the instances for which we were able to prove the optimality.

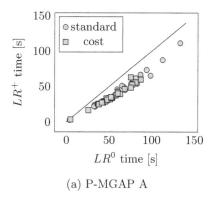
(a) P-MGAP A

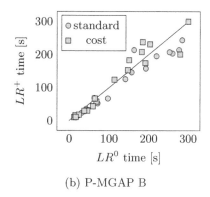
(b) P-MGAP B

Fig. 1. Time comparison of LR^0 and LR^+

finds the optimal solution of more instances than LR^0. Furthermore, LR^+ finds better solutions than LR^0 for 8 and 13 instances.

Difference Between the Problems. The main difference between the results on the dominating queens problem and the p-median problem is the average speedup. The speedup is less significant in the latter. The p-median problem is more complex than the dominating queen problem because there are more constraints and more variables. Also, those are instances randomly generated, the matrix A has no structure. In contrast with the very structured matrix A in the dominating queens problem. This might affect the convergence of the algorithm.

4.3   Determining Parameters

Experimental data helped us fix the parameters' values. Figure 2 shows the solving time and number of nodes of four instances of the p-median problem (A7, A25, B11 and B31) in function of the threshold and number of steps.

The solving time curves (Fig. 2a) follow the same trend. When the threshold lies in $[0, 0.25]$, the solving time diminishes because LR^+ starts filtering values. In $[0.25, 0.5]$, LR^+ filters even more values, but beyond 0.5 it considers too many values and wastes time. All curves share a minimum around 0.50.

In Fig. 2b, we see that when the threshold augments, the number of explored nodes decreases almost exponentially until it stops to a value. After 0.40, the number of explored nodes starts to decrease much slowly.

In Fig. 2c, we see that the solving time stops decreasing after 40 steps. The variation in time induced by the number of steps between LR^0 (the first point) and LR^+ is less than the variation induced by the threshold. It is not surprising. The threshold dictates when to trigger LR^+ and the number of steps dictates how much time is spent in LR^+. We see that for a number of steps between

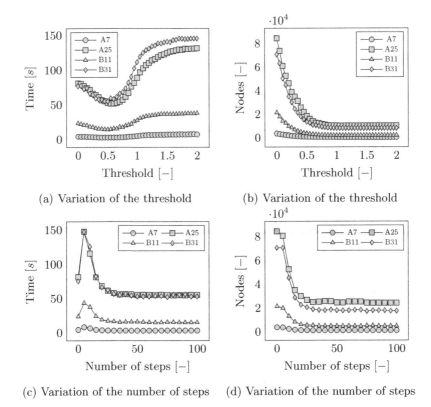

(a) Variation of the threshold   (b) Variation of the threshold

(c) Variation of the number of steps   (d) Variation of the number of steps

Fig. 2. Variation of the *threshold* and *number of steps* parameters

1 and 15, the time increases, because we are not converging fast enough to a maximum. We lose time by trying to filter values without enough steps.

In Fig. 2d, we see a tendency similar to the variation of the threshold. Again the variation, induced by the number of steps, of the number of explored nodes is less than the variation induced by the threshold. Nevertheless, we can infer that a greater number of steps helps filtering more values.

5   Conclusion

We conceived a Lagrangian relaxation based algorithm for the AtMostNValue constraint that uses local alterations of the multipliers to enhance the filtering. This new filtering algorithm leads to an average speedup of 71% on the dominating queens problem and average speedups of 33% and 8% on the instances of class A and class B of the p-median problem.

Acknowledgement. We thank Hadrien Cambazard for sharing with us his code for the AtMostNValue. It served as a thorough guide for the present work.

References

1. Beldiceanu, N.: Pruning for the minimum constraint family and for the number of distinct values constraint family. In: Principles and Practice of Constraint Programming, pp. 211–224 (2001)
2. Beldiceanu, N., Carlsson, M., Rampon, J.X.: Global constraint catalog (2010)
3. Bertsekas, D.P.: On the Goldstein-Levitin-Polyak gradient projection method. IEEE Trans. Autom. Control **21**(2), 174–184 (1976)
4. Bessiere, C., Hebrard, E., Hnich, B., Kiziltan, Z., Walsh, T.: Filtering algorithms for the nv alue constraint. Constraints **11**, 271–293 (2006)
5. Boudreault, R., Quimper, C.G.: Improved cp-based lagrangian relaxation approach with an application to the tsp. In: IJCAI, pp. 1374–1380 (2021)
6. Boyd, S., Xiao, L., Mutapcic, A.: Subgradient methods. lecture notes of EE392o, Stanford University. Autumn Quart. **2004**, 2004–2005 (2003)
7. Cambazard, H., Fages, J.G.: New filtering for atmostnvalue and its weighted variant: a Lagrangian approach. Constraints **20**(3), 362–380 (2015)
8. Fisher, M.L.: An applications oriented guide to Lagrangian relaxation. Interfaces **15**(2), 10–21 (1985)
9. Focacci, F., Lodi, A., Milano, M.: Cost-based domain filtering. In: International Conference on Principles and Practice of Constraint Programming, pp. 189–203 (1999)
10. Fumero, F.: A modified subgradient algorithm for Lagrangean relaxation. Comput. Oper. Res. **28**(1), 33–52 (2001)
11. Held, M., Wolfe, P., Crowder, H.P.: Validation of subgradient optimization. Math. Program. **6**(1), 62–88 (1974)
12. Kochetov, Y., Ivanenko, D.: Computationally difficult instances for the uncapacitated facility location problem. In: Metaheuristics: Progress as Real Problem Solvers, pp. 351–367 (2005)
13. Lemaréchal, C.: Lagrangian relaxation. In: Computational Combinatorial Optimization: Optimal or Provably Near-Optimal Solutions, pp. 112–156 (2001)
14. Petit, T., Régin, J.C., Bessiere, C.: Specific filtering algorithms for over-constrained problems. In: Principles and Practice of Constraint Programming, pp. 451–463 (2001)
15. Polyak, B.T.: Minimization of unsmooth functionals. USSR Comput. Math. Math. Phys. **9**(3), 14–29 (1969)
16. Sellmann, M.: Theoretical foundations of cp-based lagrangian relaxation. In: Principles and Practice of Constraint Programming, pp. 634–647 (2004)

Single Constant Multiplication for SAT

Hendrik Bierlee[1,2]([✉]), Jip J. Dekker[1,2], Vitaly Lagoon[3], Peter J. Stuckey[1,2],
and Guido Tack[1,2]

[1] Monash University, Wellington Road, Clayton, VIC 3800, Australia
{hendrik.bierlee,jip.dekker,peter.stuckey,guido.tack}@monash.edu
[2] OPTIMA ARC Industrial Training and Transformation Centre, Carlton, Australia
[3] Cadence Design Systems, San Jose, USA
lagoon@cadence.com

Abstract. This paper presents new methods of encoding the multiplication of a binary encoded integer variable with a constant value for Boolean Satisfiability (SAT) solvers. This problem, known as the Single Constant Multiplication (SCM) problem, is widely studied for its application in hardware design, but its techniques are currently not employed for SAT encodings. Considering the smaller and variable bit sizes and the different cost of operations in SAT, we propose further improvements to existing methods by minimizing the number of full/half adders, rather than the number of ripple carry adders. We compare our methods to simple encodings into adders, currently employed by SAT encoders, and direct Boolean encoding using logic minimization tools. We show that our methods result in improved solve-time for problems involving integer linear constraints. A library of optimal recipes for each method to encode SCM for SAT is made available as part of this publication.

Keywords: Single Constant Multiplier · Boolean Satisfiability · Encoding constraints

1 Introduction

Boolean Satisfiability (SAT) is a powerful approach to solving combinatorial problems, but in order use a SAT solver, we need to encode the problem into clauses. One fundamental constraint is the constant multiplication $y = c \times x$ of input and output integer variables x and y, and constant c. If x and y are encoded using a binary representation, any such multiplication can be achieved using a combination of additions, subtractions, and multiplications by a power of two. For example, the multiplication $5 \times x$ can be decomposed into $4 \times x + 1 \times x$. Since multiplication by a power of two in a binary encoding is a simple left-shift operation, which does not incur any cost, this decomposition effectively turns the multiplication into a sequence of additions.

The most basic decomposition for a multiplication $c \times x$ therefore simply considers the binary representation of c and introduces one addend for every 1 in this binary representation – such as in the example of $5x = 1x + 4x$ above,

B. Dilkina (Ed.): CPAIOR 2024, LNCS 14742, pp. 84–98, 2024.
https://doi.org/10.1007/978-3-031-60597-0_6

or $117x = 64x + 32x + 16x + 4x + 1x$. However, decompositions that reuse intermediate results or use a combination of additions and subtractions can result in fewer additions. For example, we could define $63x = 64x - x$, $59x = 63x - 4x$, $117x = 2 \times 59x - x$. This only requires 3 additions/subtractions, compared to the 4 in the original example.

In hardware applications such as ASIC and FPGA, it is important for the performance and cost of the application to use small circuits, and therefore there is considerable previous research on finding a circuit for a given constant that minimizes the number k of additions and subtractions. This is known as the *Single Constant Multiplication* (SCM) problem. SCM is NP-hard [7], and has an upper bound of $k \leq \lceil \log_2(c)/2 \rceil$ [3]. For many practical instances of the problem, however, k is known to be much smaller.

When encoding constant multiplication for SAT, an approach similar to SCM can be used since integer variables – like in hardware – are often encoded in a binary representation [5]. Unlike hardware, the number of bits w to represent x is not fixed, but depends on the domain of x (e.g., only $w = 2$ bits are required to represent $x \in [0, 2]$ with Boolean variables x_0, x_1).

Furthermore, the complexity of the SAT encoding of a single addition or subtraction (using a ripple carry adder [17]) varies. The required number of full (and half) adders is determined by the constant as well as the number of bits of the inputs. In fact, the addition of $x + 4x = 5x$ requires *no* adders if x has two bits x_0, x_1, since $5x$ would be represented using the existing bits x_0, x_1, x_0, x_1. If x has three bits, multiple adders are required. Consequently, instead of minimizing the number of additions/subtractions k, a better circuit for SAT minimizes the total number of full/half adders a.

Example 1. Consider two circuits for $c = 117$: $17x = 16x + x; 19x = 2x + 17x; 117x = 8 \times 17x (= 136x) - 19x$ and $63x = 64x - x; 59x = 63x - 4x; 117x = 2 \times 59x (= 118x) - x$. Both are minimal with respect to $k = 3$, but assuming $w = 4$ bits for $x \in [0, 7]$, the first requires $a = 18$ full/half adders, while the second requires $a = 31$. Yet another circuit, $3x = 2x + x; 49x = 16 \times 3x (= 48x) + x; 17x = 16x + x; 117x = 4 \times 17x (= 68x) + 49x$, requires only $a = 12$ full/half adders, even though it uses $k = 4$ additions/subtractions. For $w = 4$, $a = 12$ is optimal for $c = 117$, but for different w the best circuit changes.

Note that if we constrain all intermediates to be represented in a fixed bit width (common in hardware solutions), then this may change the answer. For example, given fixed 12 bit width for intermediates the first recipe is not applicable when x is a 5 bit number, since intermediate $136x$ is not guaranteed to fit in 12 bits, while $117x$ does. □

The main contribution of this paper is the application of the SCM methodology in the context of SAT, the extension to the basic approach that result in better SAT encodings by minimizing full/half adders, and a database with pre-computed *recipes* of the SCM problem for a range of constants and bit widths. Our database exhaustively covers the parameter space over $1 \leq w \leq 16$ and $1 \leq c \leq 2047$. We assume no fixed bit width restriction on intermediate results.

We present a set of recipes to compute $c \times x$ optimal in terms of number of additions/subtractions for the general SCM problem, independent of the size of x, assuming we can use any number of bits for intermediate results. Then, we produce a set of recipes that minimize the number of full/half adders to compute $c \times x$. These recipes differ depending on the bit size of x. The SCM SAT encoding database is available as part of the released implementation and benchmarks [8].

We evaluate the different SCM recipes in Sect. 4, where we use the pre-computed circuits to encode problems which focus on constant multiplication. We observe significant improvements in terms of the size of the final encoding and the SAT solver performance. In Sect. 5, we discuss future work and conclude our findings.

2   Preliminaries and Related Work

In this section, we give preliminaries on the *Constraint Satisfaction Problem* (CSP) and SAT problem. We then formally define the SCM problem and discuss existing approaches for solving it.

2.1   Constraint Satisfaction Problem

A CSP [15] instance, $P = (\mathcal{X}, D, \mathcal{C})$, consists of a set of variables $\mathcal{X}$, with each $x \in \mathcal{X}$ restricted to take values from some domain $D(x)$. For this paper we assume domains are ordered sets of integers, and denote by $\mathrm{lb}(x)$ and $\mathrm{ub}(x)$ the least and greatest values in $D(x)$. We will use interval notation $[l, u]$ to represent the set of integers $\{l, l+1, \ldots, u\}$. A set of constraints $\mathcal{C}$ expresses relationships between the variables. An *assignment* of a CSP instance is a mapping of variables to values. An assignment that is consistent with the domains and constraints is a *solution* to the instance. A *Constraint Optimization Problem* (COP) is a CSP with an additional objective function f that assigns a value to each assignment. An optimal solution to a minimization (maximization) COP is one that minimizes (maximizes) f.

2.2   Boolean Satisfiability

A SAT problem can be seen as a special case of a CSP, where the domain for all variables x is $D(x) \in \{0, 1\}$, representing the values *false* and *true*. A *literal* is either a Boolean variable x or its negation $\neg x$. We extend the negation operation to operate on literals, i.e., $\neg l = \neg x$ if $l = x$ and $\neg l = x$ if $l = \neg x$. We use the notation $l = v$ where l is a literal and $v \in \{0, 1\}$ to encode the appropriate form of the literal, i.e., if $v = 1$ it is equivalent to l and if $v = 0$ it is equivalent to $\neg l$. A *clause* is a disjunction of literals. In a SAT problem P, the constraints are in *Conjunctive Normal Form* (CNF), which means that $\mathcal{C}$ is a set of clauses.

A general CSP can be mapped, or encoded, into SAT, by encoding each integer variable into a set of Boolean *encoding* variables, and each constraint into a set of clauses and additional auxiliary Boolean variables. There are a number of different ways of encoding integer variables [5], such as the direct and order encodings. In this paper, we focus on the *binary* encoding.

Given an integer variable x with non-negative, possibly non-contiguous domain $D(x)$, the unsigned binary encoding method maps x to $m_x = \lfloor \log_2 \mathrm{ub}(x) \rfloor + 1$ encoding variables $[\![\mathrm{bit}(x,i)]\!], 0 \leq i < m_x$. The semantics of the encoding can be expressed by the equation $x = \sum_{i=0}^{m_x-1} 2^i \times [\![\mathrm{bit}(x,i)]\!]$. For example, for integer variables d with domain $D(d) = [0, 127]$, an encoding with $m_d = 7$ bits can be used. In an assignment with $d = 117 = 0b1110101$, we would have $[\![\mathrm{bit}(d,0)]\!] = 1$, $[\![\mathrm{bit}(d,1)]\!] = 0$, $[\![\mathrm{bit}(d,2)]\!] = 1$ and so on. If the initial domain $D(x)$ is non-contiguous, or has bounds that are not powers of 2, these have to be enforced using additional constraints (clauses). The representation can be extended in a straightforward way to support negative domain values, by assuming a two's complement encoding instead of a simple unsigned binary encoding. For the rest of the paper, we will assume non-negative domains to simplify presentation, and only discuss the general approach where required.

2.3   The Single Constant Multiplication Problem

The SCM problem can be formalized as follows. Given the multiplier target c, we are to decide a sequence of up to n equations of the form $c' \times x = 2^{s_l} \times (c_l \times x) \pm 2^{s_r} \times (c_r \times x)$. The left argument $c_l \times x$ is either a previously computed multiple or $c_l = 1$, and $s_l \geq 0$ is the left-shift of the left argument. Similarly, $c_r \times x$ is the right argument, and $s_r \geq 0$ is its left-shift.[1] The final equation should define $c \times x$.

We can thus encode a recipe for $c = c^k$ as a sequence of k equations represented by tuples $\langle c_l^i, s_l^i, c_r^i, s_r^i, \#^i, c^i \rangle$, where $c^i = (c_l^i \ll s_l^i) \#^i (c_r^i \ll s_r^i)$ such that $k \leq n$, $\#^i \in \{+, -\}$, $\{c_l^i, c_r^i\} \subseteq \{c^j \mid j < i\} \cup \{1\}$ and $s_l^i, s_r^i \geq 0$.

Example 2. The first recipe for 117 of Example 1 is defined by the sequence $[\langle 1, 4, 1, 0, +, 17 \rangle, \langle 1, 1, 17, 0, +, 19 \rangle, \langle 17, 3, 19, 0, -, 117 \rangle]$. The second recipe is defined by $[\langle 1, 6, 1, 0, -, 63 \rangle, \langle 63, 0, 1, 2, -, 59 \rangle, \langle 59, 1, 1, 0, -, 117 \rangle]$.

Usually, the goal of the SCM problem is to find the minimal k recipe for a given c (knowing that it is bounded by $k \leq \lceil \log_2(c)/2 \rceil$ [3]).

Related Work. The SCM problem has been studied extensively in the context of hardware. Many approaches for finding optimal circuits have been proposed, using graph based techniques [9], or using *Integer Linear Programming* (ILP) solvers [11], SAT solvers [2,12], as well as heuristic approaches [2]. A generalization of SCM is the MCM, which builds a circuit for multiple target constants. Compared to SCM, this enables further sharing of intermediate results between the circuits for different constants. Apart from minimizing the number of nodes

---

[1] Some versions of SCM also allows right shifts, and while these are necessary for optimal *Multiple Constant Multiplication* (MCM) solutions, we did not come across any instances of SCM where they improved the solution.

(additions and subtractions), some of these related works consider other objectives such as minimizing the required surface area, delay, or power consumption of the generated circuits. These metrics are obviously useful when generating hardware circuits, but they are not directly relevant for SAT encodings. We are not aware of approaches that minimize the number of full/half adders in the resulting circuit. This metric is useful in the context of SAT solvers, since it results in a smaller CNF in terms of the number of variables, clauses and literals, as we shall see in Sect. 4. We briefly discuss MCM for SAT as future work in Sect. 5.

Another difference with existing approaches is that the domains of intermediate results is left unbounded when solving SCM for SAT. Approaches that model hardware circuits have a fixed bit width h for all results $c' \times x$. We could modify our approach easily by fixing the number of bits for the result (and intermediates) as well. Either we would allow overflow (computing $(c \times x) \mod 2^h$), or we would enforce the (intermediate) results to always fit in h bits. However, for SAT we can consider arbitrary bit widths (up to $m_{c \times x}$ bits) for the intermediate results.

3 Applying Single Constant Multiplication to SAT

In this section, we present different approaches for solving the SCM problem. We start with well-known, simple approaches, and then introduce an encoding of SCM into COP. We discuss these approach with a particular focus on their use in SAT encodings.

3.1 A Baseline Algorithm for Encoding $y = c \times x$

In order to establish a baseline, we will first discuss a well-known, simple solution to the SCM problem that does not optimize the resulting circuit at all.

Multiplication by a constant can always be represented using a combination of shifts and additions. Suppose x is defined as a m_x width binary encoded integer. Then we know that $c \times x$ can be at most $c \times (2^{m_x} - 1)$, thus requiring $m_{c \times x} = \lceil \log(c \times (2^{m_x} - 1)) \rceil$ bits to encode.

A basic shift-and-add algorithm will construct a solution to the above SCM problem by decomposing $c \times x$ by a shifted term for every 1 in c's binary representation:

$$c \times x = \sum_{\substack{0 \leq i < m_c \\ [\![\text{bit}(c,i)]\!]=1}} 2^i x$$

The number of additions required is equal to the number of 1 s in c's binary representation minus one, or at most $\lceil \log(c) \rceil - 1$.

3.2   Using Boolean Circuit Minimization to Tackle SCM

Another option for encoding SCM into SAT is to use algorithms such as the Espresso [6] logic minimizer that directly produce small logic circuits based on truth tables or similar representations of Boolean formulae. These algorithms are remarkably powerful, and for small enough problems can produce a guaranteed minimal size circuit for a formula *without introducing auxiliary Boolean variables for intermediate results*.

Example 3. For example, we can construct a CNF encoding for $y = 117 \times x$ with $x \in [0, 15]$ by feeding the following truth table between the x and y bits into Espresso:

	x	y		x	y
0	0 0 0 0	0 0 0 0 0 0 0 0 0 0 0 0	8	1 0 0 0	0 1 1 1 0 1 0 1 0 0 0
1	0 0 0 1	0 0 0 0 1 1 1 0 1 0 1	9	1 0 0 1	1 0 0 0 0 0 1 1 1 0 1
2	0 0 1 0	0 0 0 1 1 1 0 1 0 1 0	10	1 0 1 0	1 0 0 1 0 0 1 0 0 1 0
3	0 0 1 1	0 0 1 0 1 0 1 1 1 1 1	11	1 0 1 1	1 0 1 0 0 0 0 0 1 1 1
4	0 1 0 0	0 0 1 1 1 0 1 0 1 0 0	12	1 1 0 0	1 0 1 0 1 1 1 1 1 0 0
5	0 1 0 1	0 1 0 0 1 0 0 1 0 0 1	13	1 1 0 1	1 0 1 1 1 1 1 0 0 0 1
6	0 1 1 0	0 1 0 1 0 1 1 1 1 1 0	14	1 1 1 0	1 1 0 0 1 1 0 0 1 1 0
7	0 1 1 1	0 1 1 0 0 1 1 0 0 1 1	15	1 1 1 1	1 1 0 1 1 0 1 1 0 1 1

The circuit minimization of this truth table results in a CNF encoding with 15 variables (simply representing the bits of x and y) and 84 clauses.

Unfortunately as the width w of the input variable x and the size of the constant c grows, the number of Boolean variables in the formula $y = c \times x$ quickly becomes too large for logic minimization. However, for small widths w and constants c logic minimization can generate the best SAT encoding for $y = c \times x$, as demonstrated by our experiments in Sect. 4.

3.3   Formulating SCM as a COP

We will now introduce a general method for solving the SCM problem, by formulating it as a COP.

The following proposition shows that we only need concern ourselves with generating SCM encodings for odd numbers c (since we can derive even multiples by left shifts).

Proposition 1. *Suppose* $c = 2^i c'$ *where* $c' \bmod 2 = 1$ *and* $i \geq 1$ *then we can compute* $c \times x$ *from* $c' \times x$ *without any operations.*

Given the above result we can reduce the possible forms of the equation to just three:

Proposition 2. *Given a k equation recipe to compute $c \times x, c \bmod 2 = 1$ using the general form above, then there is a k equation recipe using equations of the form* $c' = \mathit{SPLUS}(c_l, s, c_r)$: $c' = 2^s c_l + c_r$, $c' = \mathit{SMINUS}(c_l, s, c_r)$: $c' = 2^s c_l - c_r$, *and* $c' = \mathit{MINUSS}(c_l, s, c_r)$: $c' = c_l - 2^s c_r$.

Proof. Given a k length list of tuples computing c, we show by induction we can replace this by an equal length list of the new equation forms:

Suppose the first tuple not of these forms is $\langle c_l, s_l, c_r, s_r, \#, c' \rangle$. If $s_r = 0$ then we can rewrite this as either $c' = \mathtt{SPLUS}(c_l, s_l, c_r)$ if $\# = +$ or $c' = \mathtt{SMINUS}(c_l, s_l, c_r)$ if $\# = -$. If $s_l = 0$ then we can rewrite this as either $c' = \mathtt{SPLUS}(c_r, s_r, c_l)$ if $\# = +$ or $c' = \mathtt{MINUSS}(c_l, s_r, c_r)$ if $\# = -$.

Otherwise, $s_l \geq 1 \wedge s_r \geq 1$. Suppose $s_l = s_r$, then we can rewrite the tuple to either $c'' = \mathtt{SPLUS}(c_r, 0, c_l)$ if $\# = +$ or $c'' = \mathtt{SMINUS}(c_l, 0, c_r)$ if $\# = -$, and replace all later usages of c' by c'' by adding s_l to the matching shift argument. Suppose $s_l > s_r$ then we can replace the tuple by recipe $c'' = \mathtt{SPLUS}(c_l, s_l - s_r, c_r)$ if $\# = +$ and $c'' = \mathtt{SMINUS}(c_l, s_l - s_r, c_r)$ if $\# = -$, and replace later uses of c' by c'' adding s_r to the shift argument similarly. Suppose $s_l < s_r$ then we can replace the tuple by recipe $c'' = \mathtt{SPLUS}(c_r, s_r - s_l, c_l)$ if $\# = +$ and $c'' = \mathtt{MINUSS}(c_l, s_r - s_l, c_r)$ if $\# = -$, and replace later uses of c' by c'' adding s_r to the shift argument similarly.

The requirement that $c \bmod 2 = 1$ means that one of the first two cases is applicable to the last equation.   □

Example 4. Both recipes for 117 of Example 2 can be represented equivalently as $[17 = \mathtt{SPLUS}(1, 4, 1), 19 = \mathtt{SPLUS}(1, 1, 17), 117 = \mathtt{SMINUS}(17, 3, 19)]$. The second recipe is defined by $[63 = \mathtt{SMINUS}(1, 6, 1), 59 = \mathtt{MINUSS}(63, 2, 1), 117 = \mathtt{SMINUS}(59, 1, 1)]$.

A MiniZinc [13] model for solving SCM as a combinatorial optimization problem is shown in Listing 1. Note that it only makes use of the three recipes from Proposition 2. Each equation is represented by its type `ty`, with `NOP` added for an unused equation; the equation numbers defining its left (`left`) and right (`right`) arguments, and the shift amount `shift`.[2] We use equation number 0 to represent $1 \times x$. The multiplier resulting from an equation is defined by `mult`. Line 14 sets the dummy equation multiplier to 1, and enforces the final result to be the target. In lines 16–18 we enforce that all equations after `used` are NOPs, and set their other components to dummy values. Line 19 enforces that equations use only earlier defined multiples. The computation of the multiplier for each equation (lines 20–27) follows the definition, where `NOP` just returns the previous multiplier. We minimize the number of used equations. Note that the model "pre-computes" the powers of 2 in `p2` (line 13), since solvers typically propagate poorly for exponential expressions.

3.4   Encoding an SCM Recipe

Given a solution (e.g., Example 2) to the SCM problem, we can encode the shifts and adds of each equation $(c' \times x) = 2^{s_l} \times (c_l \times x) \pm 2^{s_r} \times (c_r \times x)$. We obtain the

---

[2] The careful reader will note that we limit the maximum shift using the target. Relaxing this restriction and never found better recipes (in terms of equations or adders), but we have no formal proof for it.

```
 1 int: xbits; % number of bits to represent x
 2 int: n;      % max number of shift +- equations
 3 set of int: EQ = 1..n;
 4 set of int: EQ0 = 0..n; % equation number 0 = 1x
 5 int: maxsh = ceil(log2(target)); % maximum left shift
 6 int: target;   % target multiplier
 7 enum TYPE = { SPLUS , SMINUS , MINUSS , NOP };
 8 array[EQ] of var TYPE: ty;    % type of equation
 9 array[EQ] of var EQ0: left;   % left input equation numnber
10 array[EQ] of var EQ0: right;  % right input equation number
11 array[EQ] of var 0..maxsh: shift; % shift left applied
12 array[EQ0] of var 0..infinity: mult; % multiplier value
13 array[1..maxsh+xbits] of int: p2 = [2^i | i in 1..maxsh+xbits];
14 constraint mult[0] = 1 /\ mult[n] = target;
15 var EQ0: used; % number of equations used
16 constraint forall(e in EQ)(e > used <-> ty[e] = NOP);
17 constraint forall(e in EQ)(e > used ->
18                     left[e] = 0 /\ right[e] = 0 /\ shift[e] = 1);
19 constraint forall(e in EQ)(left[e] < e /\ right[e] < e);
20 constraint forall(e in EQ)(mult[e] =
21   if ty[e] = SPLUS then
22     p2[shift[e]] * mult[left[e]] + mult[right[e]]
23   elseif ty[e] = SMINUS then
24     p2[shift[e]] * mult[left[e]] - mult[right[e]]
25   elseif ty[e] = MINUSS then
26     mult[left[e]] - p2[shift[e]] * mult[right[e]]
27   else mult[e-1] endif);
28 solve minimize used; % minimize equations
```

Listing 1. A MiniZinc model to find a recipe of at most size n for computing target × x where x is represented by xbits bits.

binary encoding of the output $z = (c' \times x)$ from the already computed binary encodings of the inputs $z_l = (c_l \times x)$ and $z_r = (c_r \times x)$.

To apply a shift $y_l = z_l \ll s$ (or equivalently, $y_l = 2^s z_l$) on the binary encoding of z_l (or to apply $y_r = z_r \ll s$), we simply extend its encoding by $[\![\text{bit}(y_l, i)]\!] = 0, 0 \leq i < s$ and $[\![\text{bit}(y_l, i)]\!] = [\![\text{bit}(z_l, i - s)]\!], s \leq i \leq m_{z_l}$ using no additional variables or clauses.

After shifting both inputs, it remains to encode an addition of the form $z = y_l + y_r$. Here, we can encode a ripple carry adder [17] which produces the output (sum) bits $[\![\text{bit}(z, i)]\!]$, and auxiliary carry bits $[\![\text{bit}(c, i)]\!]$. Inputs variables y_l, y_r might contain fixed literals due to the applied shifts, which in turn leads to output variables that are fixed, or that are equivalent to input bits, or their negations. The initial carry bit is also fixed as $[\![\text{bit}(r, 0)]\!] = 0$ (which is essentially why a half-adder is used to add $[\![\text{bit}(y_l, 0)]\!]$ and $[\![\text{bit}(y_r, 0)]\!]$). We optimize the ripple carry adder encoding to account for this.

The current sum bit $[\![\text{bit}(z, i)]\!]$ is the result of the xor operation on up to three input variables, $[\![\text{bit}(y_l, i)]\!] \oplus [\![\text{bit}(y_r, i)]\!] \oplus [\![\text{bit}(r, i)]\!]$, which we reformulate more generically as $\oplus([[\![\text{bit}(y_l, i)]\!], [\![\text{bit}(y_r, i)]\!], [\![\text{bit}(r, i)]\!]])$.

We remove any fixed variables from the xor's input, and compute the sum of their values. Then, we apply the xor on the remaining non-fixed variables, negating the output if the fixed sum is odd. Given a list l of bits, some of which are fixed, define $fs(l)$ as the sum of the fixed bits in l and $vb(l)$ the list of unfixed (variable) bits in l. We can encode the $[\![bit(z,i)]\!] = \oplus([\![bit(y_l,i)]\!], [\![bit(y_r,i)]\!], [\![bit(r,i)]\!]])$ as follows:

$$\oplus(l) = \begin{cases} fs(l) \mod 2, & \text{if } vb(l) = [] \\ b_1, & \text{if } vb(l) = [b_1] \wedge fs(l) = 0 \\ \neg b_1, & \text{if } vb(l) = [b_1] \wedge fs(l) = 1 \\ b_1 \oplus b_2 & \text{if } vb(l) = [b_1, b_2] \wedge fs(l) = 0 \\ b_1 \oplus b_2 \oplus 1 & \text{if } vb(l) = [b_1, b_2] \wedge fs(l) = 1 \\ b_1 \oplus b_2 \oplus b_3 & \text{otherwise } vb(l) = [b_1, b_2, b_3] \end{cases}$$

If $\oplus(l)$ is constant or a literal, we can just equate $[\![bit(z,i)]\!]$ with the result, requiring no encoding clauses. Otherwise, $[\![bit(z,i)]\!]$ is equated to a $\oplus$ expression, which can be encoded using a single XOR clause if supported by the SAT solver [16], or encoded with (standard) clauses in a well understood manner.

The next carry bit is set to true if at least 2 of its input variables are. In other words, $[\![bit(r,i+1)]\!]$ is true if and only if the expression $[\![bit(y_l,i)]\!] + [\![bit(y_r,i)]\!] + [\![bit(r,i)]\!] \geq 2$ is true. Again, we split off and sum the fixed variables, and compute the carry by evaluating the generic version of the at-least-2 constraint as follows:

$$\left(\sum(l) \geq 2\right) = \begin{cases} 1, & \text{if } fs(l) \geq 2 \\ 0 & \text{if } vb(l) = [] \wedge fs(l) < 2 \\ 0, & \text{if } vb(l) = [b_1] \wedge fs(l) = 0 \\ b_1, & \text{if } vb(l) = [b_1] \wedge fs(l) = 1 \\ b_1 \wedge b_2, & \text{if } vb(l) = [b_1, b_2] \wedge fs(l) = 0 \\ b_1 \vee b_2, & \text{if } vb(l) = [b_1, b_2] \wedge fs(l) = 1 \\ (b_1 \wedge b_2) \vee (b_1 \wedge b_3) \vee (b_2 \wedge b_3) & \text{otherwise } vb(l) = [b_1, b_2, b_3] \end{cases}$$

If $\sum(l) \geq 2$ is constant or a literal, we can just equate $[\![bit(r,i+1)]\!]$ with the result, requiring no encoding clauses. Otherwise, $[\![bit(r,i+1)]\!]$ is equated to the CNF expression as normal.

To handle subtraction, we use the fact that ripple-carry adders directly handle 2's complement. So, $y_l + (-y_r) = y_l + (\bar{y}_r + 1)$ where the encoding of $\bar{y}_r$ is the complement of y_r, i.e., $[\![bit(\bar{y}_r,i)]\!] = \neg[\![bit(y_r,i)]\!]$. To offset the constant 1, we can add an initial carry of 1.

3.5   Minimizing the Number of Adders

The COP formulation of Sect. 3.3 minimizes the number of equations, k. However, we show in Sect. 3.4 how the true encoding complexity for each equation is more closely related to the number of (full/half) adders. In this section, we formulate an alternative objective which minimizes the number of adders a instead.

We are now required to consider the bit width w of the input x being multiplied as an additional parameter. Consider the addition $c' \times x = 2^{s_l} \times (c_l \times x) + 2^{s_r} \times (c_r \times x)$. Assuming the left input $c_l \times x$ is represented in $m_{c_l \times x}$ bits and the right input $c_r \times x$ is represented in $m_{c_t \times x}$ bits, then the number of full/half adders naïvely required for the addition is the width of the result $m_{c' \times x}$. However, we can omit full/half adders for low bits where one argument is guaranteed 0, and if the non-zero bits never overlap we need no adders. So the number of full/half adders required is 0 if $s_l > m_{c_r \times x} + s_r$ or if $s_r > m_{c_l \times x} + s_l$, otherwise the number of adders required is $\max(m_{c_l \times x} + s_l, m_{c_r \times x} + s_r) - \max(s_l, s_r)$. For subtractions, the only bits not requiring adders are the low bits where both operands are known to be 0, so we require $m_{c' \times x} - \min(s_l, s_r)$ full/half adders.

We can similarly show that we do not need the more general recipes, we can restrict to the recipes of Proposition 2, since these also do not change the number of full/half adders required.

Proposition 3. *Given an a adder recipe to compute $c \times x, c \mod 2 = 1$ using the general form of k equations, then there is an a adder recipe using equations of the form $c' = \mathsf{SPLUS}(c_l, s, c_r)$: $c' = 2^s c_l + c_r$, $c' = \mathsf{SMINUS}(c_l, s, c_r)$: $c' = 2^s c_l - c_r$, and $c' = \mathsf{MINUSS}(c_l, s, c_r)$: $c' = c_l - 2^s c_r$.*

Proof. (Sketch) The proof follows that of Proposition 2. Note that if we shift both addends right by the same amount, then we don't change the number of adders required since the low bits do not require adders. When we replace the addition of c' by c'' shifted left by some amount s, we are simply computing the same addition since $c' = 2^s c''$ which does not change the number of adders required. □

We can modify our MiniZinc model (see Listing 1) to keep track of the number of bits required for the result of each equation, and count the number of full/half adders used for each equation, and make that the new objective. The additions to the model (and replacement objective) are shown in Listing 2. The number of bits required for the result of each equation are defined by `bits` while the number of full/half adders for each equation are given by `adders`. In lines 4–5 we compute the bit width directly from the multiple (the least power big enough to hold the maximum result). We compute the number of adders required for each recipe in lines 6–12. Finally, we minimize the sum of adders required.

Note that for minimizing the SAT model we only minimize a proxy for the size of the resulting SAT model, the number of full/half adders. We experimented with directly minimizing the number of clauses or literals (sum of size of clauses) in the SAT encoding. The resulting models were much harder to solve, we usually couldn't prove optimality, and for the small examples where we could the result

```
1 array[EQ] of var 0..infinity: bits; % number of bits
2 array[EQ] of var 0..infinity: adders; % number of adders
3 include "arg_max.mzn";
4 constraint forall(e in EQ)(bits[e] =
5   arg_max([mult[e]*(p2[xbits]-1)<=p2[i]|i in 1..maxsh+xbits ]));
6 constraint forall(e in EQ)(adders[e] =
7   if ty[e] = SPLUS then
8     if shift[e] >= bits[right[e]] then 0
9     else max(bits[left[e]],bits[right[e]]-shift[e]) endif
10  elseif ty[e] = SMINUS then bits[e]
11  elseif ty[e] = MINUSS then bits[e]
12  else 0 endif);
13 solve minimize sum(adders); % minimize adders
```

Listing 2. Additions to the MiniZinc model of Listing 1 to optimize for the number of full/half adders.

was rarely better than the SAT encoding resulting from minimizing full/half adders. It remains interesting future work to see if this more direct approach to encoding minimization could be improved.

4   Experimental Evaluation

To evaluate the effectiveness of the proposed approach, we compare the size of the encoding of $c \times x$ for a range of constants c and bit widths for x in Sect. 4.1. In Sect. 4.2 we apply our approach to a realistic benchmark, comparing both the encoding size and solve times.

4.1   Constructing and Analyzing the SCM Databases

In this section, we compare the different SCM approaches in terms of the size of their respective encodings. For each bit width $2 \leq w \leq 16$, we average the number of variables, clauses, and literals that are generated when encoding $c \times x$ over $1 \leq c \leq 2047$. The four methods are base (the baseline from Sect. 3.1), espresso (using logic minimization with Espresso 2.3 from Sect. 3.2), min-k (minimizing the number of ripple carry adders) and min-a (minimizing the total number of adders). It becomes increasingly hard for Espresso to compute CNF for all constants as the number of input bits grows, so for espresso we are limited to $w \leq 12$. The results are shown in Fig. 1.

It is clear that the naive decomposition of base produces by far the most variables. In contrast, min-k achieves the same multiplication using fewer ripple carry adders. Importantly, min-a has even fewer variables than min-k, indicating that minimizing the number of adders serves as a proxy for avoiding additional variables. The effect is stronger for smaller bit widths, presumably because then there is more opportunity for completely free additions where one input is shifted by the other input's bit width. The number of variables for espresso is always minimal, since no additional variables are ever constructed. In terms of clauses and literals, we see a linear relationship between the other methods other than espresso, which generates an exponential number of clauses and literals, min-a better than min-k better than base.

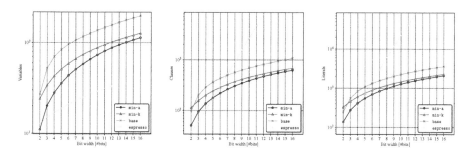

Fig. 1. Encoding size of $c \times x$ using different SCM approaches

4.2 Solving Multidimensional Bounded Knapsack Problem

To evaluate the different approaches, we adapt the approach from [10] to generate hard *Multidimensional Bounded Knapsack Problem* (MBKP) instances. In MBKP, we decide for N item types how many x_i to pack (up to B) such that a minimum profit $\sum_{i=1}^{N} x_i p_i \geq P$ is reached. Each item is restricted by M dimensions of weight with $\bigwedge_{i=1}^{M} \sum_{j=1}^{N} x_j w_{i,j} \leq W_i$. Instances can be generated by first generating C coefficient sets for $w_{i,j}$ and p_i, sampling uniformly from $[1, Q]$. Then, for each coefficient set, we choose S capacity sets for $1 \leq s \leq S$ with a capacity factor $f = \frac{s}{S}$: generating $W_i = f \sum_{j=1}^{N} B w_{i,j}, 1 \leq i \leq M$, and $P = (1 - f) \sum_{j=1}^{N} B p_j$. As f increases, the constraints become less strict, and instances turn from unsatisfiable to satisfiable. To avoid an excess of trivial instances on the lower and higher ends of f, we normalize f to be within $0.2 \leq f \leq 0.8$. One instance set is generated for each $B = 2^w - 1, 4 \leq w \leq 16$, with parameters $N = 15$, $M = 100$, $S = 100$, $C = 3$.

In the SAT encoding of MBKP each integer variable x_i is binary encoded in the least number of bits required to represent its full domain. Each linear constraint of the form $\sum_{j=1}^{N} w_{i,j} x_j \leq W_i$ is encoded by first constructing $y_j = w_{i,j} \times x_j$ using the SCM method, and then computing $y_0 = \sum_{j=1}^{N} y_j$ by repeated use of ripple-carry adders, and finally constraining $y_0 \leq W_i$ using a lexicographic constraint.

To provide another control method, we encode the benchmarks using Picat-SAT (version 3.5) [18], a state-of-the-art SAT encoder that uses the binary encoding. The `picat` encodes SCM similar to `base`, but with additional ad-hoc optimizations [19]. Note that for these problems with large coefficients and domain sizes alternate encoding approaches, e.g., domain or order encodings are non-competitive.

All encodings produced in our experiments are solved using the same binary of the SAT solver CaDiCal (version 1.9.1) [4] with default parameters, 8 GB memory limit and a time limit of 180 s. A PAR2 penalty applies to timeouts.

In Fig. 2, we compare the encoding sizes of the MBKP, similar to the results from Sect. 4.1. `picat` matches `min-a` in the number of variables. However, for clauses and especially for literals, larger bit widths result in larger encodings, closer to `base`. For unknown reasons, `picat` shows a temporary spike in clauses and literals at $w = 5$ (for some constraints, `picat` also relies on Espresso; which it perhaps uses in this case).

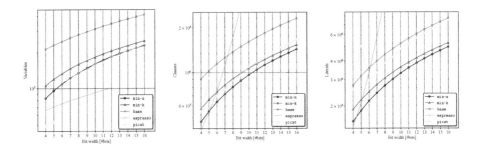

Fig. 2. Encoding size of MBKP using different SCM approaches

Comparing solve times in Fig. 3, we observe that `espresso` performs best up to a bit width of 10, after which its performance drops off drastically. Even if pre-computing larger input bit widths would be tractable for `espresso`, we can extrapolate that an encoding without auxiliary variables loses its effectiveness. In contrast, `min-k` and `min-a` remain relatively constant, even though the instances grow exponentially in size. Furthermore, `min-a` performs better than `min-k`, which shows the benefit of minimizing adders over ripple carry adders, even if some recipes have sub optimal length (such as in Example 1). Finally, `picat` performs much better than `base` for smaller bit widths but similar for larger bit widths, suggesting that `picat` is able to apply certain encoding optimizations in the former case.

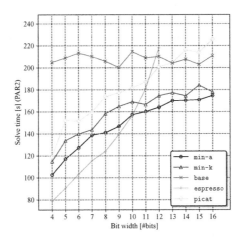

Fig. 3. Solve time comparison for MBKP

5   Conclusion and Future Work

In conclusion, we have shown how to tackle and apply SCM specifically for SAT in order to encode linear constraints. Encoding linear constraints is a key challenge for SAT-based solvers [1]. To our knowledge, this is the first work that adapts SCM circuits to encode linear constraints for SAT. Since ripple carry adders are not uniformly encoded in the presence of partially fixed inputs, this gives us the opportunity to develop different optimal circuits compared to the traditional SCM problem. In the experimental evaluation, we have seen how this approach significantly improves both the encoding size as well as the solver's performance. Finally, we have made the pre-computed SCM encodings freely available.

In future work, we aim to extend our approach by applying MCM to SAT, which – unlike SCM – cannot be comprehensively pre-computed. Instead, we would first have to collect the target constants (e.g., through common sub-expression elimination [14]) from a given instance and then dynamically solve the MCM problem at encode time. This poses an interesting trade-off between encode and solve time, but has the potential for additional improvements.

Acknowledgements. This research was partially funded by the Australian Government through the Australian Research Council Industrial Transformation Training Centre in Optimisation Technologies, Integrated Methodologies, and Applications (OPTIMA), Project ID IC200100009.

References

1. Abío, I., Mayer-Eichberger, V., Stuckey, P.J.: Encoding Linear Constraints into SAT. CoRR **abs/2005.02073** (2020). https://arxiv.org/abs/2005.02073
2. Aksoy, L., Flores, P.F., Monteiro, J.: Exact and approximate algorithms for the filter design optimization problem. IEEE Trans. Signal Process. **63**(1), 142–154 (2015). https://doi.org/10.1109/TSP.2014.2366713
3. Avizienis, A.: Signed-digit number representations for fast parallel arithmetic. IRE Trans. Electron. Comput. **EC-10**(3), 389–400 (1961). https://doi.org/10.1109/TEC.1961.5219227
4. Biere, A., Fazekas, K., Fleury, M., Heisinger, M.: CaDiCaL, Kissat, Paracooba, Plingeling and Treengeling Entering the SAT Competition 2020. In: Balyo, T., Froleyks, N., Heule, M., Iser, M., Järvisalo, M., Suda, M. (eds.) Proc. of SAT Competition 2020 – Solver and Benchmark Descriptions. Department of Computer Science Report Series B, vol. B-2020-1, pp. 51–53. University of Helsinki (2020)
5. Biere, A., Heule, M., van Maaren, H., Walsh, T. (eds.): Handbook of Satisfiability, 2 edn. IOS Press (2021). google-Books-ID: dUAvEAAAQBAJ
6. Brayton, R.K., Hachtel, G.D., McMullen, C.T., Sangiovanni-Vincentelli, A.L.: Logic minimization algorithms for VLSI synthesis. In: The Kluwer International Series in Engineering and Computer Science, vol. 2. Springer, Heidelberg (1984). https://doi.org/10.1007/978-1-4613-2821-6
7. Cappello, P., Steiglitz, K.: Some complexity issues in digital signal processing. IEEE Trans. Acoust. Speech Signal Process. **32**(5), 1037–1041 (1984)

8. Dekker, J.J., Bierlee, H.: Pindakaas: CPAIOR-24 (2024). https://doi.org/10.5281/zenodo.10851856

9. Gustafsson, O.: Towards optimal multiple constant multiplication: a hypergraph approach. In: 42nd Asilomar Conference on Signals, Systems and Computers, ACSSC 2008, Pacific Grove, CA, USA, 26–29 October 2008, pp. 1805–1809. IEEE (2008). https://doi.org/10.1109/ACSSC.2008.5074738

10. Han, B., Leblet, J., Simon, G.: Hard multidimensional multiple choice knapsack problems, an Empirical Study. Comput. Oper. Res. **37**(1), 172–181 (2010). https://doi.org/10.1016/j.cor.2009.04.006

11. Kumm, M.: Optimal constant multiplication using integer linear programming. IEEE Trans. Circuits Syst. II Express Briefs **65-II**(5), 567–571 (2018).https://doi.org/10.1109/TCSII.2018.2823780

12. Ma, S., Ampadu, P.: Optimal SAT-based minimum adder synthesis of linear transformations. In: Lee, H., Geiger, R.L. (eds.) 62nd IEEE International Midwest Symposium on Circuits and Systems, MWSCAS 2019, Dallas, TX, USA, 4–7 August 2019, pp. 335–338. IEEE (2019). https://doi.org/10.1109/MWSCAS.2019.8885033

13. Nethercote, N., Stuckey, P., Becket, R., Brand, S., Duck, G., Tack, G.: MiniZinc: towards a standard CP modelling language. In: Bessiere, C. (ed.) CP 2007. LNCS, vol. 4741, pp. 529–543. Springer, Heidelberg (2007). https://doi.org/10.1007/978-3-540-74970-7_38

14. Nightingale, P., Spracklen, P., Miguel, I.: Automatically improving SAT encoding of constraint problems through common subexpression elimination in savile row. In: Pesant, G. (ed.) CP 2015. LNCS, vol. 9255, pp. 330–340. Springer, Heidelberg (2015). https://doi.org/10.1007/978-3-319-23219-5_23

15. Rossi, F., van Beek, P., Walsh, T. (eds.): Handbook of Constraint Programming, Foundations of Artificial Intelligence, vol. 2. Elsevier (2006). https://www.sciencedirect.com/science/bookseries/15746526/2

16. Soos, M., Nohl, K., Castelluccia, C.: Extending SAT solvers to cryptographic problems. In: Kullmann, O. (ed.) SAT 2009. LNCS, vol. 5584, pp. 244–257. Springer, Heidelberg (2009). https://doi.org/10.1007/978-3-642-02777-2_24

17. Warners, J.P.: A linear-time transformation of linear inequalities into conjunctive normal form. Inf. Process. Lett. **68**(2), 63–69 (1998). https://doi.org/10.1016/S0020-0190(98)00144-6

18. Zhou, N., Kjellerstrand, H.: The Picat-SAT Compiler. In: Gavanelli, M., Reppy, J.H. (eds.) PADL 2016. LNCS, vol. 9585, pp. 48–62. Springer, Heidelberg (2016). https://doi.org/10.1007/978-3-319-28228-2_4

19. Zhou, N., Kjellerstrand, H.: Optimizing SAT encodings for arithmetic constraints. In: Beck, J.C. (ed.) CP 2017. LNCS, vol. 10416, pp. 671–686. Springer, Heidelberg (2017). https://doi.org/10.1007/978-3-319-66158-2_43

Towards a Generic Representation of Combinatorial Problems for Learning-Based Approaches

Léo Boisvert[1]([✉]), Hélène Verhaeghe[2][iD], and Quentin Cappart[1][iD]

[1] Polytechnique Montréal, Montreal, Canada
{leo.boisvert,quentin.cappart}@polymtl.ca
[2] DTAI, KU Leuven, Leuven, Belgium
helene.verhaeghe@kuleuven.be

Abstract. In recent years, there has been a growing interest in using learning-based approaches for solving combinatorial problems, either in an end-to-end manner or in conjunction with traditional optimization algorithms. In both scenarios, the challenge lies in encoding the targeted combinatorial problems into a structure compatible with the learning algorithm. Many existing works have proposed problem-specific representations, often in the form of a graph, to leverage the advantages of *graph neural networks*. However, these approaches lack generality, as the representation cannot be easily transferred from one combinatorial problem to another one. While some attempts have been made to bridge this gap, they still offer a partial generality only. In response to this challenge, this paper advocates for progress toward a fully generic representation of combinatorial problems for learning-based approaches. The approach we propose involves constructing a graph by breaking down any constraint of a combinatorial problem into an abstract syntax tree and expressing relationships (e.g., a variable involved in a constraint) through the edges. Furthermore, we introduce a graph neural network architecture capable of efficiently learning from this representation. The tool provided operates on combinatorial problems expressed in the XCSP3 format, handling all the constraints available in the 2023 mini-track competition. Experimental results on four combinatorial problems demonstrate that our architecture achieves performance comparable to dedicated architectures while maintaining generality. Our code and trained models are publicly available at https://github.com/corail-research/learning-generic-csp.

1 Introduction

Combinatorial optimization has drawn the attention of computer scientists since the discipline emerged. Combinatorial problems, such as the *traveling salesperson problem* or the *Boolean satisfiability* have been the focus of many decades of research in the computer science community. We can now solve large problems of these kinds efficiently with exact methods [2] and heuristics [14]. Heuristics are handcrafted procedures that crystallize expert knowledge and intuition about the structure of the problems they are applied to. While heuristics have found success and applications for the resolution of combinatorial problems either as a direct-solving process or integrated into a search procedure, the rise of deep learning in many different fields [4,7,18,24] has attracted the

© The Author(s), under exclusive license to Springer Nature Switzerland AG 2024
B. Dilkina (Ed.): CPAIOR 2024, LNCS 14742, pp. 99–108, 2024.
https://doi.org/10.1007/978-3-031-60597-0_7

attention of researchers [5]. Among deep learning architectures, *graph neural networks* (GNNs) [29] have proven to be a powerful and flexible tool for solving combinatorial problems. However, as identified by Cappart et al. (2023) [8], it is still cumbersome to integrate GNN and machine learning into existing solving processes for practitioners. One reason is that a dedicated architecture must be designed and trained for each combinatorial problem. In addition to potentially expensive computing resources for training, this also requires the existence of a large and labeled training set.

Building such a problem-specific graph representation has been the preferred choice of most related approaches, such as NeuroSAT [30] leveraging an encoding for SAT formulas, or approaches tightly linked to the traveling salesperson problem [16,27]. These approaches suffer from a lack of generality as the architecture could not be trivially exported from one combinatorial problem to another one (e.g., the representation in NeuroSAT cannot be used for encoding an instance of the traveling salesperson problem). Few attempts have been realized to bridge this gap, but they still provide only a partial genericity. For instance, Gasse et al. (2019) [13] proposed a bipartite graph linking variables and constraints when a variable is involved in a given constraint. However, this approach only encodes binary mixed-integer programs. Chalumeau et al. (2021) [10] later introduced a tripartite graph where variables, values, and constraints are specific types of vertices. This approach also lacks genericity as the method requires retraining when the number of variables changes. To our knowledge, the last attempt in this direction has been realized by Marty et al. (2023) [23] who leveraged a tripartite graph allowing decorating each vertex type with dedicated features. Although any combinatorial problem can theoretically be encoded with this framework, some information is lost with the encoding. For instance, it was not clear how the constraint $3x_1 \leq 4x_2$, could be differentiated from the constraint $2x_1 \leq 5x_2$. On the one hand, both constraints can be represented as an *inequality*, but we lose information about the variables' coefficients. On the other hand, both constraints can be encoded as two distinct relationships, but in this case, we lose the fact that we have an inequality in both cases. More formally, their encoding function was not *injective*. Different instances could have the same encoding with no option to differentiate them. Besides, the experiments proposed only targeted relatively pure problems (*maximum independent set*, *maximum cut*, and *graph coloring*). Similar limitations are observed in the approach of Tönshoff et al. (2022) [31].

Based on this context, this paper progresses towards a fully generic representation of combinatorial problems for learning-based approaches. Intuitively, our idea is to break down any constraint of a problem instance as an *abstract syntax tree* and connect similar items (e.g., the same variables or constraints) through an edge. Then, we introduce a GNN architecture able to leverage this graph. To demonstrate the genericity of this approach, our architecture directly operates on instances expressed with the XCSP3 format [6] and can handle all the constraints available in the 2023 mini-track competition. Experiments are carried out on four problems (featuring standard intension, and global constraints such as ALLDIFFERENT [28], TABLE [11], NEGATIVETABLE [33], ELEMENT and SUM) and aims to predict the satisfiability of the decision version of combinatorial problems. The results show that our generic architecture gives performances close to problem-specific architectures and outperforms the tripartite graph of Marty et al. (2023) [23].

2   Encoding Combinatorial Problem Instances as a Graph

Formally, a combinatorial problem instance $\mathcal{P}$ is defined as a tuple $\langle X, D(X), C, O \rangle$, where X is the set of variables, $D(X)$ is the set of domains, C is the set of constraints, and O $(X \rightarrow \mathbb{R})$ is an objective function. A valid solution is an assignment of each variable to a value of its domain such that every constraint is satisfied. The optimal solution is a valid solution such that no other solution has a better value for the objective. Our goal is to build a function $\Phi : \langle X, D(X), C, O \rangle \mapsto \mathcal{G}(V, f, E)$, where $\mathcal{G}$ is a graph and V, f, E are its vertex, vertex features and arc sets, respectively. We want this function to be *injective*, i.e., an encoding refers to at most one combinatorial problem instance. We propose to do so by introducing an encoding consisting in a heterogeneous and undirected graph featuring 5 types of vertices: *variables* (VAR), *constraints* (CST), *values* (VAL), *operators* (OPE), and *model* (MOD). The idea is to split each constraint as a sequence of elementary operations, to merge vertices representing the same variable or value, and connect together all the relationships. An illustration of this encoding is proposed in Fig. 1 with a running example. Intuitively, this process is akin to building the abstract syntax tree of a program. Formally, the encoding gives a graph $\mathcal{G}(V, f, E)$ with $V = V_{\mathrm{VAR}} \cup V_{\mathrm{CST}} \cup V_{\mathrm{VAL}} \cup V_{\mathrm{OPE}} \cup V_{\mathrm{MOD}}$ is a set containing the five types of vertices, $f = f_{\mathrm{VAR}} \cup f_{\mathrm{CST}} \cup f_{\mathrm{VAL}} \cup f_{\mathrm{OPE}} \cup f_{\mathrm{MOD}}$ is the set of specific features attached to each vertex, and E is the set of edges connecting vertices. Each type is defined as follows.

Values (V_{VAL}). A *value-vertex* is introduced for each constant appearing in $\mathcal{P}$. Such values can appear inside either à domain or a constraint. All the values are distinct, i.e., they are represented by a unique vertex. The type of the value (integer, real, etc.) is added as a feature (f_{VAL}) to each *value-vertex*, using a one-hot encoding.

Variables (V_{VAR}). A *variable-vertex* is introduced for each variable appearing in $\mathcal{P}$. A vertex related to a variable $x \in X$ is connected through an edge to each value v inside the domain $D(x)$. Such as the *value-vertices*, all variables are represented by a unique vertex. The type of the variable (boolean, integer, set, etc.) is added as a feature (f_{VAR}) using a one-hot encoding.

Constraints (V_{CST}). A *constraint-vertex* is introduced for each constraint appearing in $\mathcal{P}$. A vertex related to a constraint $c \in C$ is connected through an edge to each value v that is present in the constraint. The type of the constraint (*inequality*, *allDifferent*, *table*, etc.) is added as a feature (f_{CST}) using a one-hot encoding.

Operators (V_{OPE}). Tightly related to constraints, *operator-vertices* are used to unveil the combinatorial structure of constraints. Specifically, operators represent modifications or restrictions on variables involved in constraints (e.g., arithmetic operators, logical operators, views, etc.). An *operator-vertex* is added each time a variable $x \in X$ is modified by such operations. The vertex is connected to the vertices related to the impacted variables. The type of the operator ($+$, $\times$, $\wedge$, $\vee$, etc.) is added as a feature (f_{OPE}) using a one-hot encoding. If the operator uses a numerical value to modify a variable, this value is used as a feature (f_{OPE}) as well.

Model (V_{MOD}). There is only one *model-vertex* per graph. This vertex is connected to all *constraint-vertices* and all *variable-vertices* involved in the objective function. Its semantics is to gather additional information about the instance (e.g., the direction of the optimization) by means of its feature vector (f_{MOD}).

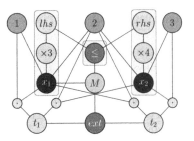

$$\max x_1$$
$$\text{s.t. } 3x_1 \leq 4x_2$$
$$\text{TABLE}([x_1, x_2], [(1, 2), (2, 3)])$$
$$x_1 \in \{1, 2\}$$
$$x_2 \in \{2, 3\}$$

Fig. 1. Encoding of a combinatorial problem instance presented as a running example. There are 3 *value-vertices* depicted (in green) and 2 *variable-vertices* (in red). As x_1 contains values 1 and 2 in its domain, they are connected with an edge, and similarly for the domain of x_2. There are 2 *constraint-vertices* (in blue), one for the inequality ($\leq$) and one for the TABLE constraint (**ext**). The figure's gray area illustrates the constraint $3x_1 \leq 4x_2$, highlighted by operators (in orange) and showing a multiplication ($\times$, with feature 3) of x_1 on the right-side (**rhs**) and another ($\times$, with feature 4) of x_2 on the left-side (**lhs**). The **rhs** and **lhs** operators clarify equation sides, essential for distinguishing between $3x_1 \leq 4x_2$ and $3x_1 \geq 4x_2$, and link to the associated constraint (e.g., inequality $\leq$). The constraint TABLE($[x_1, x_2], [(1, 2), (2, 3)]$) is expressed in a similar way. It involves two tuples t_1 and t_2. Finally, the *model-vertex* (in yellow) is connected to the two constraints, and to variable x_1, as it is part of the objective function. (Color figure online)

As a final remark, this encoding can be used to represent any combinatorial problem instance in a unique way. To do so, a parser for each constraint, describing how a constraint has to be split with the involved operators and variables, must be implemented. We currently support all the constraints formalized in XCSP3-CORE modeling format [6] and used for the mini-solver tracks of the XCSP-2023 competition: *binary intension, table, negative table, short table, element* and *sum*. Our repository contains the required documentation to build a graph from an instance in the XCSP3-CORE format with a description of all the features considered.

3   Learning from the Encoding with a Graph Neural Network

To achieve the next step and *learn* from this representation, we designed a tailored *graph neural network* (GNN) architecture to leverage this encoding. A GNN is a specialized neural architecture designed to compute a latent representation (known as an *embedding*) for each node of a given graph [29]. This process involves iteratively aggregating information from neighboring nodes. Each aggregation step is denoted as a *layer* of the GNN and incorporates learnable weights. There are various ways to perform this aggregation, leading to different variants of GNNs documented in the literature [20,25,32]. The model is differentiable and can then be trained with gradient descent methods.

Let $\mathcal{G}(V, f, E)$ be the graph encoding previously obtained, and let $h_{t,v}^{[i]} \in \mathbb{R}^p$ be a p-dimensional vector representation of a vertex $v \in V_t$ (t refers to a vertex type from $\mathcal{T} : \{\text{VAR, VAL, CST, OPE, MOD}\}$) at iteration $i \in \{0, \dots, I\}$. The inference process of a GNN consists in computing the next representations ($h_{t,v}^{[i+1]}$) from the previous ones for each vertex. This operation is commonly referred to as *message passing*. First, we set $h_{t,v}^{[0]} = f_{t,v}$ for each type, where $f_{t,v}$ is the vector of features related to vertex $v \in V_t$. Then, the representations at each iteration are obtained with LAYERNORM [3] and LSTM layers [15]. The whole inference process is formalized in Algorithm 1. First, the initial embedding of each vertex is set to its feature (line 5) and the hidden states of the LSTM are initialized to 0 (line 6), as commonly done. Then, I steps of message-passing are carried out (main loop). At each iteration i, a *message* $(\mu_{t_1,v}^{[i]})$ is obtained for each vertex. The computation is done in three steps (line 8): (1) the embedding of each neighbor of a given type is summed up, (2) the resulting value is fed to a standard multi-layer perceptron (MLP$_{t_1,t_2}^{in}$), note that there is a specific module for each edge type, (3) the messages related to each type are concatenated together ($\bigoplus$) to obtain the global message for each vertex. Notation $\mathcal{N}_{t_2}(v)$ refers to the set of neighbors of vertex v of type t_2. The result is then given as input to an LSTM cell (line 9, one cell for each type) and is used to obtain the embedding at the next layer. We note that each LSTM has its internal state $(\gamma_{t,v}^{[i]})$ updated. At the end of the loop, each vertex has a specific embedding $(h_{t,v}^{[I]})$. After the last iteration, we compute the vertex-type dependent output by passing $h_{t,v}^{[I]}$ through a standard multi-layer perceptron MLP$_t^{out}$ (line 10). This produces the output embeddings for all nodes, which are then averaged (line 11). Finally, the sigmoid function (σ) is used to obtain an output between 0 and 1.

Algorithm 1: Inference process of our GNN architecture.

1 ▷ **Pre:** $\mathcal{G}(V_{\text{VAR,CST,VAL,OPE,MOD}}, f_{\text{VAR,CST,VAL,OPE,MOD}}, E)$ is the graph encoding.

2 ▷ **Pre:** $\mathcal{T} : \{\text{VAR, VAL, CST, OPE, MOD}\}$ is the set of *vertex-types*.

3 ▷ **Pre:** I is the number of iterations of the GNN.

4

5 $h_{t,v}^{[0]} := f_{t,v}$ $\quad \forall v \in V_t, \forall t \in \mathcal{T}$

6 $\gamma_{t,v}^{[0]} := 0$ $\quad \forall v \in V_t, \forall t \in \mathcal{T}$

7 **for** i **from** 1 **to** I **do**

8 $\quad \mu_{t_1,v}^{[i]} := \bigoplus_{t_2 \in \mathcal{T}} \text{MLP}_{t_1,t_2}^{in} \left(\sum_{u \in \mathcal{N}_{t_2}(v)} h_{t_2,u}^{[i-1]} \right)$ $\forall v \in V_{t_1}, \forall t_1 \in \mathcal{T}$

9 $\quad h_{t,v}^{[i]}, \gamma_{t,v}^{[i]} := \text{LSTM}_t \left(\mu_{t,v}^{[i]}, \gamma_{t,v}^{[i-1]} \right)$ $\forall v \in V_t, \forall t \in \mathcal{T}$

10 $\nu_{t,v} := \text{MLP}_t^{out} \left(h_{t,v}^{[I]} \right)$ $\forall v \in V_t, \forall t \in \mathcal{T}$

11 $\hat{y} := \sigma \left(\frac{1}{|\mathcal{T}| \times |V|} \sum_{t \in \mathcal{T}} \sum_{v \in V_t} \nu_{t,v} \right)$

12 **return** $\hat{y}$

---

4 Experiments

This section evaluates our approach on combinatorial task, focusing on four problems: Boolean satisfiability (SAT), graph coloring (COL), knapsack (KNAP), and the traveling salesperson problem (TSP) with TSP-EXT (table constraint) and TSP-ELEM (element constraint) models. We trained models on the decision version of the problems, asking if a solution exists with costs under a target k. When there is no objective function, we have a plain constrained satisfaction problem (e.g., SAT). The aim is not to find the solution but to determine its existence, aligning with recent studies [19,21,27,30]. We compared our approach with problem-specific architectures and the tripartite graph of Marty et al. (2023) [23]. For the latter, we extracted their graph representation and used it in our GNN. The evaluation metric considered is the accuracy in correctly predicting the answer to the decision problem. Details on experimental protocols and implementations follow.

Boolean Satisfiability. Instances are generated with the random generator of Selsam et al. (2018) [30]. Briefly, the generator builds random pairs of SAT instances of n variables by adding new random clauses until the problem becomes unsatisfiable. Once the problem becomes unsatisfiable, it changes the sign of the first literal of the problem, rendering it satisfiable. On average, clauses have an arity of 8. Both the satisfiable and unsatisfiable instances are included in the dataset. Still following Selsam et al. (2018), we built a dataset containing millions of instances having 10 to 40 literals. The SAT-dependent architecture considered in our analysis has been also introduced in the same paper. We used a training set of size 3,980,000 and a validation set of size 20,000.

Traveling Salesperson Problem. Instances are generated with the random generator of Prates et al. (2018) [27]. The generation consists in (1) creating n points in a $(\sqrt{2}/2 \times \sqrt{2}/2)$ square, (2) building the distance matrix with the Euclidean distance, and (3) solving them using the Concorde solver [1] to obtain the optimal tour cost C^*. Two instances are then created: a feasible one and an unfeasible one with target costs of $1.02C^*$ and $0.98C^*$, respectively. We build two TSP models: a first one where the distance constraints are expressed with an extension constraint (TSP-EXT), and a second one with an element constraint (TSP-ELEM). The motivation is to analyze the impact of the model on the resulting graph encoding and the performances. Still following Prates et al. (2018), we built the dataset with a number of cities n sampled uniformly from 20 to 40. The TSP-dependent architecture considered in our analysis has been also introduced in the same paper. We used a training set of size 850,000 and a validation set of size 50,000.

Graph Coloring. Instances are generated following Lemos et al. (2019) [19]. This generator builds graphs with 40 to 60 vertices. For each graph, a single edge is added such that the k-colorability is altered. The instances are produced in pairs: one where the optimal value is k and another one where it is higher. Our encoding leverages a standard model of the graph coloring featuring binary intension ($\neq$) constraints. The COL-dependent architecture has been also introduced in the same paper. We used a training set of size 140,000 and a validation set of size 10,000.

Knapsack. We built instances containing 20 to 40 items and solved them to optimality to find the optimal value V^*. Then, we created two instances, a feasible one and an unfeasible one with a target cost of $1.02V^*$ and $0.98V^*$, respectively. Our encoding leverages a standard knapsack model featuring a SUM constraint. The KNAPSACK-specific model was based on the model of Liu et al. (2020) [21]. We used a training set of size 950,000 and a validation set of size 50,000.

Implementation Details. All models were trained with PyTorch [26] and PyTorch-Geometric [12] on a single Nvidia V100 32 GB GPU for up to 4 days or until convergence. We selected the model having the best performance on the validation set. To make the comparisons between specific and generic graphs fair, we trained all our models using a single GPU, and we tuned each model by varying the number of hidden units in the MLP and LSTM layers. Concerning the problem-specific architectures, we reused to same hyperparameters as described in their original publications. All models are trained with Adam optimizer [17] coupled with a learning rate scheduler and a weight decay of 10^{-8}. The main hyperparameters used for our different models are detailed in the accompanying code. All our models are expressed using XCSP3 formalism. We implemented a parser to build the graph from this representation. Our code and trained models are publicly available.

4.1 Results: Accuracy of the Approaches

Table 1 summarizes the accuracy in predicting the correct answer on the validation set for each problem and baseline. Interestingly, we observe that our approach achieves similar or close performances as the problem-specific architectures for all the problems. We see it as a promising result, as our approach, thanks to its genericity, can be directly used for all the problems without designing a new dedicated architecture. On the other hand, the approach of Marty et al. (2023) [23], fails to achieve similar results except for the graph coloring. This is because this representation does not preserve the combinatorial structure of complex constraints (COL has only constraints like $x_1 \neq x_2$).

Table 1. Prediction accuracy on holdout test set

Architecture	SAT	TSP-EXT	TSP-ELEM	COL	KNAP
Problem-specific	94.3%	**96.3%**		77.0 %	**98.8%**
Tripartite [23]	50.0%	50.0%		**84.6%**	50.0%
Ours	**94.4%**	84.5%	91.4%	84.4%	97.9%

While TSP-ELEM achieves results close to the TSP-specific approach, TSP-EXT falls short significantly. This highlights the importance of using an appropriate combinatorial model for the encoding. Specifically, the encoding TSP-ELEM has a size of 1841 vertices and 13042 edges, while TSP-EXT yields a graph of 5661 vertices and 28322 edges, for a same instance of 40 cities. This is consequently larger encoding, which is not desirable as it makes the model harder to train.

4.2 Analysis: Generalization to Larger Instances

Figure 2 shows the generalization ability of the previous model, without retraining, on new instances with 60, 80 and 100 variables (5000 instances for each size). We observe

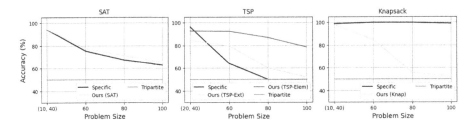

Fig. 2. Analysis of the generalization ability on larger instances.

that our generic representation provides a better generalization than the problem-specific architecture for SAT and TSP. Notably, TSP-ELEM offers a far better generalization than TSP-EXT, confirming the impact of the input model and the size of the graph. Interestingly, we observe a strong generalization ability of the specific architecture for the knapsack. Our preliminary analysis indicates that this is achieved thanks to a GNN aggregation function based on a weighted sum. Understanding in detail the root causes of this generalization is part of our future work.

4.3 Discussion: Limitations and Challenges

Although the empirical results show the promise of this generic architecture, some limitations must be addressed. First, the training time required to obtain such results is prohibitive, although we have considered only relatively small instances so far. This is mainly because our encoding generates large graphs. This opens the door to integrating compression methods, dedicated to shrinking the size of the encoding without losing information. A parallel can be done with the SMARTTABLE [22] constraints, which encodes tables more compactly. It also highlights the importance of having a good input model, yielding a small embedding (e.g.,TSP-ELEM versus TSP-EXT). Besides, the current task is restricted to solving the decision version of problems with an end-to-end approach. The next step will be to integrate this architecture in a full-fledged solver, such as what has been proposed by Gasse et al. (2019) for binary mixed-integer programs [13], and by Cappart et al. (2021) for constraint programming [9]. Finally, the generalization ability of a model trained for a specific problem (e.g., on the TSP) to a similar one (e.g., TSP with time windows) is an interesting aspect to investigate.

5 Conclusion and Perspectives

This paper introduced a first version of a generic procedure to encode arbitrary combinatorial problem instances into a graph for learning-based approaches. The encoding proposed is injective, meaning that each encoding can be obtained by only one instance. Besides, a tailored graph neural network has been proposed to learn from this encoding. Experimental results showed that our approach could achieve similar results as problem-specific architectures, without requiring the need to hand-craft a dedicated representation. All the constraints involved in the 2023 mini-track competition of XCSP3 are currently supported. Adding new constraints only requires the implementation of a parser. Our next steps are to challenge the approach on bigger and more complex problems, and on other combinatorial tasks (e.g., learning branching heuristics).

Acknowledgement. This research has been mainly funded thanks to a NSERC Discovery Grant (Canada) held by Quentin Cappart.This research received funding from the European Union's Horizon 2020 research and innovation program under grant agreement No 101070149, project Tuples.

References

1. Applegate, D., Bixby, R., Chvatal, V., Cook, W.: Concorde TSP solver (2006)
2. Applegate, D.L., et al.: Certification of an optimal TSP tour through 85,900 cities. Oper. Res. Lett. **37**(1), 11–15 (2009). https://doi.org/10.1016/j.orl.2008.09.006
3. Ba, J.L., Kiros, J.R., Hinton, G.E.: Layer normalization. arXiv preprint arXiv:1607.06450 (2016)
4. Bahdanau, D., Cho, K.H., Bengio, Y.: Neural machine translation by jointly learning to align and translate. In: 3rd International Conference on Learning Representations, ICLR 2015 (2015)
5. Bengio, Y., Lodi, A., Prouvost, A.: Machine learning for combinatorial optimization: a methodological tour d'horizon. Eur. J. Oper. Res. **290**(2), 405–421 (2021). https://doi.org/10.1016/j.ejor.2020.07.063
6. Boussemart, F., Lecoutre, C., Audemard, G., Piette, C.: XCSP3-core: a format for representing constraint satisfaction/optimization problems. arXiv preprint arXiv:2009.00514 (2020)
7. Brown, T., et al.: Language models are few-shot learners. Adv. Neural Inf. Process. Syst. **33**, 1877–1901 (2020)
8. Cappart, Q., Chételat, D., Khalil, E.B., Lodi, A., Morris, C., Velickovic, P.: Combinatorial optimization and reasoning with graph neural networks. J. Mach. Learn. Res. **24**(130), 1–61 (2023)
9. Cappart, Q., Moisan, T., Rousseau, L.M., Prémont-Schwarz, I., Cire, A.A.: Combining reinforcement learning and constraint programming for combinatorial optimization. In: Proceedings of the AAAI Conference on Artificial Intelligence, vol. 35, pp. 3677–3687 (2021)
10. Chalumeau, F., Coulon, I., Cappart, Q., Rousseau, L.-M.: SeaPearl: a constraint programming solver guided by reinforcement learning. In: Stuckey, P.J. (ed.) CPAIOR 2021. LNCS, vol. 12735, pp. 392–409. Springer, Cham (2021). https://doi.org/10.1007/978-3-030-78230-6_25
11. Demeulenaere, J., et al.: Compact-Table: efficiently filtering table constraints with reversible sparse bit-sets. In: Rueher, M. (ed.) CP 2016. LNCS, vol. 9892, pp. 207–223. Springer, Cham (2016). https://doi.org/10.1007/978-3-319-44953-1_14
12. Fey, M., Lenssen, J.E.: Fast graph representation learning with PyTorch geometric. In: ICLR Workshop on Representation Learning on Graphs and Manifolds (2019)
13. Gasse, M., Chételat, D., Ferroni, N., Charlin, L., Lodi, A.: Exact combinatorial optimization with graph convolutional neural networks, vol. 32 (2019)
14. Helsgaun, K.: An Extension of the Lin-Kernighan-Helsgaun TSP Solver for Constrained Traveling Salesman and Vehicle Routing Problems: Technical report. Roskilde Universitet, December 2017
15. Hochreiter, S., Schmidhuber, J.: Long short-term memory. Neural Comput. **9**(8), 1735–1780 (1997)

16. Joshi, C.K., Cappart, Q., Rousseau, L.M., Laurent, T.: Learning the travelling salesperson problem requires rethinking generalization. Constraints 27(1–2), 70–98 (2022)
17. Kingma, D.P., Ba, J.: Adam: a method for stochastic optimization. arXiv preprint arXiv:1412.6980 (2014)
18. Krizhevsky, A., Sutskever, I., Hinton, G.E.: Imagenet classification with deep convolutional neural networks. Adv. Neural Inf. Process. Syst. 25 (2012)
19. Lemos, H., Prates, M., Avelar, P., Lamb, L.: Graph colouring meets deep learning: Effective graph neural network models for combinatorial problems. In: 2019 IEEE 31st International Conference on Tools with Artificial Intelligence (ICTAI), pp. 879–885. IEEE (2019)
20. Li, Y., Zemel, R., Brockschmidt, M., Tarlow, D.: Gated graph sequence neural networks. In: International Conference on Learning Representations (2016)
21. Liu, M., Zhang, F., Huang, P., Niu, S., Ma, F., Zhang, J.: Learning the satisfiability of Pseudo-Boolean problem with graph neural networks. In: Simonis, H. (ed.) CP 2020. LNCS, vol. 12333, pp. 885–898. Springer, Cham (2020). https://doi.org/10.1007/978-3-030-58475-7_51
22. Mairy, J.-B., Deville, Y., Lecoutre, C.: The smart table constraint. In: Michel, L. (ed.) CPAIOR 2015. LNCS, vol. 9075, pp. 271–287. Springer, Cham (2015). https://doi.org/10.1007/978-3-319-18008-3_19
23. Marty, T., François, T., Tessier, P., Gautier, L., Rousseau, L.M., Cappart, Q.: Learning a generic value-selection heuristic inside a constraint programming solver. In: 29th International Conference on Principles and Practice of Constraint Programming (2023)
24. Mnih, V., et al.: Human-level control through deep reinforcement learning. Nature 518(7540), 529–533 (2015)
25. Monti, F., Boscaini, D., Masci, J., Rodola, E., Svoboda, J., Bronstein, M.M.: Geometric deep learning on graphs and manifolds using mixture model CNNs. In: Proceedings of the IEEE Conference on Computer Vision and Pattern Recognition, pp. 5115–5124 (2017)
26. Paszke, A., et al.: Pytorch: an imperative style, high-performance deep learning library. Adv. Neural Inf. Process. Syst. 32 (2019)
27. Prates, M., Avelar, P.H., Lemos, H., Lamb, L.C., Vardi, M.Y.: Learning to solve np-complete problems: a graph neural network for decision TSP. In: Proceedings of the AAAI Conference on Artificial Intelligence, vol. 33, pp. 4731–4738 (2019)
28. Régin, J.C.: A filtering algorithm for constraints of difference in CSPs. In: AAAI, vol. 94, pp. 362–367 (1994)
29. Scarselli, F., Gori, M., Tsoi, A.C., Hagenbuchner, M., Monfardini, G.: The graph neural network model. IEEE Trans. Neural Netw. 20(1), 61–80 (2008)
30. Selsam, D., Lamm, M., Bünz, B., Liang, P., de Moura, L., Dill, D.L.: Learning a SAT solver from single-bit supervision. In: International Conference on Learning Representations (2019)
31. Tönshoff, J., Kisin, B., Lindner, J., Grohe, M.: One model, any CSP: graph neural networks as fast global search heuristics for constraint satisfaction. arXiv preprint arXiv:2208.10227 (2022)
32. Veličković, P., Cucurull, G., Casanova, A., Romero, A., Lió, P., Bengio, Y.: Graph attention networks. In: International Conference on Learning Representations (2018)
33. Verhaeghe, H., Lecoutre, C., Schaus, P.: Extending compact-table to negative and short tables. In: Proceedings of the AAAI Conference on Artificial Intelligence, vol. 31 (2017)

Accelerating Continuous Variable Coherent Ising Machines via Momentum

Robin A. Brown[1,2,3(✉)], Davide Venturelli[2,3], Marco Pavone[1], and David E. Bernal Neira[2,3,4]

[1] Autonomous Systems Laboratory, Stanford University, Stanford, USA
rabrown1@stanford.edu
[2] USRA Research Institute for Advanced Computer Science (RIACS), Mountain View, USA
[3] NASA Quantum Artificial Intelligence Laboratory (QuAIL), Mountain View, USA
[4] Davidson School of Chemical Engineering, Purdue University, West Lafayette, USA

Abstract. The Coherent Ising Machine (CIM) is a non-conventional architecture that takes inspiration from physical annealing processes to solve Ising problems heuristically. Its dynamics are naturally continuous and described by a set of ordinary differential equations that have been proven to be useful for the optimization of continuous variables non-convex quadratic optimization problems. The dynamics of such Continuous Variable CIMs (CV-CIM) encourage optimization via optical pulses whose amplitudes are determined by the negative gradient of the objective; however, standard gradient descent is known to be trapped by local minima and hampered by poor problem conditioning. In this work, we propose to modify the CV-CIM dynamics using more sophisticated pulse injections based on tried-and-true optimization techniques such as momentum and Adam. Through numerical experiments, we show that the momentum and Adam updates can significantly speed up the CV-CIM's convergence and improve sample diversity over the original CV-CIM dynamics. We also find that the Adam-CV-CIM's performance is more stable as a function of feedback strength, especially on poorly conditioned instances, resulting in an algorithm that is more robust, reliable, and easily tunable. More broadly, we identify the CIM dynamical framework as a fertile opportunity for exploring the intersection of classical optimization and modern analog computing.

Keywords: Ising Model · Momentum · Analog Computing

1 Introduction

Optimization is a powerful and intuitive framework for expressing and reasoning about the desirability of certain decisions subject to constraints. With diverse applications ranging from supply chain management, vehicle routing, machine learning, and chip design, it is difficult to overstate the ubiquity of optimization.

B. Dilkina (Ed.): CPAIOR 2024, LNCS 14742, pp. 109–126, 2024.
https://doi.org/10.1007/978-3-031-60597-0_8

In many contexts, casting a particular decision as an optimization problem–deriving mathematical expressions for the costs and constraints–permits the off-the-shelf application of commercial solvers as a satisfactory, general-purpose resolution to the problem. However, even if a problem complies with a particular solver's specifications, practical constraints, such as memory or time, may bottleneck the size of problems that can be addressed or the quality of the solution obtainable. While computational power has increased exponentially in the past few decades, according to Moore's law, if it stagnates these bottlenecks will persist, barring radical breakthroughs in computer engineering.

Quantum computing has emerged as an attractive and high-profile alternative to von Neumann computers based on its promise to accelerate specific computational tasks. The past few years have seen significant commercial interest in quantum computing, leading to the rapid development of prototype quantum computers based on superconducting qubits [21], photonics [36], neutral atoms [15], and trapped ions [5], encoding the gate-based model that underpins much of the theoretical work on quantum complexity theory. However, due to the significant constant overhead induced by error correction, it is unlikely that quantum computing will result in practical speed-ups for optimization, notably when the best classical algorithms can be parallelized on specialized hardware, such as graphics processing units (GPUs) [1]. A competing paradigm to the gate-based model are quantum computers that can only solve particular forms of problems, or "primitives", including quantum annealers (which solve Ising problems) [14], and Rydberg atom arrays (that encode the maximum independent set) [11]. While these are theoretically expressive primitives due to being NP-complete, the number of additional variables and connections required to construct a reduction from other NP-complete problems to one of these forms may significantly reduce the size of problems addressable on near-term hardware, despite only being polynomial overheads [22,25]. It is against this backdrop that there has been a recent surge of interest in non-quantum, yet non-conventional computing architectures for optimization based on coupled lasers [31], memristors [8], polaritons [2,18], and optical parametric oscillators [26,27].

From an algorithmic perspective, the development of these architectures has concurrently ushered in research on hybrid algorithms that explore how non-conventional and classical computers should complement each other [9]. These algorithms are typically developed with specific hardware abstractions in mind, so their practical impact will depend heavily on which technologies prevail in a highly dynamic field. In this paper, we pose a complementary question of how classical optimization algorithms can inform the development of non-conventional architectures. Despite their theoretical limitations, classical optimization techniques have achieved impressive empirical success, even for difficult, non-convex optimization problems. Rather than discarding the innovations developed in the context of classical optimization, we contend that these ideas should be incorporated into future computing architectures. Crucially, much of the discussion of these technologies has been couched in the language of physics and optics, making it largely inscrutable for the optimization community and obscuring the opportunities to incorporate well-known optimization techniques.

This paper focuses on one such device, the Coherent Ising Machine (CIM), which takes inspiration from the principles of thermalization to search for the ground state of Ising problems heuristically. It was first prototyped in 2014 as a network of four optical parametric oscillators (OPOs) coupled by optical delay lines [26]. Since then, the technology has rapidly developed, with prototype devices reaching sizes of 100,000 spins [16]. One significant deviation in today's architecture versus the original architecture is the inclusion of a measurement-feedback circuit implemented with a field programmable gate array (FPGA) [27]. Not only has this development enabled scalable all-to-all connections between the OPOs, it has also enabled the implementation of more sophisticated feedback methods, which have primarily been focused on mitigating the "amplitude heterogeneity" problem observed in early instantiations of the device [24]. This paper focuses on the CIM because the measurement-feedback circuit allows for modular modifications to the feedback term without otherwise affecting the operational principles of the device, making it a fertile opportunity for exploring the intersection of classical optimization and analog computing.

Contribution: Our contribution hinges on the insight that all existing variations of the CIM rely on gradient descent to encourage the optimization of a desired objective. In this paper, we contend that the feedback term should be modularly swapped with more sophisticated optimization techniques, particularly the accelerated and adaptive optimizers that are tried and true for large-scale non-convex optimization, such as neural network training. We conduct an extensive numerical evaluation of the proposed modifications to evaluate the anticipated performance gains realizable in a physical device.

Organization: In Sect. 2, we will first define our notation and terminology. We will then present a high-level overview of the CIM's operational principles, along with relevant developments, and conclude the section with a discussion of the momentum and Adam optimizers. In Sect. 3, we will present the Gaussian-state model of the CIM's dynamics, highlighting opportunities to incorporate other optimization techniques. We will also discuss the specific application of the CIM to box-constrained quadratic programs, which we later use for benchmarking. In Sect. 4, we present numerical experiments evaluating convergence speed, diversity of solutions, and parameter sensitivity of the modified CIM against its original dynamics.

2   Background and Related Work

In this section, we will first present our notation and terminology. We will then provide a high-level overview of the CIM's operational principles, discussing its history and recent developments designed to improve optimization performance. Finally, we will briefly introduce momentum and the Adam optimizer, overviewing some of their desirable properties, particularly for non-convex optimization.

2.1   Notation and Terminology

In this paper, we work with vectors and matrices defined over the real numbers and reserve lowercase letters for vectors and uppercase letters for matrices. We will also follow the convention that a vector $x \in \mathbb{R}^n$ is to be treated as a column vector, i.e., equivalent to a matrix of dimension $n \times 1$. For a matrix M, we use $M_{i,j}$ to denote the entry in the ith row and jth column, $M_{i,*}$ denotes the entire ith row, and $M_{*,j}$ denotes the entire jth column. An Ising problem is an optimization problem of the form: $\min_{x \in \{-1,1\}^n} f(x) := \sum_{i,j} Q_{i,j} x_i x_j + \sum_i c_i x_i$, where $Q \in \mathbb{R}^{n \times n}$, $c \in \mathbb{R}^n$ are real coefficients and $x_i \in \{-1,1\}$ are discrete variables to be optimized over. We will use the convention that the discrete variables are ± 1, also known as *spin* variables.

2.2   Coherent Ising Machines

The Coherent Ising Machine (CIM) is an optical network of optical parametric oscillators (OPOs) that was originally designed to heuristically search for the solution ("ground state") of Ising problems. At a high level, the CIM relaxes the discrete variables to continuous variables and augments the Ising objective, $f(x)$, with a "double-well potential" of the form $\phi(x, a) := \sum_i \frac{1}{4} x_i^4 - \frac{a}{2} x_i^2$. The OPOs, whose amplitudes represent the variables of the Ising problem, are encouraged to optimize the Ising objective by evolving approximately in the negative gradient direction of the augmented objective. At the same time, the parameter a (the "pump term") is gradually increased ("annealed") during the device's run. This has the effect of deepening the wells of the potential and encouraging the OPOs to take on the same magnitude. While this closely resembles the standard penalty method (i.e., approximating constraints with a penalty term in the objective), it also has the effect of gradually reshaping the energy landscape from a trivial form to one that encodes the desired objective. Moreover, the wells of the potential are at $\pm\sqrt{a}$, so the mapping from OPO amplitude to binary variables also varies while the pump term is annealed.

The double-well potential is physically implemented using a pumped laser and a periodically-poled lithium niobate (PPLN) crystal [26]. Early prototypes encoded the objective with optical delay lines and only accommodated objectives with linear gradients (i.e., quadratic functions), allowing the CIM to naturally encode Ising problems, as its name suggests. Despite the simplicity of their functional form, Ising problems are a rich class of problems encompassing Karp's NP-complete problems [25], and NP-hard problems such as job-shop scheduling [38], vehicle routing [13], and community detection [28], to name a few.

While significant progress has recently been made towards deriving a theory of the (fully classical) CIM's performance on Sherington-Kirkpatrick spin glass problems [41], a complete theory characterizing its performance on generic optimization problems still remains elusive. Currently, the CIM is primarily benchmarked by simulating a quantitative model of its behavior on different applications, such as MIMO detection [20] and compressed sensing [12]. Even though this is a widely accepted approach, no single model of the CIMs dynamics exists. Instead, different models have been constructed with varying degrees of fidelity when modeling the quantum-mechanical effects.

In parallel with the theoretical developments, the hardware of the CIM has matured considerably from the early prototypes [16]. Instead of implementing couplings between spins using optical delay lines, couplings are now implemented using a measurement-feedback circuit designed to allow for scalable all-to-all connections between OPOs [27]. This circuit consists of a measurement process (optical homodyne detection) to determine the amplitudes of the OPOs, and a feedback term based on these measurements computed on a field-programmable gate array (FPGA). In this work, we propose to leverage the flexibility afforded to us by the measurement-feedback circuit to modify the feedback computation.

While modifying the feedback term has been explored previously, efforts have primarily been targeted toward heuristic resolutions to the "amplitude hetero-geneity" problem. This occurs when the OPO amplitudes are not exactly $\pm\sqrt{a}$, breaking the mapping between the CIM and Ising objectives. From an opti-mization perspective, this can be seen as the infeasibility that may occur when replacing a constraint with a penalty function. One such modification is the amplitude heterogeneity correction proposed in [23], which introduced an auxil-iary error correction variable that exponentially increases/decreases if the spin amplitudes are greater/less than $\sqrt{a}$. These error variables modulate the relative weights of the Ising objective and double-well potential and increase the latter when the spins have different amplitudes. Later work extended this idea to addi-tionally modulate the pump term, a, to encourage the CIM to continue exploring the solution space within a single run rather than converging to a single solution [17]. A fully optical implementation of the error correction mechanism was later developed in [34], thus overcoming some of the downsides of having an electronic component in the loop.

The focus on resolving the amplitude-heterogeneity problem has primarily been motivated by problems whose complexity stems from optimizing over dis-crete domains. This starkly contrasts the many applied works that artificially discretize continuous variables to force them into the Ising form [3,10,30,39]. This encoding dramatically increases the number of OPOs mapped to a sin-gle variable and may result in a significant skew in the coupling coefficients. Recognizing the expressive power of *continuous* optimization problems, [19] recently proposed embracing the naturally continuous nature of the OPOs to solve box-constrained quadratic programs (BoxQPs), whose complexity stems from non-convexity rather than discreteness–this modified device is called the coherent continuous variable machine (CCVM). The CCVM does not funda-mentally change the operational principles of the CIM but instead interprets the OPO amplitudes as their native continuous values rather than projecting them to binary values. While this primarily impacts how the resulting solution is inter-preted, it can materially affect the dynamics of the device, as the feedback term is also computed with the continuous values rather than their binary projection. This modification was further supported by [4], who showed that BoxQPs are an algorithmic primitive that enables copositive optimization, a broad class of optimization problems that includes Ising problems.

2.3   Accelerated and Adaptive Optimizers

Given an optimization problem, $\min_x f(x)$, gradient descent is one of the most primitive and fundamental algorithms one could use to solve it. It formalizes the intuition that the optimization variable should be iteratively updated in the direction of the greatest decrease in the objective. The variables are updated according to $x^{(t+1)} = x^{(t)} - \gamma_t \nabla f(x^{(t)})$, where $\gamma_t > 0$ are iteration-specific step-sizes. Stochastic gradient methods replace $\nabla f(x^{(t)})$ with a cheap, unbiased estimate, and are among the most popular methods for large-scale optimization due to their broad applicability, simplicity, and efficiency.

One scenario known to cause slow convergence for gradient descent is when the objective function is poorly-conditioned. Intuitively, the conditioning of an optimization problem refers to the disparity in sensitivity of the objective function along different directions. A well-conditioned problem is one where the objective function has roughly the same sensitivity in all directions. In contrast, in a poorly conditioned problem, the objective is far more sensitive along some directions than others. In the latter case, gradient descent exhibits oscillatory behavior along the directions where the objective is most sensitive while making little progress along directions of low sensitivity.

Momentum methods, named for the analogy to mechanical systems, have been shown to improve convergence for poorly conditioned problems by updating the variable with an exponentially decreasing weighted sum of all past gradients. Mathematically, the updates are of the form

$$g^{(t+1)} = \theta g^{(t)} + (1-\theta)\nabla f(x^{(t)}), \tag{1}$$

$$x^{(t+1)} = x^{(t)} - \gamma g^{(t+1)}, \tag{2}$$

where $\theta \in [0,1)$ is a damping parameter. In poorly conditioned problems, this has the effect of smoothing out oscillations in the high sensitivity direction while adding up contributions along the low sensitivity direction, thereby improving convergence speed [33]. It is also effective for helping iterates escape saddle-points and local minima, which can trap ordinary gradient descent methods.

In principle, if poor conditioning is a result of inconsistent scaling between variables, it could be combated by using different step sizes per variable. However, this introduces additional challenges, as the step sizes must be individually chosen for each variable. Instead, the Adam optimizer uses estimates of the first and second moments of the gradient to adapt the step size for each variable automatically. The complete algorithm is presented in Algorithm 1. Adam is well-known for its empirical success on large-scale, non-convex optimization problems, such as neural network training, for which it is often chosen as the default optimizer.

Algorithm 1: Adam optimizer updates

Input: $f(\cdot)$ function to be optimized; Parameters $\beta_1, \beta_2 \in [0,1)$;
Initial variable vector $x^{(0)}$
Output: $x^{(t)}$ that approximately minimizes $f(x)$
$t \leftarrow 0, \quad m^{(0)} \leftarrow 0, \quad v^{(0)} \leftarrow 0$;
while $x^{(t)}$ *has not converged* **do**

$\quad g^{(t)} \leftarrow \nabla_x f(x^{(t)})$;
$\quad m^{(t)} \leftarrow \beta_1 m^{(t-1)} + (1 - \beta_1) g^{(t)}$;
$\quad v^{(t)} \leftarrow \beta_2 v^{(t-1)} + (1 - \beta_2)(g^{(t)})^2$;
$\quad \hat{m}^{(t)} \leftarrow \frac{m^{(t)}}{1 - \beta_1^t}$;
$\quad \hat{v}^{(t)} \leftarrow \frac{v^{(t)}}{1 - \beta_2^t}$;
$\quad x^{(t)} \leftarrow x^{(t-1)} - \gamma \frac{\hat{m}^{(t)}}{\sqrt{\hat{v}^{(t)}} + \epsilon}$;
$\quad t \leftarrow t + 1$
end
return $x^{(t)}$;

---

3 Dynamics of the CIM

In this section, we will present the Gaussian-state model of the CIM's dynamics, which will be the basis of our numerical experiments. While this is not a fully quantum treatment of the device, as it only considers up to second-order quantum correlations, it has been found to be consistent with experimental CIMs and models important phenomena such as squeezing/anti-squeezing, measurement uncertainty, and backaction, while maintaining the computational tractability necessary to simulate large-scale systems [29]. Under this model, it is assumed that each OPO pulse amplitude is a Gaussian distribution with mean μ_i and variance σ_i that evolves according to the following dynamics [17]:

$$\mu(t + \Delta t) = \mu(t) + \left[(-(1 + j_t) + p_t - g^2 \mu(t)^2) \mu(t) - \lambda \partial f(\tilde{\mu}(t)) \right] \Delta t \\ + \left[\sqrt{j_t}(\sigma(t) - 1/2) \right] Z(t) \sqrt{\Delta t}, \tag{3a}$$

$$\sigma(t + \Delta t) = \sigma(t) + \left[2(-(1 + j_t) + p_t - 3g^2 \mu(t)^2) \sigma(t) - 2j_t(\sigma(t) - 1/2)^2 \\ + ((1 + j_t) + 2g^2 \mu(t)^2) \right] \Delta t, \tag{3b}$$

$$Z_i(t) \sim \mathcal{N}(0,1), \qquad \tilde{\mu}(t) = \mu(t) + \frac{Z(t)}{\sqrt{4 j_t \Delta t}}. \tag{3c}$$

The remaining quantities in the above equations represent the normalized continuous measurement strength, j_t, a normalized non-linearity coefficient, g, the pump strength, p_t, and feedback magnitude, λ (which is analogous to the step-size in optimization algorithms). While these quantities are governed by the physical parameters of the device, in this work, we treat them abstractly as hyperparameters that can be optimized to improve the performance of the CIM. Following the treatment in [19], the hyperparameters j_t and p_t are indexed by t to emphasize that they are modulated over the run of the device.

In Eq. (3a), the term $(-(1 + j_t) + p_t - g^2\mu(t)^2)\mu(t)$ is the gradient of a double-well potential with minima at $\pm\sqrt{p_t-(1+j_t)}/g$. At each iteration ("round trip"), the mean-field amplitude is measured using optical homodyne detection. The measured amplitude, presented in Eq. (3c), may differ from its true value by a finite uncertainty depending on the measurement strength. The measurement also induces a shift in the mean-field amplitude, represented by the term $\sqrt{\bar{j}_t}(\sigma(t) - 1/2)Z(t)\sqrt{\Delta t}$, through backaction. The feedback term, $-\lambda\partial f(\tilde{\mu}(t))$, is computed based on the measured amplitudes. Crucially, for measurement-feedback variants of the CIM, this term is computed on a field programmable gate array (FPGA) rather than being implemented in optics and thus can be modified modularly without otherwise altering the hardware.

Abstractly, this is represented by replacing the term $\partial f(\tilde{\mu}(t))$ in Eq. (3a) with an arbitrary update $\Phi(\{\tilde{\mu}(t)\}_t)$. Because updates that are too complex or compute-intensive may undermine the resource bottlenecks motivating the CIM in the first place, we aim to design Φ to enhance the optimization performance of the system defined by Eq. (3) while maintaining comparable complexity to the feedback computations in the amplitude-heterogeneity-correction (AHC) and chaotic-amplitude-control (CAC) variants of the CIM. In both variants, the complexity of the update is dominated by the complexity of computing the gradients, $O(n^3)$. At the same time, additional computations induce an additional overhead of $O(n)$, where n is the dimension of the decision variable x. Specifically, the momentum and Adam updates from Eq. (1) and Algorithm 1, respectively, satisfy these desiderata. In the remainder of the paper, we will refer to the CIM with momentum or Adam updates as momentum-CV-CIM and Adam-CV-CIM, respectively. The original CIM variant will be referred to as GD-CV-CIM to emphasize that the updates are based on gradient descent.

3.1 Application to Box-Constrained Quadratic Programs

In this work, we will benchmark the proposed modifications of the continuous CIM on box-constrained quadratic programs (BoxQP). These are optimization problems of the form:

$$\begin{aligned} \underset{x\in\mathbb{R}^n}{\text{minimize}} \quad & x^\top Q x + c^\top x \\ \text{subject to} \quad & \mathbf{0} \leq x \leq \mathbf{1}, \end{aligned} \qquad \text{(BoxQP)}$$

where there are no restrictions on $Q \in \mathbb{R}^{n\times n}$ or $c \in \mathbb{R}^n$. Equation (BoxQP) has been shown to be NP-hard if Q is negative semi-definite or indefinite [32]. Furthermore, Eq. (BoxQP) is an expressive primitive that enables solving a broader class of mixed-binary quadratic programs, including Ising problems [4]. In this work, we seek to evaluate whether momentum or Adam updates improve the convergence speed of the CIM dynamics, reducing the time needed to obtain each sample from the device. Evaluating the CIM based on the continuous values of the OPO amplitudes will allow us to detect mechanistic improvements to convergence without the confounding and discontinuous process of projecting OPO amplitudes to binary variables based on their sign.

While the double-well potential induces a bias for the amplitudes $\pm\sqrt{a}$ (in Eq. (3a), $a = p_T - (1+j_T)/g^2$) we will map the entire domain of the OPOs, $[-\sqrt{a}, \sqrt{a}]^n$, to the domain of Eq. (BoxQP). This is given by the transformation,

$$x_i = \max\left(\min\left(\frac{1}{2}\left(\frac{\mu_i}{\sqrt{a}} + 1\right), 1\right), 0\right). \tag{4}$$

Intuitively, this re-scales and shifts the range $[-\sqrt{a}, \sqrt{a}]$ to $[0, 1]$ and projects any infeasible variables to the feasible domain. We note that the double-well potential only softly confines the OPOs to this domain (akin to penalty functions), so the OPOs may exceed the feasible region. It also introduces an extraneous quadratic term due to its inception as an Ising solver rather than a continuous optimization solver. While we expect this to degrade the performance of the resulting solver, our objective is to assess the proposed modifications on already realized instantiations of the CIM and thus should be reflected by the model.

4   Experiments

In Sect. 3, we presented a model of the CIM's dynamics, highlighting an opportunity to incorporate more sophisticated optimization techniques. In this section, we aim to assess whether such substitutions have any performance benefit for the CIM. The performance of quantum/quantum-inspired devices is often evaluated on the time-to-target metric [35]. The time-to-target metric with s confidence represents the number of repetitions to sample a state that achieves a particular "target" with probability s, multiplied by the time for a single run of the algorithm, T_{single}, i.e.,

$$\text{TTT}_s = T_{single} \frac{\log(1-s)}{\log(1-\hat{p}_{succ})}. \tag{5}$$

In the context of continuous optimization, the targets typically correspond to achieving a particular relative optimality gap, i.e., $\text{gap}_f(x) := \frac{f(x)-f(x^*)}{|f(x^*)|} \leq \epsilon$.

The time-to-target metric incorporates two figures of merit: the time to convergence of a single run of the algorithm and the distribution to which the samples from the algorithm converge. If a particular method improves this metric, it can be due to any combination of these two mechanisms. First, the samples can converge to their final state more quickly, resulting in a decrease in the multiplier T_{single}. Secondly, the method may change the distribution that the samples converge to, effectively changing the probability of success $\hat{p}_{succ}$. Our numerical experiments are intended to evaluate the performance of the modified CIM against the original variant along these two metrics. In addition, we will evaluate the robustness of the different variants as a function of the feedback strength. This is one component of the effort required for a practitioner to successfully apply the CIM and its variants on an unseen problem instance; this is an important, yet often neglected, factor in the practical applicability of a solver.

In this section, we report results on the standard benchmark instances from [6,7,37], available at https://github.com/sburer/BoxQP_instances. All code required to replicate our experiments is available at https://github.com/SECQU OIA/continuous-CIM.

4.1   Time to Optimality Gaps

The first experiment evaluates the relative convergence speed of the modified CIMs versus the original dynamics. For each problem instance, we evaluate the relative optimality gaps for the Xth percentile, with $X = [5, 10, 25, 50]$, throughout the CIM's evolution and determine the best gap achieved by the Xth percentile of both variants (hereafter referred to as the target gap). We then evaluate the roundtrip where each variant achieves the target gap for the first time. We record the ratio of the faster variant's roundtrip value to that of the slower variant, along with which one was faster. Figure 1 plots how this metric is calculated for a single problem instance. Intuitively, the roundtrip fraction represents the relative improvement in T_{single}.

We note that the roundtrip fraction metric does not capture whether a sample achieves its minimum optimality gap only transiently or if the dynamics stabilize around that value. Consequently, this metric makes the most sense in contexts where the optimality gaps are close to decreasing monotonically. In this paper, we leave the OPO bounds to their "natural" values of $[-\sqrt{a}, \sqrt{a}]$, with $a = \frac{p_T - (1 + j_T)}{g^2}$.[1] Setting the bounds to be significantly smaller than those encouraged by the double-well potential means that the OPO amplitudes are primarily evolving outside the feasible domain. This induces fast transients in optimality gaps when a particular variable can be updated from one side of the feasible domain to another within a single iteration. Because the roundtrip fraction metric is designed to measure differences in mechanistic convergence to a solution, the bounds are left to the natural domain of the OPOs, instead of the aggressive clipping proposed in [19].

Figure 2 plots the histogram of roundtrip fractions, with the top row comparing the Adam-CV-CIM to the GD-CV-CIM and the bottom row comparing the momentum-CV-CIM to the GD-CV-CIM. We find that both the momentum-CV-CIM and Adam-CV-CIM consistently reach the target gap before the GD-CV-CIM does. However, for Adam-CV-CIM, the roundtrip fractions are normally distributed around 0.5 (meaning the Adam-CV-CIM is ~2 times faster), while for the momentum-CV-CIM, the roundtrip fractions are more uniformly distributed across the entire range of values with a slight peak in the 0–0.1 bucket (≥ 10 times faster than the GD-CV-CIM). The difference between the momentum-CV-CIM

---

[1] The remaining parameters are given by: $T = 15000$, $dt = 0.0025$, $g = 0.01$, $\lambda = 550$, $p_t = 2.5\frac{t}{T}$, $j_t = 25\exp\left(-3\frac{t}{T}\right)$.

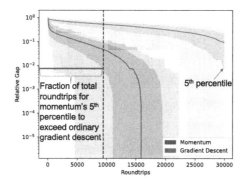

Fig. 1. This figure shows how the roundtrip ratio is computed for a single problem instance. We determine the best relative optimality gap achieved by the 5th percentile of both CIM variants (8e−3 at the end of GD-CV-CIM's evolution) and compare the roundtrip numbers where the 5th percentile first reaches this gap (∼9500 for momentum, 30000 for gradient descent). The roundtrip ratio is approximately 0.32, and we indicate which variant achieves the target gap faster.

and ordinary CIM is most pronounced for the top percentiles of samples, while it is less pronounced for the 50th percentile (i.e., median of samples). In contrast, the relative performance of the Adam-CV-CIM is more consistent across percentiles.

4.2 Sample Distribution

Secondly, we sought to understand whether there was a difference in the distributions found by the modified CIMs. Figure 3 depicts violin plots of the relative optimality gaps for three representative instances. In all instances, we find that the Adam-CV-CIM and momentum-CV-CIM achieve a greater diversity of samples with multi-modal distributions. In contrast, for the spar040-040-3 and spar080-050-3 instances, all samples from the GD-CV-CIM achieve the same optimality gap; they concentrate on the global optimum for spar040-040-3, but only achieve a relative optimality gap of 0.003275 for spar080-050-3. For the instance spar125-025-1, all variants result in multi-modal distributions, which share peaks in optimality gaps. The momentum-CV-CIM has three peaks in its distribution with relative optimality gaps of 0, 0.007, and 0.018. The GD-CV-CIM shares two of these peaks at 0 and 0.018, while the momentum-CV-CIM shares the peaks at 0 and 0.007.

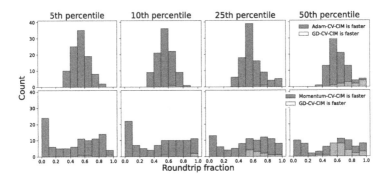

Fig. 2. This figure plots the histogram of roundtrip fractions, computed according to the process illustrated in Fig. 1. The top row compares the Adam-CV-CIM to the GD-CV-CIM, while the bottom row compares the momentum-CV-CIM to the GD-CV-CIM. The data points are categorized depending on which CIM feedback variant converged faster.

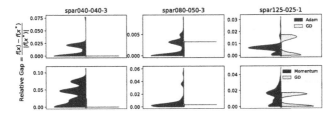

Fig. 3. This figure plots the relative optimality gaps for three representative instances. In all instances, the Adam-CV-CIM and momentum-CV-CIM achieve a greater diversity of samples with multi-modal distributions. For `spar040-040-3`, all samples from GD-CV-CIM concentrate on the global optimum, while for `spar080-050-3`, all samples concentrate on a suboptimal solution. For `spar125-025-1`, all variants result in multi-modal distributions with some shared modes between solvers.

We aggregate these metrics over all problem instances by computing the standard deviation in the relative optimality gap achieved by each sample. Figure 4 plots the histogram of optimality gap standard deviations for all CIM variants. We find that the momentum-CV-CIM has the highest standard deviations, with values roughly double that of the Adam-CV-CIM. In contrast, consistent with the examples depicted in Fig. 3, the GD-CV-CIM has very low variation in optimality gaps, with most instances having no variation. We note that the relationship between the degree of variation and solver performance is not monotonic. In the context of Eq. (5), if all samples converge to the same value, $\hat{p}_{\text{succ}}$ will be either 0 or 1, depending on the desired target. This means that the solver will either achieve the target with a single sample or not at all. While this may improve the time-to-target for problems that are particularly amenable to the solver, it will result in a complete failure for problems that are not. This is contrary to the perspective of the CIM as a sampler, from which one may expect a diverse sample of high-quality, albeit sub-optimal, solutions. Instead, the solver should strike a delicate balance between achieving a good diversity of samples while still concentrating on those with good objective value.

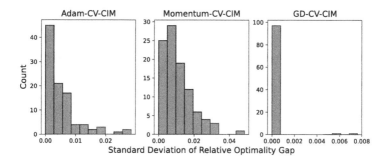

Fig. 4. This figure plots the histogram of optimality gap standard deviations for all CIM variants. The momentum-CV-CIM has the highest standard deviations, with values roughly double that of the Adam-CV-CIM. The GD-CV-CIM has very low variation in optimality gaps, with most instances having no variation.

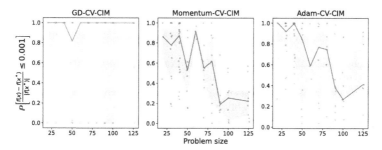

Fig. 5. This plot shows the likelihood of samples having a relative optimality gap $\leq$ 0.1%. Adam-CV-CIM and momentum-CV-CIM consistently achieved this, while GD-CV-CIM often met the target, except in 17 instances out of 99.

Figure 5 plots the probability that each variant finds samples with a relative optimality gap of at most 0.1%. On the standard benchmarks, using default parameters, both the Adam-CV-CIM and momentum-CV-CIM had non-zero probabilities of finding a solution within a 0.1% gap on all instances. In contrast, the GD-CV-CIM frequently had all samples converging within a 0.1% gap, however in 17 (out of 99 total) instances, the GD-CV-CIM failed to find any solutions within a 0.1% gap.

4.3 Sensitivity to Feedback Strength

In Sects. 4.1 and 4.2, we found that all variants of the CIM had reasonable performance on the standard benchmarks, even using default parameters. However, one commonly neglected factor is the degree to which solver performance

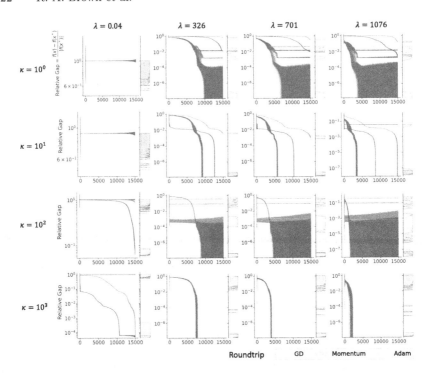

Fig. 6. Each plot in the grid illustrates the convergence in relative optimality gaps over roundtrips for varying combinations of κ and feedback strengths, λ. We plot the kernel-density estimate of the best optimality gaps on the right axis of each subplot to illustrate their distributions. Notably, the GD and momentum CV-CIMs show limited consistency in successful feedback strengths with respect to problem size, while the Adam-CV-CIM exhibits more consistent dynamics across instances of varying κ.

depends on its hyperparameters. Indeed, performance metrics are often reported based on hyperparameters that have been meticulously hand-tuned, typically in an opaque process that obscures the effort required for a practitioner to apply the solver to a new problem. In this section, we introduce a new benchmark set designed to test the CIM variants on problems with poor conditioning. At a high level, these instances are generated to have a non-convex landscape and are additionally parameterized by a constant $\kappa \in \mathbb{R}_{++}$, that dictates the skew when re-scaling the objective along different axes.[2]

In this section, we evaluate the sensitivity of each of the variants to the feedback strength, λ in Eq. (3a), analogous to step-size in classical optimization. Figure 6 plots the convergence in relative optimality gaps for representative

---

[2] Specifically, instances are constructed as $M = D(\kappa)U\Sigma(n)U^\top D(\kappa)$, where $U \in \mathbb{R}^{n \times n}$ is a random orthogonal matrix, $\Sigma(n) \in \mathbb{R}^{n \times n}$ is a diagonal matrix with $\Sigma(n)_{ii} = 1$ if $i \leq \frac{n}{2}$ and -1 otherwise (to induce a non-convex landscape), and $D(\kappa) = \mathrm{diag}(1, 1 + \frac{\kappa-1}{n}, \ldots, \kappa)$.

instances of dimension 20 with $\kappa \in [1, 10, 1000, 1000]$ across feedback strengths of $\lambda \in [0.04, 326, 701, 1076]$. We plot the kernel-density estimate of the best gap per sample on the right axes. We note that absence of samples from an individual plot indicates that the sample diverged, and the best gap is computed until the point of divergence. We find that there is limited consistency in successful feedback strengths as a function of problem size for the GD-CV-CIM and momentum-CV-CIM; the step sizes required for converge on instances with small κ lead to divergent behavior when κ is large. We expect that, much like classical optimization, problem conditioning is a primary factor dictating stable feedback strengths. While setting the feedback strength to be too small hampers the convergence of the Adam-CV-CIM, its dynamics tend to be consistent as a function of feedback strength, even across instances of varying κ. This is likely because Adam regularizes the momentum updates using the second-moment estimates, and is scale-invariant as a result. Moreover, the learning rate automatically adapted per variable, allowing Adam to explicitly combat the poor problem scaling when κ is large.

Surprisingly, momentum is more unstable than gradient descent, while it is typically known for improving stability for poorly conditioned instances. This is likely because the CIM relies solely on the double-well potential to enforce the constraints, and they are not otherwise incorporated in the feedback calculation. As a consequence, the momentum update continues to add up the contribution from the gradients, even if they lead to constraints violation. This suggest that the feedback term should also take constraints into account, similarly to the approach taken by the AHC-CIM.

The purpose of this experiment is not to suggest that there are no values of λ or re-parameterizations of the instances where the performance of the GD-CV-CIM and momentum-CV-CIM are comparable to that of the Adam-CV-CIM. In fact, a diagonal pre-conditioner that undoes the skew-inducing step in the instance construction would suffice. However, parameter-tuning can be tedious process that may result in disparate results depending on the expertise and persistence of the user, particularly when differences in problem characteristics prevent extrapolating good parameter settings from one set of instances to another. While it is unlikely to make these solvers fully parameter-free, a solver that is robust, reliable, and user-friendly will be accommodating of some user-error in the parameter settings. These qualities are part of the reason for Adam's wide adoption in deep-learning, and we believe that adoption of non-conventional optimization architectures will follow similar desiderata.

5   Discussion

In this work, we introduced the CIM and the Gaussian-state model of its dynamics, highlighting clear opportunities to incorporate more sophisticated optimization techniques in the feedback term. We benchmarked variants with the Adam and momentum updates, replacing the standard feedback term. We found that both the Adam-CV-CIM and momentum-CV-CIM greatly improved convergence speed over the GD-CV-CIM. While they did not consistently improve the probability of reaching specific optimality gap targets within our benchmarks, particularly when all samples from the GD-CV-CIM converged to the global optimum, they generally resulted in a greater diversity of samples. We also found that the performance of the Adam-CV-CIM was significantly less sensitive to the feedback strength parameter (analogous to step size in classical optimization), resulting in one fewer parameter that needs to be meticulously tuned.

Many of the same benefits that Adam and momentum confer to classical optimization also translate to the CIM. While this is unsurprising when the exposition of the CIM is couched in the language of optimization, it typically is not, obscuring the connections to classical optimization theory. Despite the limitations of classical optimization algorithms running on von Neumann computers, they have undoubtedly achieved impressive empirical success, including on difficult, non-convex optimization problems. Even if non-conventional computing architectures are intended to address the limitations of von Neumann computers, we contend that the lessons learned from classical optimization theory should be re-purposed where appropriate. For example, some challenges identified by the numerical experiments include: (1) better incorporation of the constraints in the feedback computation, which could potentially be addressed with a Lagrangian-inspired update, and (2) standardization of problem inputs to ease parameter tuning, which could take inspiration from preconditioning.

This raises the question of how one should trade off the complexity of $\Phi(\{\tilde{\mu}(t)\}_t)$ and the performance of the solver as a whole. One can envision increasingly complicated feedback terms, $\Phi(\{\tilde{\mu}(t)\}_t)$, that start to resemble higher-order optimization algorithms at the expense of more computational effort. We expect a Pareto frontier of feedback terms with different trade-offs between time/energy-per-iteration and the number of iterations required to reach a desired optimization target. The optimal feedback term for a particular scenario will likely depend on resource constraints, such as wall clock time or energy usage.

In this work, we consider a discrete-time perspective of the dynamical system for ease of simulation; however, the continuous-time dynamical systems perspective has recently had significant impacts on optimization theory [40]. Exploring the continuous-time models of optimization algorithms will likely result in tighter analogs with the continuous-time models often used to describe non-conventional architecture, which we leave to future work.

Acknowledgements. This work was supported by NSF CCF (grant #1918549), NASA Academic Mission Services (contract NNA16BD14C – funded under SAA2-403506). R.B. acknowledges support from the NASA/USRA Feynman Quantum Academy Internship program.

References

1. Babbush, R., McClean, J.R., Newman, M., Gidney, C., Boixo, S., Neven, H.: Focus beyond quadratic speedups for error-corrected quantum advantage. PRX Quantum **2**(1), 010103 (2021)
2. Berloff, N.G., et al.: Realizing the classical XY Hamiltonian in polariton simulators. Nat. Mater. **16**(11), 1120–1126 (2017)
3. Borle, A., Lomonaco, S.J.: Analyzing the quantum annealing approach for solving linear least squares problems. In: Das, G.K., Mandal, P.S., Mukhopadhyaya, K., Nakano, S. (eds.) WALCOM 2019. LNCS, vol. 11355, pp. 289–301. Springer, Cham (2019). https://doi.org/10.1007/978-3-030-10564-8_23
4. Brown, R., Bernal Neira, D.E., Venturelli, D., Pavone, M.: A copositive framework for analysis of hybrid ising-classical algorithms. arXiv preprint arXiv:2207.13630 (2022)
5. Bruzewicz, C.D., Chiaverini, J., McConnell, R., Sage, J.M.: Trapped-ion quantum computing: progress and challenges. Appl. Phys. Rev. **6**(2) (2019)
6. Burer, S.: Optimizing a polyhedral-semidefinite relaxation of completely positive programs. Math. Program. Comput. **2**(1), 1–19 (2010)
7. Burer, S., Vandenbussche, D.: Globally solving box-constrained nonconvex quadratic programs with semidefinite-based finite branch-and-bound. Comput. Optim. Appl. **43**(2), 181–195 (2009)
8. Cai, F., et al.: Power-efficient combinatorial optimization using intrinsic noise in memristor hopfield neural networks. Nat. Electron. **3**(7), 409–418 (2020)
9. Callison, A., Chancellor, N.: Hybrid quantum-classical algorithms in the noisy intermediate-scale quantum era and beyond. Phys. Rev. A **106**(1), 010101 (2022)
10. Chang, C.C., Gambhir, A., Humble, T.S., Sota, S.: Quantum annealing for systems of polynomial equations. Sci. Rep. **9**(1), 10258 (2019)
11. Ebadi, S., et al.: Quantum optimization of maximum independent set using rydberg atom arrays. Science **376**(6598), 1209–1215 (2022)
12. Gunathilaka, M.D.S.H., et al.: Effective implementation of l_0-regularised compressed sensing with chaotic-amplitude-controlled coherent ising machines. arXiv preprint arXiv:2302.12523 (2023)
13. Harwood, S., Gambella, C., Trenev, D., Simonetto, A., Bernal, D., Greenberg, D.: Formulating and solving routing problems on quantum computers. IEEE Trans. Quantum Eng. **2**, 1–17 (2021)
14. Hauke, P., Katzgraber, H.G., Lechner, W., Nishimori, H., Oliver, W.D.: Perspectives of quantum annealing: methods and implementations. Rep. Prog. Phys. **83**(5), 054401 (2020)
15. Henriet, L., et al.: Quantum computing with neutral atoms. Quantum **4**, 327 (2020)
16. Honjo, T., et al.: 100,000-spin coherent Ising machine. Sci. Adv. **7**(40), eabh0952 (2021)
17. Kako, S., Leleu, T., Inui, Y., Khoyratee, F., Reifenstein, S., Yamamoto, Y.: Coherent Ising machines with error correction feedback. Adv. Quantum Technol. **3**(11), 2000045 (2020)
18. Kalinin, K.P., Amo, A., Bloch, J., Berloff, N.G.: Polaritonic xy-ising machine. Nanophotonics **9**(13), 4127–4138 (2020)
19. Khosravi, F., Yildiz, U., Scherer, A., Ronagh, P.: Non-convex quadratic programming using coherent optical networks. arXiv preprint arXiv:2209.04415 (2022)
20. Kim, M., Mandrà, S., Venturelli, D., Jamieson, K.: Physics-inspired heuristics for soft MIMO detection in 5G new radio and beyond. In: Proceedings of the 27th Annual International Conference on Mobile Computing and Networking, pp. 42–55 (2021)

21. Kjaergaard, M., et al.: Superconducting qubits: current state of play. Annu. Rev. Condens. Matter Phys. **11**, 369–395 (2020)
22. Kleinberg, J., Tardos, E.: Algorithm Design. Addison-Wesley Longman Publishing Co., Inc., USA (2005)
23. Leleu, T., Yamamoto, Y., McMahon, P.L., Aihara, K.: Destabilization of local minima in analog spin systems by correction of amplitude heterogeneity. Phys. Rev. Lett. **122**(4), 040607 (2019)
24. Leleu, T., Yamamoto, Y., Utsunomiya, S., Aihara, K.: Combinatorial optimization using dynamical phase transitions in driven-dissipative systems. Phys. Rev. E **95**(2), 022118 (2017)
25. Lucas, A.: Ising formulations of many NP problems. Front. Phys. 5 (2014)
26. Marandi, A., Wang, Z., Takata, K., Byer, R.L., Yamamoto, Y.: Network of time-multiplexed optical parametric oscillators as a coherent ising machine. Nat. Photonics **8**(12), 937–942 (2014)
27. McMahon, P.L., et al.: A fully programmable 100-spin coherent ising machine with all-to-all connections. Science **354**(6312), 614–617 (2016)
28. Negre, C.F., Ushijima-Mwesigwa, H., Mniszewski, S.M.: Detecting multiple communities using quantum annealing on the D-Wave system. PLoS ONE **15**(2), e0227538 (2020)
29. Ng, E., Onodera, T., Kako, S., McMahon, P.L., Mabuchi, H., Yamamoto, Y.: Efficient sampling of ground and low-energy ising spin configurations with a coherent ising machine. Phys. Rev. Res. **4**(1), 013009 (2022)
30. Ottaviani, D., Amendola, A.: Low rank non-negative matrix factorization with d-wave 2000q. arXiv preprint arXiv:1808.08721 (2018)
31. Pal, V., Mahler, S., Tradonsky, C., Friesem, A.A., Davidson, N.: Rapid fair sampling of the X Y spin Hamiltonian with a laser simulator. Phys. Rev. Res. **2**(3), 033008 (2020)
32. Pardalos, P.M., Vavasis, S.A.: Quadratic programming with one negative eigenvalue is np-hard. J. Glob. Optim. **1**(1), 15–22 (1991)
33. Qian, N.: On the momentum term in gradient descent learning algorithms. Neural Netw. **12**(1), 145–151 (1999)
34. Reifenstein, S., Kako, S., Khoyratee, F., Leleu, T., Yamamoto, Y.: Coherent ising machines with optical error correction circuits. Adv. Quantum Technol. **4**(11), 2100077 (2021)
35. Rønnow, T.F., et al.: Defining and detecting quantum speedup. Science **345**(6195), 420–424 (2014)
36. Slussarenko, S., Pryde, G.J.: Photonic quantum information processing: a concise review. Appl. Phys. Rev. **6**(4) (2019)
37. Vandenbussche, D., Nemhauser, G.: A branch-and-cut algorithm for nonconvex quadratic programs with box constraints. Math. Program. **102**(3), 559–575 (2005)
38. Venturelli, D., Marchand, D., Rojo, G.: Job shop scheduling solver based on quantum annealing. In: Proc. of ICAPS-16 Workshop on Constraint Satisfaction Techniques for Planning and Scheduling (COPLAS), pp. 25–34 (2016)
39. Willsch, D., Willsch, M., De Raedt, H., Michielsen, K.: Support vector machines on the d-wave quantum annealer. Comput. Phys. Commun. **248**, 107006 (2020)
40. Wilson, A.C., Recht, B., Jordan, M.I.: A lyapunov analysis of accelerated methods in optimization. J. Mach. Learn. Res. **22**(1), 5040–5073 (2021)
41. Yamamura, A., Mabuchi, H., Ganguli, S.: Geometric landscape annealing as an optimization principle underlying the coherent ising machine. arXiv preprint arXiv:2309.08119 (2023)

Decision-Focused Predictions via Pessimistic Bilevel Optimization: A Computational Study

Víctor Bucarey[1,4]([✉]) [ID], Sophia Calderón[2] [ID], Gonzalo Muñoz[3,4] [ID],
and Frédéric Semet[5] [ID]

[1] Engineering Sciences Institute, Universidad de O'Higgins, Rancagua, Chile
`victor.bucarey@uoh.cl`
[2] Engineering School, Universidad de O'Higgins, Rancagua, Chile
[3] Industrial Engineering Department, Universidad de Chile, Santiago, Chile
[4] Instituto Sistemas Complejos de Ingeniería (ISCI), Santiago, Chile
[5] Univ. Lille, CNRS, Centrale Lille, Inria, UMR 9189 - CRIStAL, Lille, France

Abstract. Dealing with uncertainty in optimization parameters is an important and longstanding challenge. Typically, uncertain parameters are predicted accurately, and then a deterministic optimization problem is solved. However, the decisions produced by this so-called *predict-then-optimize* procedure can be highly sensitive to uncertain parameters. In this work, we contribute to recent efforts in producing *decision-focused* predictions, i.e., to build predictive models constructed to minimize a *regret* measure on the decisions taken with them. We formulate the exact expected regret minimization as a pessimistic bilevel optimization model and then we reformulate it as a non-convex quadratic problem. Finally, we show various computational techniques to achieve tractability. We report extensive computational results on shortest-path instances with uncertain cost vectors. Our results indicate that our approach can improve training performance over the approach of Elmachtoub and Grigas (2022), a state-of-the-art method for decision-focused learning.

Keywords: predict and optimize · pessimistic bilevel optimization · non-convex quadratics

1 Introduction

Decision-making processes often involve uncertain parameters. Commonly, a two-stage approach is employed: firstly, training a machine learning (ML) model to estimate the uncertain input, and secondly, using this estimate to tackle the decision task. This approach overlooks how inaccurate predictions can lead to decisions of poor quality, which has sparked the recent interest in *decision-focused*

Supported by the Chilean National Agency of Research and Development through Fondecyt grants 1231522 and 11220864 and ANID PIA/PUENTE AFB230002.

learning (DFL): here, prediction and optimization tasks are integrated into a single model. The typical measure for a prediction quality in this setting is the *regret*: the excess of cost in the task caused by prediction errors. We refer the readers to the recent survey [6] for details about the algorithmic approaches and models dealing with DFL.

In this work, we center on studying the mathematical object behind the expected regret minimization in the context of DFL, in order to develop better predictions and algorithms. The contributions of this work are the following: 1) we formulate the problem of minimizing the regret as a pessimistic bilevel optimization problem; 2) we reformulate the bilevel pessimistic problem as a non-convex quadratically-constrained quadratic program (QCQP), which can be tackled by current optimization technology for moderate sizes; 3) we propose heuristics to improve the solution procedure for the quadratic reformulation; 4) we conduct a comprehensive computational study on shortest path instances.

1.1   Problem Setting

We consider a nominal optimization problem with a linear objective function:

$$P(c): \quad z^*(c) = \min_{v \in V} c^\top v \tag{1}$$

In this work, we restrict V to be a non-empty polytope, i.e. a non-empty bounded polyhedron. For a given c, we define $V^*(c)$ as the set of optimal solutions to (1).

In our setting, $c \in \mathbb{R}^n$ is unknown, but we have access to a dataset $\mathcal{D} = \{(x^i, c^i)\}_{i=1}^N$ with historical observations of c and correlated feature vectors $x \in \mathbb{R}^K$. Given these observations, we consider a parametric predictive model $m(\omega, x)$ for the costs. For a fixed set of parameters ω, we can empirically measure the sensitivity of the decisions given by the predictions using an average *regret*:

$$\max_v \quad \frac{1}{N} \sum_{i \in [N]} \left(c^{i\top} v^i - z^*(c^i) \right), \quad \text{s.t.} \quad v^i \in V^*(m(\omega, x^i)) \quad \forall i \in [N] \tag{2}$$

Here, we are comparing the *true* optimal value $z^*(c^i)$ with the value $c^{i\top} v^i$, which is the "true" cost of a solution that is optimal for the prediction $m(\omega, x^i)$. To find the values that minimize the regret (2), we must solve the following pessimistic bilevel optimization problem.

$$\min_\omega \max_{v^i \in V^*(m(\omega, x^i))} \quad \frac{1}{N} \sum_{i \in [N]} \left(c^{i\top} v^i - z^*(c^i) \right) \tag{3}$$

In (3), three optimization problems are involved: i. the lower-level problem optimizing $m(\omega, x^i)^\top v$ over V; ii. over all the possible optimal solutions of the latter, take the one with the worst regret. This corresponds to the pessimistic version of the bilevel formulation. iii. Minimize the pessimistic regret using ω as a variable.

In what follows, we use the notation $\hat{c}^i(\omega) := m(\omega, x^i)$.

2   A Non-convex Quadratic Reformulation

In this section, we will apply duality arguments, supported by the assumption that V in problem (1) is non-empty and bounded. Under this assumption, we can assume (1) has the following form:

$$\min_{v} \quad \{c^\top v \, : \, Av \geq b, \, 0 \leq v \leq 1\} \tag{4}$$

Our predicted costs have the form $m(\omega, x)$ for some feature vector x and parameters ω (to be determined), and the terms $z^*(c^i)$ are constant. Thus, an equivalent formulation of our (pessimistic) regret-minimization problem is:

$$\min_{\omega} \max_{v} \quad \frac{1}{N} \sum_{i \in [N]} (c^i)^\top v^i$$
$$\text{s.t.} \quad v^i \in \arg\min \{m(\omega, x^i)^\top \tilde{v}^i \, : \, A\tilde{v}^i \geq b, \, 0 \leq \tilde{v}^i \leq 1\}. \tag{5}$$

2.1   Duality Arguments

A common approach to solving optimistic bilevel problems involving convex lower-level problems is to reformulate them by replacing the lower-level problems with their KKT conditions (see [4]). Here, we follow the same type of argument twice to achieve a single-level reformulation of the pessimistic problem (5).

Since the feasible region of the lower-level problem in (5) is a non-empty polytope, which is unaffected by ω, we can apply LP duality and reformulate it as:

$$\min_{\omega} \max_{v, \rho, \alpha} \quad \frac{1}{N} \sum_{i \in [N]} (c^i)^\top v^i \tag{6a}$$

$$\text{s.t.} \quad Av^i \geq b \qquad\qquad \forall i \in [N] \tag{6b}$$
$$0 \leq v^i \leq 1 \qquad\qquad \forall i \in [N] \tag{6c}$$
$$A^\top \rho^i + \alpha^i \leq m(\omega, x^i) \qquad\qquad \forall i \in [N] \tag{6d}$$
$$\rho^i \geq 0, \, \alpha^i \leq 0 \qquad\qquad \forall i \in [N] \tag{6e}$$
$$m(\omega, x^i)^\top v^i \leq b^\top \rho^i + \mathbf{1}^\top \alpha^i \qquad\qquad \forall i \in [N] \tag{6f}$$

In this formulation (6b)–(6c) impose primal feasibility, (6d)–(6e) impose dual feasibility, and (6f) imposes strong duality[1].

The inner maximization problem of (6) is an LP, which is feasible for every value of ω (it is a primal-dual system which always has a solution). Moreover, since the objective function in (6a) only involves the v variables which are bounded, we know strong duality holds and we can take a dual again. This yields the following reformulation of (5).

---

[1] Strong duality is typically written as an equality constraint; however, the $\geq$ inequality always holds due to weak duality.

$$\min_{\omega} \min_{\mu,\theta,\delta,\gamma} \sum_{i \in [N]} \left(b^\top \mu^i + \mathbf{1}^\top \theta^i + m(\omega, x^i)^\top \delta^i \right) \tag{7a}$$

$$\text{s.t.} \quad A^\top \mu^i + m(\omega, x^i)\gamma^i + \theta^i \geq \frac{1}{N} c^i \qquad \forall i \in [N] \tag{7b}$$

$$A\delta^i - b\gamma^i \geq 0 \qquad \forall i \in [N] \tag{7c}$$

$$- \gamma^i \mathbf{1} + \delta^i \leq 0 \qquad \forall i \in [N] \tag{7d}$$

$$\delta^i, \gamma^i, \theta^i \geq 0, \ \mu^i \leq 0 \qquad \forall i \in [N] \tag{7e}$$

This is a single-level, non-convex quadratic problem.

2.2   Shortest Path as a Bounded Linear Program

In this work, and motivated by [2], we consider the shortest path problem with unknown cost vectors. It is well known that this problem can be formulated as a linear program using a totally unimodular constraint matrix. However, the feasible region may not be bounded, as the underlying graph may have negative cycles. To apply our framework, we need to assume no negative cycles exist for *every* possible prediction $m(\omega, x)$. For this reason, we make the following assumption.

Assumption 1. *The underlying graph G defining the shortest path is acyclic.*

Under this assumption, we can safely formulate the shortest path problem as (4), and every extreme point solution will be a binary vector indicating the shortest path. Additionally, the inner maximization problem in (6) always has a binary optimal solution, as its feasible region is an extended formulation of the optimal face of the lower-level problem. From this discussion, we can guarantee that under Assumption 1, formulation (7) is valid for the shortest path problem with uncertain costs.

3   Solution Methods

In this work, we focus on the case when $m(\omega, x)$ is a linear model. For this reason, we will use as baselines the ordinary least square estimator for the linear regression (LR) and the SPO+ loss function presented in [2].

Local Search. The intermediate reformulation presented in (6) can be seen as an unrestricted optimization problem of the form $\min_\omega F(\omega)$. Here, $F(\omega)$ is a function that for each ω returns the optimal value of the inner maximization problem in (6). As mentioned above, for each ω, $F(\omega)$ is a feasible linear problem. We propose the following local-search-based heuristic: Given an initial incumbent solution ω_0, we randomly generate T new solutions in a neighborhood of ω_0, evaluate the regret for each, and update the incumbent solution with the one with the smallest regret. We repeat these steps during L iterations. The procedure is detailed in Algorithm 1. In our experiments we used $\epsilon = 1$, $T = 5$ and $L = 20$.

Algorithm 1. Local-search based algorithm

1: **Input** Training data, Starting model parameters ω_0.
2: **Hyperparameters**: size of neighbourhood ϵ, sample size T, maximum number of iterations L.
3: Solve $F(\omega_0)$
4: **for** $i = 0, \ldots, N$ **do**
5:     Sample T parameters in the neighbourhood of ω_i: $\omega_t \leftarrow \omega_i + \epsilon \cdot \mathcal{N}(0, 1)$
6:     Compute $F(\omega_t)$ $\forall t \in \{1, \ldots, T\}$
7:     Update $\omega_{i+1} \leftarrow \arg\min_{t=1,\ldots,T} F(\omega_t)$
8: **end for**

Penalization. Based on the ideas of [1], we propose the following related formulation: we fix the variables γ^i in (7) to have all the same fixed value κ. This transforms (7b) into a linear constraint. This parameter κ is set before optimization and can be seen as a hyperparameter of the optimization problem. This formulation corresponds to a restricted version, corresponding to a slice of the formulation (7). Moreover, by adopting this approach, we solve an optimization program with quadratic non-convex objective and linear constraints. In our experiments we used $\kappa = 0.1$.

We remark that this approach results in a problem that is not a traditional penalization (which typically yields relaxations) but rather a restriction of the problem. The name "penalization", which is used in [1], comes from a derivation that follows a similar approach to the one described in Sect. 2. The difference lies in penalizing (6f) using κ before taking the dual a second time.

4 Computational Experiments

4.1 Computational Set-Up

Data generation. We adapt the data generation described in [2,9]. We consider a directed acyclic graph consisting of a grid of 5×5 nodes. The data consists of $\{(x^i, c^i)\}_{i=1}^N$ generated synthetically as follows. We consider values of $N = \{50, 100, 200\}$ split 70% for training and 30% for testing. Feature vectors are sampled from a standard normal distribution. We generate cost vectors by fixing the parameters ω, representing the true underlying linear model, and then using the following formula:

$$c_a^i = \left[\frac{1}{3.5^{\mathrm{Deg}}} \left(\frac{1}{\sqrt{K}} \left(\sum_{k=1}^K \omega_k x_{ak}^i \right) + 3 \right)^{\mathrm{Deg}} + 1 \right] \cdot \varepsilon$$

where i is the observation, a is the arc, *Deg* represents the model misspecification, and ε is sampled from a uniform distribution in $[0.5, 1.5]$.

Algorithms. In our experiments, we consider the following sequence of steps to generate decision-focused predictions:

1. Generate an initial solution using a common linear regression (LR) or the SPO+ (S) method described in the previous section.
2. Improve the previous solution using Algorithm 1 (L).
3. Solve the Penalized model (P) described in Sect. 3 or (7) (E) using Gurobi [3] with the previous solution as a warm start.

We note that each step is computationally more expensive than the previous. In fact, Step 1 involves solving an LP (in the case of SPO+) or a simple linear system (in the case of LR); Step 2 is a fixed number of LPs; and Step 3 is a non-convex quadratic problem. For the latter, we use a one-hour limit.

This generates four different combinations that we test below, with their names indicating the sequence: S-L-P, S-L-E, LR-L-P, and LR-L-E. To prevent the solver from generating solutions with large coefficients in the non-convex QCQPs, and since the lower-level optimization problem is invariant to scalings of the objective, we added arbitrary bounds to the values of ω. Any bound is valid, but we avoided small numbers to prevent numerical instabilities.

Additionally, to improve the performance of Gurobi, we included the following valid inequality to (7):

$$\sum_{i \in [N]} \left(b^\top \mu^i + \mathbf{1}^\top \theta^i + m(\omega, x^i)^\top \delta^i \right) \geq \frac{1}{N} \sum_{i \in [N]} z^*(c^i).$$

This is a dual cut-off constraint. Its left hand side takes the same value as $\frac{1}{N} \sum_{i \in N}(c^i)^\top v^i$ in (5), and thus, by optimality of z^*, the inequality holds. This simple inequality provided considerable improvements in Gurobi's performance.

4.2   Results

In this section, we compare the decision-focused predictions we obtained with the aforementioned methods based on the regrets they achieve. To display values of the regrets of the same magnitudes, and following [2], instead of reporting the regret directly (as defined in (2)) we use the *normalized* regret defined as $\sum_{i \in [N]} \text{Regret}(m(\omega, x^i), c^i) / \sum_{i \in [N]} z^*(c^i)$, where $\text{Regret}(m(\omega, x^i), c^i)$ is defined as in (2), but for a single observation i. Henceforth, when we reference the regret, we mean this normalized version.

In Table 1, we show the results obtained in the training set when running the four sequences described before. We note that the penalized problem may reject a warm start due to its nature, since we are strictly fixing the dual variables. This can result in the penalized problem producing a worse solution. On the contrary, the exact model always accepts the warm start and can only improve the solutions. Table 1 shows that while SPO+ and LR have small regret, they can still be considerably improved. In some cases, the regret decreases by 50% or more. One notable and surprising result is that the penalized method performs better than the exact method in general, even though it is a restricted version of it. In a few cases, though, the penalized method worsened the regret; we note that the two cases where the penalized method has its worse performance are

Table 1. Experiments showing regret improvements over SPO+ and LR in the training set. Columns 1–3 indicate the parameters used to generate the instance. The column labeled SPO (LR resp.) shows the regret obtained by the SPO+ (LR resp.). Columns 5–7 (9–11 resp.) show the regret change compared to SPO (LR resp.). Entries in boldface indicate the lowest regret.

N	Deg	Noise	S	S-L	S-L-E	S-L-P	LR	LR-L	LR-L-E	LR-L-P
50	2	0.0	0.004	0.0%	0.0%	**−50.0%**	0.006	−50.0%	−50.0%	**−66.7%**
50	2	0.5	0.094	−20.2%	−20.2%	**−20.2%**	0.087	**−10.3%**	**−10.3%**	−10.3%
50	8	0.0	0.016	0.0%	−12.5%	**−12.5%**	0.027	−7.4%	−7.4%	**−48.1%**
50	8	0.5	0.123	−14.6%	−14.6%	**−26.0%**	0.12	−19.2%	−21.7%	**−24.2%**
50	16	0.0	1.767	0.0%	0.0%	**−69.3%**	2.728	0.0%	0.0%	**−67.5%**
50	16	0.5	0.734	−0.8%	−0.8%	**−16.8%**	**0.735**	0.0%	0.0%	+336.9%
100	2	0.0	**0.001**	0.0%	0.0%	0.0%	0.002	0.0%	0.0%	**−50.0%**
100	2	0.5	0.121	−7.4%	−7.4%	**−9.9%**	0.126	**−14.3%**	**−14.3%**	−13.5%
100	8	0.0	0.029	0.0%	−3.4%	**−10.3%**	0.038	−15.8%	−18.4%	**−31.6%**
100	8	0.5	0.116	0.0%	0.0%	**−7.8%**	0.117	0.0%	0.0%	**−8.5%**
100	16	0.0	0.251	−2.4%	−2.4%	**−9.6%**	0.257	**−11.7%**	**−11.7%**	**−11.7%**
100	16	0.5	0.574	**−15.3%**	**−15.3%**	+27.5%	1.296	**−60.9%**	**−60.9%**	−50.6%
200	2	0.0	**0.002**	0.0%	0.0%	0.0%	**0.002**	0.0%	0.0%	0.0%
200	2	0.5	0.116	0.0%	0.0%	**−4.3%**	0.12	0.0%	0.0%	**−7.5%**
200	8	0.0	0.031	0.0%	0.0%	**−12.9%**	0.031	0.0%	0.0%	**−12.9%**
200	8	0.5	0.161	−4.3%	−4.3%	**−6.2%**	0.162	**−6.8%**	**−6.8%**	**−6.8%**
200	16	0.0	0.141	−39.0%	−39.0%	**−39.7%**	0.138	−34.8%	−34.8%	−34.8%
200	16	0.5	0.442	−12.0%	−12.0%	**−15.8%**	0.44	−10.7%	−10.7%	**−15.5%**

high-degree cases. Overall, the results for both starting points (SPO+ and LR) greatly align, although there are interesting differences. For example, for LR, the local search method provides the best solution more often than for SPO+.

In Table 2, we perform the analogous analysis for the test set. Now, in the case of SPO+, the results are more heterogeneous: the extra steps after SPO+ worsen the test performance more often. However, most instances are improved (11 out of 18), and the improvements are quite significant in some cases. These results both advocate for a good generalization in many of the SPO+ predictors and show that improvements are possible in many cases.

In the cases where LR was the starting point, in 12 out of 18 instances, LR provided the best performance, and our methods were not able to improve these solutions. It is quite remarkable that in so many cases, no improvements were found. This suggests that the "improvements" after LR lead to over-fitting. However, we need to be careful before concluding that LR provides the best overall test performance. If one compares both starting points in Table 2, one can see that only in 6 cases LR provides a better regret than SPO+. Therefore, regarding regret, the solutions starting from SPO+ performed better in the test set. Interestingly, the LR solutions are, in some sense, worse, but they are not improved by our methods. And on the contrary, the SPO+ solutions are good, and our approach can further improve them.

Table 2. Experiments showing regret improvements over SPO+ and LR in the test set. Columns 1–3 indicate the parameters used to generate the instance. The column labeled SPO (LR resp.) shows the regret obtained by the SPO+ (LR resp.). Columns 5–7 (9–11 resp.) show the regret change compared to SPO (LR resp.). Entries in boldface indicate the lowest regret for each starting point.

N	Deg	Noise	S	S-L	S-L-E	S-L-P	LR	LR-L	LR-L-E	LR-L-P
50	2	0.0	**0.006**	0.0%	0.0%	0.0%	**0.007**	0.0%	0.0%	+28.6%
50	2	0.5	0.142	−10.6%	−10.6%	−10.6%	**0.154**	+10.4%	+10.4%	+10.4%
50	8	0.0	0.012	0.0%	−16.7%	−33.3%	0.024	+33.3%	+33.3%	−66.7%
50	8	0.5	0.17	−14.7%	−14.7%	−14.1%	**0.112**	+30.4%	+30.4%	+30.4%
50	16	0.0	0.581	0.0%	0.0%	−71.6%	0.659	0.0%	0.0%	−52.2%
50	16	0.5	0.292	0.0%	0.0%	−12.0%	**0.282**	0.0%	+3.5%	+3235.5%
100	2	0.0	**0.001**	0.0%	0.0%	+100.0%	**0.002**	0.0%	0.0%	0.0%
100	2	0.5	0.116	−19.0%	−19.0%	−19.0%	0.152	−33.6%	−33.6%	−37.5%
100	8	0.0	**0.058**	0.0%	+1.7%	+13.8%	**0.061**	+9.8%	+4.9%	+8.2%
100	8	0.5	**0.113**	0.0%	0.0%	+15.0%	**0.118**	0.0%	0.0%	+10.2%
100	16	0.0	**0.418**	+20.6%	+20.6%	+0.2%	**0.415**	+4.8%	+0.7%	+1.0%
100	16	0.5	0.951	−7.2%	−7.2%	−5.8%	0.88	−48.6%	−45.7%	−9.4%
200	2	0.0	0.002	0.0%	0.0%	−50.0%	**0.003**	0.0%	0.0%	0.0%
200	2	0.5	**0.117**	0.0%	0.0%	+10.3%	**0.122**	0.0%	0.0%	+3.3%
200	8	0.0	0.052	0.0%	0.0%	−1.9%	**0.051**	0.0%	0.0%	0.0%
200	8	0.5	**0.147**	+10.9%	+10.9%	+11.6%	**0.146**	+11.6%	+11.6%	+11.6%
200	16	0.0	0.076	−2.6%	−2.6%	0.0%	0.077	−2.6%	−2.6%	−2.6%
200	16	0.5	0.335	+1.8%	+1.8%	−4.2%	0.344	+15.7%	+15.7%	−6.7%

5 Conclusions and Future Work

In this work, we have shown a framework that can generate high-quality decision-focused predictions. Using known methods in tandem with a local search algorithm and a non-convex quadratic reformulation, we can provide predictions with a good performance in training and test sets for challenging shortest path instances with unknown cost vectors.

Our current and future work involves the following challenges. Firstly, improving scalability. So far, we have tried grids of up to 5×5 nodes and 200 observations. We would like to scale our approach further in these two dimensions. Secondly, improving solver performance. The non-convex quadratic formulations rarely finish with a provably optimal solution. We suspect many of these solutions are optimal, but the solver has issues finding a matching dual bound to prove so. We want to derive custom valid inequalities that can improve these bounds and thus accelerate the convergence of the non-convex solver. Thirdly, testing our methods in other nominal problems where this methodology can be applied, for instance, bipartite matchings, and spanning trees. Finally, comparing our methods with other state-of-the-art approaches in the literature, such as [5,7,8,10].

References

1. Aboussoror, A., Mansouri, A.: Weak linear bilevel programming problems: existence of solutions via a penalty method. J. Math. Anal. Appl. **304**(1), 399–408 (2005)
2. Elmachtoub, A., Grigas, P.: Smart predict, then optimize. Manag. Sci. **68**(1), 9–26 (2022)
3. Gurobi Optimization, LLC: Gurobi Optimizer Reference Manual (2023). https://www.gurobi.com
4. Kleinert, T., Labbé, M., Ljubić, I., Schmidt, M.: A survey on mixed-integer programming techniques in bilevel optimization. EURO J. Comput. Optim. **9**, 100007 (2021)
5. Mandi, J., Bucarey, V., Tchomba, M.M.K., Guns, T.: Decision-focused learning: through the lens of learning to rank. In: International Conference on Machine Learning, pp. 14935–14947. PMLR (2022)
6. Mandi, J., et al.: Decision-focused learning: foundations, state of the art, benchmark and future opportunities. arXiv preprint arXiv:2307.13565 (2023)
7. Mulamba, M., Mandi, J., Diligenti, M., Lombardi, M., Lopez, V.B., Guns, T.: Contrastive losses and solution caching for predict-and-optimize. In: 30th International Joint Conference on Artificial Intelligence (IJCAI-21): IJCAI-21, p. 2833. International Joint Conferences on Artificial Intelligence (2021)
8. Niepert, M., Minervini, P., Franceschi, L.: Implicit MLE: backpropagating through discrete exponential family distributions. In: Ranzato, M., Beygelzimer, A., Dauphin, Y., Liang, P., Vaughan, J.W. (eds.) Advances in Neural Information Processing Systems, vol. 34, pp. 14567–14579. Curran Associates, Inc. (2021)
9. Tang, B., Khalil, E.B.: PyEPO: a pytorch-based end-to-end predict-then-optimize library for linear and integer programming. arXiv preprint arXiv:2206.14234 (2022)
10. Wilder, B., Dilkina, B., Tambe, M.: Melding the data-decisions pipeline: decision-focused learning for combinatorial optimization. In: Proceedings of the AAAI Conference on Artificial Intelligence, vol. 33, pp. 1658–1665 (2019)

Bi-objective Discrete Graphical Model Optimization

Samuel Buchet[✉], David Allouche, Simon de Givry[✉], and Thomas Schiex[✉]

Universite Fédérale de Toulouse, ANITI, INRAE, UR 875, 31326 Toulouse, France
{samuel.buchet,david.allouche,simon.de-givry,thomas.schiex}@inrae.fr

Abstract. Discrete Graphical Models (GMs) are widely used in Artificial Intelligence to describe complex systems through a joint function of interest. Probabilistic GMs such as Markov Random Fields (MRFs) define a joint non-normalized probability distribution while deterministic GMs such as Cost Function Networks (CFNs) define a joint cost function. A typical query on GMs consists in finding the joint state that optimizes this joint function, a problem denoted as the *Maximum a Posteriori* or Weighted Constraint Satisfaction Problem respectively.

In practice, more than one function of interest may need to be optimized at the same time. In this paper, we develop a two-phase scalarization method for solving bi-objective discrete graphical model optimization, with the aim of computing a set of non-dominated solutions—the Pareto frontier—representing different compromises between two GM-defined objectives. For this purpose, we introduce a dedicated higher-order constraint, which bounds the value of one GM-defined objective while minimizing another GM on the same variables. Discrete GM optimization is NP-hard, and its bi-objective variants are even harder. We show how existing GM global lower and upper bounds can be exploited to provide anytime bounds of the exact Pareto frontier. We benchmark it on various instances from Operations Research and Probabilistic Graphical Models.

Keywords: bi-objective combinatorial optimization · graphical model · constraint optimization · cost function network · exact methods

1 Introduction

Many real problems require considering more than one objective. In protein design [2], one may look for the most probable (minimum energy) amino acid sequence given a target structure that also minimizes the complexity of the sequence [43], or which also optimizes its probability given another structure [42]. The same interest for multi-objective optimization exists in drug design [27,31]. In the uncapacitated warehouse location problem, it is desirable to minimize the setup costs of all facilities while minimizing the serving costs from these facilities [25]. In frequency assignment problems, one typically wants to minimize the number of different frequencies used but also their span (the maximum difference in frequencies) [10]. In all these cases, one objective can be compactly

B. Dilkina (Ed.): CPAIOR 2024, LNCS 14742, pp. 136–152, 2024.
https://doi.org/10.1007/978-3-031-60597-0_10

represented as one Graphical Model [14]. But the existence of multiple objectives makes discrete optimization challenging [20]. Even classical polytime problems such as the shortest path or assignment problems become NP-hard (to find a non-dominated solution) or even possibly #P-hard, when a set of states having non-dominated costs (the so-called Pareto frontier) is sought [16].

The hardness of these problems means that they often cannot be exhaustively solved. In a single objective minimization case, most algorithms will deliver an incumbent solution and a global lower bound, providing a desirable *a posteriori* optimality gap. In this paper, we are interested in providing similar services when two objectives, each defined by a Graphical Model, need to be simultaneously optimized under hard constraints. To exploit the efficiency of existing GM solvers [3], we reduce the problem of computing non-dominated states to a series of single GM optimization problems.

Our approach lies in the family of two-phase methods [16,40,41], where a first phase identifies all the solutions that can be found by solving a linear combination of both objectives (called *supported* non-dominated states). This phase reduces to a dichotomic series of optimization of a single objective combining the two original objectives linearly [4] that can be directly solved by any discrete GM optimization solver. To identify non-supported solutions that may exist inside the convex envelope of supported solutions, a second phase requires to optimize one objective while the other is bounded, an approach usually denoted as the ε-constraint method [16]. In our algorithm, for a proper anytime behavior, every single objective resolution is bounded in CPU time, as is the maximum number of non-dominated states produced. In its first phase, the algorithm produces both incumbent solutions defining an upper-bounding set and a set of lower-bounding half-spaces. When this first phase finishes, the second phase enumerates non-supported solutions as well as additional lower-bounding rectangular regions. These two sets offer both incumbent solutions (in the upper bounding set) as well as an optimality gap that can be represented graphically and computed as a ratio of surfaces.

2   Background and Notations

Definition 1. *A Bi-Objective GM (BO-GM) N is a tuple $(\boldsymbol{X}, \boldsymbol{D}, \boldsymbol{F} = \boldsymbol{F_1} \bigcup \boldsymbol{F_2})$ where $\boldsymbol{X}$ is the set of variables, $\boldsymbol{D}$ the finite domains of each variable and $\boldsymbol{F}$ the set of potential functions. Each function $f_{\boldsymbol{S}} \in \boldsymbol{F}$ associates a cost c to every assignment of the variables in its scope $\boldsymbol{S} \subseteq \boldsymbol{X}$. $\boldsymbol{F_1}$ and $\boldsymbol{F_2}$ define the two joint functions (objectives) $F_i(\boldsymbol{x}) = \sum_{f_{\boldsymbol{S}} \in F_i} f_{\boldsymbol{S}}(\boldsymbol{x}|_{\boldsymbol{S}})$ where $\boldsymbol{x}$ is a complete assignment of $\boldsymbol{X}$ and $\boldsymbol{x}|_{\boldsymbol{S}}$ denotes the projection of $\boldsymbol{x}$ to variables $\boldsymbol{S}$.*

We assume that the functions $f_{\boldsymbol{S}} \in \boldsymbol{F}$ may take infinite values, representing forbidden states. In probabilistic GMs such as Markov Random Fields, a joint function $F_i(\boldsymbol{x})$ is used to define a joint probability distribution $p_i(\boldsymbol{X} = \boldsymbol{x}) \propto \exp(-F_i(\boldsymbol{x}))$ and the cost $F_i(\boldsymbol{x})$ is called an *energy*. Minimizing $F_i(\boldsymbol{x})$ is equivalent to maximizing $p_i(\boldsymbol{x})$ (infinite energies define zero probabilities).

For any state x, we denote by $\ddot{F}(x)$ the pair of costs $(F_1(x), F_2(x))$. Given two pairs of costs or bi-costs $c = (c_1, c_2)$ and $c' = (c'_1, c'_2)$, we say that c weakly dominates c', denoted as $c \preccurlyeq c'$, iff $\forall i, c_i \leq c'_i$. c dominates c', denoted as $c \prec c'$ iff $c \preccurlyeq c'$ and $\exists i, c_i < c'_i$. This can be extended to states where $x \prec x'$ iff $\ddot{F}(x) \prec \ddot{F}(x')$. We denote by P the set of all states with non-dominated bi-costs or efficient assignments. The image $\ddot{F}(P)$ is a set of non-dominated bi-costs of states called the Pareto frontier. An efficient state x such that $\ddot{F}(x) = (c_1, c_2)$ is said to be supported if $\exists \lambda_1, \lambda_2 \in \mathbb{R}^{+2}$ such that $\forall c' \in \ddot{F}(P), \lambda_1 c_1 + \lambda_2 c_2 \leq \lambda_1 c'_1 + \lambda_2 c'_2$. The bi-costs of supported efficient assignments are known to lie on the frontier of the convex envelope of $F(P)$. They are said to be extreme when located at extreme points of the convex envelope. On the other hand, non-supported efficient assignments have costs in the interior of this convex envelope.

The Bi-Objective GM Optimization Problem (BO-GMO) is to find a set E of efficient states such that $\ddot{F}(E) = \ddot{F}(P)$. In the worst case, the set $\ddot{F}(P)$ (and therefore E) can have a size that grows exponentially with the number of variables. Finding a single efficient state is NP-hard while counting the number of elements in $\ddot{F}(P)$ is #P-hard [14,16]. Exploring the complete Pareto front is therefore challenging and quickly infeasible. However, as in single GM optimization, the Pareto front can be approximated by a lower and upper bound. The global ideal cost, defined as the pair $\iota = (\min_{x \in P} F_1(x), \min_{x \in P} F_2(x))$ defines a bi-cost that weakly dominates the bi-cost of any state.

More generally, we define a lower-bound set L as any closed subset of $\mathbb{R}^2$ such that all states have a cost that is weakly dominated by some element of L and is dominated by some element of the interior of L. By definition, the union of two lower-bound sets is also a lower-bound set. In this paper, we consider lower-bound sets defined as the union of simple polygonal regions (half-spaces or rectangles). An upper bound set is a set U of bi-costs such that any efficient state weakly dominates some element of U.

3   A Two-Phase Method for Bi-objective GM Optimization

Multi-objective optimization problems can be reduced to a series of single-objective optimization problems, using so-called scalarization techniques. As our method follows the two-phase scheme, it relies on two types of scalarizations. The overall search algorithm is described as Algorithm 1 and its different phases are outlined in Fig. 1 and presented in detail below.

3.1   First Phase

In the first phase, we focus on linear scalarization, which consists of aggregating objectives into a weighted sum. This technique ensures that the optimal solution of the resulting problem is a supported efficient solution for the original problem [16]. Moreover, as the linearly scalarized problem corresponds precisely to a GM, it can be solved by ready-to-use solvers. We assume we have a discrete

GM optimizer which is called by Solve $(N, \top, t)$ where N is a GM, $\top$ is a global upper bound (the solver will only report solutions of cost lower than $\top$) and t is a time-limit. In all cases, the solver returns a pair $(\boldsymbol{x}, lb)$ where $\boldsymbol{x}$ is the best state identified in the time budget and lb is a global lower bound on the joint function defined by N. The difference in cost between lb and the cost of $\boldsymbol{x}$ defines a possibly non-zero optimality gap. When there is provably no solutions, the solver returns $\boldsymbol{x} = \varnothing$ and $lb = \top$.

Definition 2. *Given a BO-GM* $(\boldsymbol{X}, \boldsymbol{D}, \boldsymbol{F} = \boldsymbol{F_1} \bigcup \boldsymbol{F_2})$ *and two multipliers* $\lambda_1, \lambda_2 \in \mathbb{R}^2$, *its* (λ_1, λ_2)-*scalarized GM is a GM* $(\boldsymbol{X}, \boldsymbol{D}, \boldsymbol{F} = \lambda_1 \boldsymbol{F_1} + \lambda_2 \boldsymbol{F_2})$ *with a single objective* $F = \lambda_1 \sum_{fs \in F_1} fs + \lambda_2 \sum_{fs \in F_2} fs$.

When a linear scalarization of a given GM is solved, the single-objective lower bound returned by the solver defines a half-space lower bound set over feasible bi-costs, guaranteed to contain no feasible bi-cost. Let $\lambda_1, \lambda_2 \in \mathbb{R}^2$ and lb be a lower bound on the optimum of the (λ_1, λ_2)-scalarized GM, we know that $lb \leq \lambda_1 F_1(\boldsymbol{x}) + \lambda_2 F_2(\boldsymbol{x})$ and the half-space $\lambda_1 c_1 + \lambda_2 c_2 \leq lb$ defines a lower bound set.

To identify supported efficient states, we use the dichotomic enumeration approach [4]. We maintain a heap Q_1 of lexicographically sorted pairs of supported states. Q_1 is initialized with a pair $(\boldsymbol{x_1^*}, \boldsymbol{x_2^*})$ containing extreme states that respectively minimize F_1 and F_2. At each iteration, we extract a pair of supported states $(\boldsymbol{x_1}, \boldsymbol{x_2})$ and check if a new improved (extreme) supported state between $\boldsymbol{x_1}$ and $\boldsymbol{x_2}$ exists (see Fig. 1(a)) by solving the (λ_1, λ_2)-scalarization of the BO-GM for $\lambda_1 = F_2(\boldsymbol{x_1}) - F_2(\boldsymbol{x_2})$ and $\lambda_2 = F_1(\boldsymbol{x_2}) - F_1(\boldsymbol{x_1})$ with an upper bound of $\lambda_1 F_1(\boldsymbol{x_1}) + \lambda_2 F_2(\boldsymbol{x_1})$. To focus the exploration on the sparsest regions of the Pareto frontier, we choose $(\boldsymbol{x_1}, \boldsymbol{x_2})$ such that $\|F(\boldsymbol{x_1}) - F(\boldsymbol{x_2})\|$ is maximum. An optimal solution $\boldsymbol{x}$ of the problem, if any, results in new supported pairs $(\boldsymbol{x_1}, \boldsymbol{x})$ and $(\boldsymbol{x}, \boldsymbol{x_2})$ added in Q_1 for further exploration (Fig. 1(b)). Whether the new state is efficient or not, it is added in $\boldsymbol{U}$ as part of the upper bound set. However, when there is provably no solution, $(\boldsymbol{x_1}, \boldsymbol{x_2})$ is stored for phase 2. In all cases, the lower bound l found together with the multipliers λ_1, λ_2 represent a lower bounding half-space which is stored as a triple $\langle \lambda_1, \lambda_2, l \rangle$ in a growing list $\boldsymbol{L_1}$ (Fig. 1(c)). The algorithm proceeds until Q_1 is emptied. If all sub-problems have been solved to optimality, the convex envelope of $\boldsymbol{P}$ is determined.

3.2   Second Phase

The linear scalarization method, however, cannot identify non-supported efficient states and may potentially miss supported non-extreme efficient solutions. In two-phase algorithms, starting from the pairs of supported efficient states pushed in $\boldsymbol{Q_2}$ during the first phase, the second phase explores all pairs to identify possibly non-supported states that may exist in between. This requires restricting the search to a polygonal region where we can optimize one objective (say F_1) and bound the other objective F_2 to reside between $F_2(\boldsymbol{x_2})$ and $F_2(\boldsymbol{x_1})$. Additionally, an upper bound requires F_1 not to be worse than $F_1(\boldsymbol{x_2})$. Contrarily to the first phase, a lower bound l on the optimum of this constrained problem

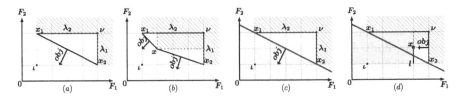

Fig. 1. (a, left) In phase 1, given two efficient solutions (x_1, x_2), $obj = \lambda_1 F_1 + \lambda_2 F_2$ is minimized. (b, center-left) If a supported efficient solution x is found, one recursively considers two new subproblems given by (x_1, x) and (x, x_2). (c, center-right) if no solution is found, the half-space region depicted in light cyan defines a lower-bounding space. (d, right) In phase 2, after (c), objective F_1 is minimized with a bounding constraint on F_2 (*i.e.*, efficient solutions are searched inside the rectangle given by x_1, x_2 and their nadir point ν). Any lower bound l of this problem defines a forbidden rectangle region (light gray). The union of half-space and rectangle regions defines a lower bounding set. If an efficient solution x is found, then the method explores the subproblem given by (x, x_2). Each solution found defines a region of dominated solutions. (Color figure online)

does not define a lower bounding half-space. It instead excludes all bi-costs in the interior of the rectangle defined by $0 \leq c_1 \leq l$ and $F_2(x_2) \leq c_2 \leq F_2(x_1)$. This region is added as a triple $(l, F_2(x_2), F_2(x_1))$ to the set L_2 of second-phase lower-bounding regions. The new solution x, if any, is added to U. Moreover, similarly to phase 1, when optimality has been proven, the remaining interval pair (x, x_2) is inserted in the list Q_2 of pairs of efficient states for exhaustive search (Fig. 1(d)).

In the end, the union of the lower-bounding regions in L defines a lower-bounding polygon. Similarly, each upper-bound set defined by a bi-cost $\ddot{F}(u)$ with $u \in U$ excludes a top-right quadrant with a corner in $\ddot{F}(u)$ and their union defines an upper-bounding polygon. All our polygons are quadrangles, and their union can be computed in time $O(n \log n)$ where n is the number of regions [33, Chapter 7]. The surfaces s_L and s_U of the lower and upper bounding regions can be computed using the Shoelace formula [9]. Together with the overall surface s_o of the rectangle defined by the global ideal point with coordinates $(F_1(x_1^*), F_2(x_2^*))$ and the opposite $(F_1(x_2^*), F_2(x_1^*))$ they define a Pareto optimality gap as $1 - \frac{s_L + s_U}{s_o}$. Eventually, the states in U with a bi-cost that lies on the frontier defined by s_L are known to be efficient.

Theorem 1. *Algorithm 1 is correct and has a bounded time complexity.*

Proof. Correctness follows from the exactly identified lower bounding regions, found by an exact solving method (Solve) in phases 1 and 2. The algorithm outputs a guaranteed Pareto optimality gap in time bounded by $m \times t$ (input parameters m and t).

Function TwoPhase(N, t, m)

    $L_1 := \varnothing;\ L_2 := \varnothing;\ U := \varnothing;\ Q_1 := \varnothing;\ Q_2 := \varnothing;$

    /* Solve F_1 within time t, store upper & lower bound   */

1    $\boldsymbol{x_1^*}, l_1 := \mathsf{Solve}((X, D, F_1), \infty, t);$

2    $L_1.\mathsf{push}(\langle 1, 0, l_1 \rangle);$

3    $U.\mathsf{push}(\boldsymbol{x_1^*});$

    /* Solve F_2 within time t, store upper & lower bound   */

4    $\boldsymbol{x_2^*}, l_2 := \mathsf{Solve}((X, D, F_2), \infty, t);$

5    $L_1.\mathsf{push}(\langle 0, 1, l_2 \rangle);$

6    $U.\mathsf{push}(\boldsymbol{x_2^*});$

    /* Proceed if F_1 and F_2 have been solved to optimality   */

    if $\boldsymbol{x_1^*} \neq \varnothing \wedge \boldsymbol{x_2^*} \neq \varnothing \wedge l_1 = F_1(\boldsymbol{x_1^*}) \wedge l_2 = F_2(\boldsymbol{x_2^*})$ **then**

    | $Q_1 := \{(\boldsymbol{x_1^*}, \boldsymbol{x_2^*})\}$

7   **while** $Q_1 \neq \varnothing$ **and** $|L_1| \leq m$ **do**

        /* Bisect by (λ_1, λ_2)-scalarization and store bounds   */

        $(x_1, x_2) := \mathsf{pop\text{-}best}(Q_1);$

        $\lambda_1 := F_2(x_1) - F_2(x_2);\ \lambda_2 := F_1(x_2) - F_1(x_1);$

        $\top := \lambda_1 F_1(x_1) + \lambda_2 F_2(x_1)$;

8        $x, l := \mathsf{Solve}((X, D, (\lambda_1 F_1 + \lambda_2 F_2)), \top, t);$

9        $L_1.\mathsf{push}(\langle \lambda_1, \lambda_2, l \rangle)$;

10       $U.\mathsf{push}(x)$;

        /* Push in Q_1 only if solved to optimality   */

        if $x \neq \varnothing$ **and** $l = \lambda_1 F_1(x) + \lambda_2 F_2(x)$ **then**

11 | $Q_1.\mathsf{push}((x_1, x))$; $Q_1.\mathsf{push}((x, x_2))$;

        /* Push in Q_2 when there is provably no solution   */

        if $x = \varnothing$ **and** $l = \top$ **then**

12 | $Q_2.\mathsf{push}((x_1, x_2))$

    /* Phase 2 if all phase 1 problems solved to optimality   */

13   **if** $|Q_2| \neq |U| - 1$ **then return** L_1, L_2, U **while** $Q_2 \neq \varnothing$ **and** $|L_1| + |L_2| \leq m$ **do**

    $(x_1, x_2) := \mathsf{pop\text{-}best}(Q_2);$

    /* Optimize F_1 with a bounding constraint on F_2 */

15    $x, l := \mathsf{Solve}((X, D, F_1 \bigcup (F_2(x_2) + 1 \leq F_2 < F_2(x_1))), F_1(x_2), t)$;

16    $L_2.\mathsf{push}(\langle l, F_2(x_2), F_2(x_1) \rangle)$;

17    $U.\mathsf{push}(x)$;

18   **if** $x \neq \varnothing$ **and** $l = F_1(x)$ **then** $Q_2.\mathsf{push}((x, x_2))$

    return L_1, L_2, U

Algorithm 1: Two-phase method. N is the BO-GM to optimize, t is a maximum CPU-time for any execution of **Solve** and m the maximum number of calls to **Solve**.

4   The Higher-Order GM Bounding Constraint

To optimize $\boldsymbol{F_1}$ while constraining $\boldsymbol{F_2}$, a dedicated bounding constraint needs to be added to the discrete GM solver. The corresponding higher-order GM bounding *hard* constraint takes as input a GM $N = (\boldsymbol{X}, \boldsymbol{D}, \boldsymbol{F})$ and two costs lb and ub such that $lb < ub$. It enforces $lb \leq F(\boldsymbol{x}) < ub$ for all assignments $\boldsymbol{x}$ of $\boldsymbol{X}$.

To enforce this constraint during branch and bound search, since solving N would be NP-hard, we rely on polynomial time convergent message passing algorithms, also known as equivalence preserving soft local consistencies [13,14]. We use soft local consistencies such as EDAC [13,19]. When enforced on a GM N, soft arc consistencies produce a lower-bound $lb_N(x_i, a)$ on the joint cost F if variable x_i is assigned state a, for every state a of every variable x_i. To enforce $F(\boldsymbol{x}) < ub$, one can simply prune any state a for any variable x_i such that $lb_N(x_i, a) \geq ub$. Since these lower bounds eventually become exact on fully assigned GMs, this guarantees that no solution violating $F(\boldsymbol{x}) < ub$ will ever be produced while pruning branches as soon as lower bounds allow for.

To enforce $lb \leq F(\boldsymbol{x})$, we build the GM denoted as $-N = (\boldsymbol{X}, \boldsymbol{D}, -\boldsymbol{F})$ simply by taking the opposite *finite* costs in every cost function, *i.e.*, $\forall f_S \in \boldsymbol{F}$ of N, we have $f'_S \in -\boldsymbol{F}$ in $-N$ such that $\forall \boldsymbol{x} : f'_S(\boldsymbol{x}) = -f_S(\boldsymbol{x})$ if $f_S(\boldsymbol{x}) < \infty$ else $f'_S(\boldsymbol{x}) = \infty$. To enforce $lb \leq F(\boldsymbol{x})$, we prune a state a for a variable x_i as soon as the lower bound $lb_{-N}(x_i, a) \geq -lb$. For the same reason as above, this guarantees that no solution violating $lb \leq F(\boldsymbol{x})$ will ever be produced while pruning branches.

Interestingly, if N does not contain any infinite cost ($\forall f_S \in \boldsymbol{F}, f_S(\boldsymbol{x}) < \infty$), if the current lower bound of N is greater than lb and the current lower bound of $-N$ is strictly greater than $-ub$ then the higher-order GM bounding constraint is always satisfied and can be ignored. This is checked after every variable assignment during branch and bound search. Notice that any state removal in the subproblem N (resp. $-N$) is informed/channeled in $-N$ (resp. N) and in the main problem optimizing F_1 and *vice versa*. By doing so, we synchronize domains between the different GMs sharing the same variables.

When adding the higher-order GM bounding constraint, after having tried to decompose all non-unary cost functions into a sum of unary cost functions [17], we check if N contains only unary cost functions. In this case, the objective is linear and the higher-order GM bounding constraint can be replaced by two generalized linear constraints, without introducing any extra variables [30].

For variable ordering heuristics, we use the weighted-degree heuristic $\frac{dom}{wdeg}$ [8]. For each higher-order GM bounding constraint, we maintain counters of conflicts for all the non-unary cost functions inside the sub-problems N and $-N$. Their sum gives a conflict weight associated with every variable in the scope of the higher-order GM bounding constraint, used to compute the $\frac{dom}{wdeg}$ heuristic.

It is important to note, as indicated in [23], that mono-objective preprocessing techniques that do not preserve all optimal solutions cannot be used safely in such a multi-objective optimization. This is the case for bounded variable elimination [26] and dominance rules [18] which need to be deactivated when the higher-order GM bounding constraint is processed.

5   Related Works

While systematic multi-objective optimization has been well explored in (integer) Linear Programming [16], there is far less algorithmic work for GMs. The C-semiring framework with a partial order [7] can in principle represent such problems, but algorithms are restricted to min-max instead of min-sum problems in MRFs and CFNs and no implementation is available. More closely to what we propose, MO-MBE [35] is a multi-objective mini-bucket-elimination based higher-order constraint that computes a set of non-dominated solutions. However, the algorithm requires initial upper bounds on the optimum and has space complexity in $O(ed^{z-1} \prod_{j=1}^{p-1}(ub_j))$ and time complexity in $O(ed^z \prod_{j=1}^{p-1}(ub_j^2))$ with e cost functions, p objectives, maximum bucket size z and initial upper bounds ub_i. This limits the approach to problems with small known initial upper bounds ub_j. The algorithm is not incremental and is used as a constraint in a constraint programming context. In the same context, dedicated Pareto constraints have been developed using either a support-based algorithm or a multi-valued decision diagram to filter dominated solutions in a multi-objective branch and bound [21,28]. Instead, our two-phase approach is based on single-objective B&B and the bounding constraint helps to reason about costs.

The introduction of a bounding constraint on a function defined by a Graphical Model, similar to our higher-order GM bounding constraint, has been explored for stochastic GMs [34] but the proposed bounds are not used for exact bi-objective search.

Compared to the original two-phase method [40,41], instead of depth-first search, we used (hybrid) best-first branch and bound methods that offer anytime global lower bounds [1], allowing to produce an anytime Pareto front gap on difficult instances.[1] The main novelty of our approach is to provide an anytime Pareto front bounding within a generic two-phase method for (stochastic) graphical models for the first time. The higher-order bounding constraint for discrete GM optimization is another contribution.

6   Computational Experiments

We implemented the two-phase method and the higher-order GM bounding constraint in C++ using the GM optimization solver `toulbar2`, winner of several medals in constraint programming (XCSP3 2022 and 2023) and probabilistic graphical model (UAI'2022) competitions.[2,3,4,5] We used the Hybrid-Best First Search branch and bound method [1] with default parameters (function Solve in Algorithm 1). Additional preprocessing techniques (Virtual Arc Consistency [13] and Virtual Pairwise Consistency [29]) were also considered.

As a baseline competitive approach, we also implemented the two-phase method replacing `toulbar2` by the state-of-the-art commercial integer linear programming solver `cplex` (version 22.1.1.0). We used default parameters except for

---

[1] In phase 2, we did not use a combination of objectives as in Test 3 with u3 bound [41] but used only one objective to optimize and prune search nodes (see Fig 1.d *obj* arrow). A comparison of these two bounding approaches remains to be done.

[2] https://forgemia.inra.fr/samuel.buchet/tb2_twophase in *release* branch.

[3] https://github.com/toulbar2/toulbar2 in *master* branch from version 1.2.1 including a dedicated linear constraint propagation method [30].

[4] https://xcsp.org/competitions.

[5] https://uaicompetition.github.io/uci-2022.

tolerance MIP gaps set to zero to ensure a complete search. The two objectives were linearized using the *local polytope* [13] formulation (*aka* the tuple encoding in [22], providing tighter bounds than a direct linear encoding):

$$\min \sum_{\substack{f_{\{x_i\}} \in \mathbf{F} \\ a \in \mathbf{D}_i}} f_{\{x_i\}}(a) x_{i:a} + \sum_{\substack{f_{\mathbf{S}} \in \mathbf{F}, |\mathbf{S}| > 1 \\ \tau \in \ell(\mathbf{S})}} f_{\mathbf{S}}(\tau) y_{\mathbf{S}:\tau} \tag{1}$$

$$\text{s.t. } \forall x_i \in \mathbf{X}, \quad \sum_{a \in \mathbf{D}_i} x_{i:a} = 1 \tag{2}$$

$$\forall f_{\mathbf{S}} \in \mathbf{F}, |\mathbf{S}| > 1, x_i \in \mathbf{S}, a \in \mathbf{D}_i, \sum_{\substack{\tau \in \ell(\mathbf{S}) \\ \tau | \{x_i\} = a}} y_{\mathbf{S}:\tau} = x_{i:a} \tag{3}$$

where $x_{i:a}$ is a Boolean 0/1 variable taking value 1 if $x_i = a$, similarly, $y_{\mathbf{S}:\tau}$ is a non-negative continuous variable taking value 1 if tuple $\tau \in \ell(\mathbf{S})$ is chosen ($\ell(\mathbf{S})$ representing the Cartesian product of domains in $\mathbf{S}$). The objective function (1) minimizes the sum of the linear and nonlinear cost functions in one criterion ($\mathbf{F_1}$ or $\mathbf{F_2}$), while constraints (2) enforce that each variable must be assigned to exactly one value and constraints (3) enforce that the assignments of variables and tuples are compatible.[6] After linearization, expressing the bounding constraint on the second criterion becomes trivial. We just have to replace the objective function (1) with two linear constraints using the lower and upper bounds defined in TwoPhase (line 15).

Experiments were run on a single core of an Intel Xeon E5-2680 v3 2.5 GHz processor with 128 GB of RAM, using Linux Debian 6.1.52-1 operating system. The CPU time limit of Solve is $t = 30$ s per call (except for Protein where $t = 300$ and SetCover where $t = 3,600$) with a maximum number of Solve calls $m = 1,000$ (except for Knapsack where $m = 2,000$). The total CPU time limit is 1 h.

6.1   Benchmarks

We experimented on six benchmarks. We took four existing benchmarks from the multi-objective literature in Operations Research. We added two benchmarks coming from Graphical Models. The first five have at least one linear objective function ($\mathbf{F_1}$ or $\mathbf{F_2}$), allowing to replace the GM bounding constraint with linear constraints. The last benchmark has nonlinear objective functions and constraints. All benchmarks, sources, and results are made publicly available.[7]

---

[6] Further simplifications are made on the model, as done in [22]. First, original GM variables with a domain size of 2 are translated into a single Boolean 0/1 variable and no constraint (2). Secondly, a more-compact direct encoding of hard constraints is used. For every forbidden tuple τ in $f_{\mathbf{S}}$, we have $\sum_{x_i \in \mathbf{S}} (1 - x_{i:\tau|\{x_i\}}) \geq 1$.

[7] https://forgemia.inra.fr/samuel.buchet/tb2_twophase (release branch).

Bi-objective Vertex Cover Problem. In the weighted vertex cover problem (VertexCover), the aim is to select a subset of vertices with a minimum total weight in order to cover at least one extremity of every edge in a given graph. We followed the same protocol as in [36], generating random graphs with N vertices having two associated random costs $c_j^i \in [0, C], i \in \{1, 2\}$ for every vertex $j \in \{1, \ldots, N\}$, and E edges (randomly selected among $\frac{N(N-1)}{2}$ possible ones). In the GM, there is a 0/1 variable x_j for every vertex j and two cost functions $f_j^i, i \in \{1, 2\}$ with $f_j^i(0) = 0, f_j^i(1) = c_j^i$, representing the 2 objectives. For every edge (k, l), a hard constraint enforces that x_k or x_l is equal to one. We tested on 25 samples for every parameter combination of $N \in \{60, 70, 80, 90\}$, $E \in \{95, 250, 500, 950\}$, and $C = 4$, resulting in 400 bi-objective instances.

Bi-objective Set Cover Problem. A related problem is the weighted set cover problem (SetCover). Here, the aim is to select a subset of elements with minimum total weight such that all the sets of a given list are covered by at least one of the selected elements. We took 120 randomly-generated instances from [24] composed of 5 instances for each combination of the following parameters: number of elements $N \in \{100, 150, 200\}$, number of sets $M \in \{20, 40, 60, 80\}$, fixed set cardinality $C \in \{5, 10\}$.[8][9] The integer cost values were chosen uniformly at random in the range $[1, 100]$. In the GM, there is a 0/1 variable x_j for every element j and two cost functions $f_j^i, i \in \{1, 2\}$ with $f_j^i(0) = 0, f_j^i(1) = c_j^i$, representing the 2 objectives. A hard constraint/clause is used to represent the fact that at least one element is selected in every set.

Bi-objective Knapsack Problem. The Knapsack problem is to select a subset of items such that the total weight of the selected items is less than a given capacity and the total profit associated to the items is maximized. Following [41], we generated 20 bi-objective instances with $N = 300$ items; weights w_j and profits p_j^i (uncorrelated) being randomly chosen in $[1, 100]$, and the capacity $W = \frac{\sum_{j=1}^{N} w_j}{2}$. In the GM, there is a 0/1 variable x_j for every item and two cost functions $f_j^i, i \in \{1, 2\}$ with $f_j^i(0) = p_j^i, f_j^i(1) = 0$, representing the 2 objectives in minimization. A hard linear constraint is added to enforce the capacity constraint.

Bi-objective Warehouse Location Problem. The goal of the uncapacitated warehouse or facility location problem (Warehouse) is to open a subset of N warehouses to fulfill the demands of M stores. Each store must be served by one warehouse. Following [5], in every objective $i \in \{1, 2\}$, we add a random cost $o_j^i \in [C, 2C]$ for opening warehouse j. In the first objective, the cost c_{jk}^i for serving store k by warehouse j is the Manhattan distance between k and l, stores

---

[8] https://bitbucket.org/coreo-group/bioptsat.
[9] The current version of `toulbar2` could not tackle large cardinality sets occurring in the fixed element probability set benchmark (SetCovering-EP) of [24].

and warehouses being randomly placed in a square of side C. In the 2nd objective, the corresponding cost c_{jk}^2 is randomly chosen in $[1, C]$. In the GM, there are N 0/1 variables for the warehouses and M variables with domain size N indicating which warehouse serves each store. Hard constraints ensure consistency between the two sets of variables [19]. Bi-objective linear cost functions associated with warehouses (resp. stores) are f_j^i with $f_j^i(0) = 0, f_j^i(1) = o_j^i$ (resp. f_k^i with $f_k^i(j) = c_{jk}^i$). We tested on 20 samples for $N = 6, M = 30$.

Bi-objective Computational Protein Design Problem. Our main motivation is to solve the bi-criteria problem (Protein) of designing a protein with minimum energy and a minimum number of different amino acid types, as done heuristically in [43]. We used a data set of 109 protein backbones from [32,38,43] for benchmarking. The first objective is defined by a Deep Learned decomposable protein design scoring function [15] and the second one is the number of different values used (a straightforward decomposition of NValue [6] as a GM using one extra Boolean variable B_a per value a, constrained to take value 1—with associated cost 1—if one variable uses a). This benchmark contains 109 instances with $n \in [42, 100]$ amino acid positions, $d = 20$ domain values, and $e \in [643, 2275]$ (non-linear) cost functions.

Bi-objective UAI'2022 Benchmark. Last, we experimented with UAI'2022 Final Evaluation MMAP benchmark (UAI2022).[10] From the initial 100 instances, we kept those that were solved by toulbar2 in less than 600 s. As in [34], the second objective of each instance is obtained by adding Gaussian noise $\mathcal{N}(\mu = 0, \sigma = 0.1)$ to the original functions (in the exponential/probability domain), truncating to non-negative numbers.[11] After the removal of instances with no solution, we obtained 26 instances with $n \in [225, 1997]$ variables, $d \in [2, 21]$ domain size, $e \in [578, 3334]$ cost functions per objective, and max. arity 5.

6.2   Experimental Results

We present a summary of our comparative results in Table 1.[12] The comparison measure is the number of instances completely solved per class of benchmark. The solving task is to find a single representative state for each point of the exact Pareto front. To break ties when all the instances are solved, we use the average CPU-time.

---

[10] https://www.ics.uci.edu/~dechter/uaicompetition/2022/FinalBenchmarks/MMAP. zip.

[11] Zero probabilities lead to forbidden tuples after the required $-\log(\cdot)$ transform.

[12] Detailed results are available in Supplementary Materials. See https://forgemia.inra. fr/samuel.buchet/tb2_twophase/-/tree/release/results/supplementals.pdf.

Table 1. For each of the 6 benchmark classes, number of solved instances per method (in parenthesis, average CPU-time in seconds when all instances have been solved). Tested methods are `toulbar2`: two-phase method using `toulbar2`; `tb2 no kp`: two-phase method using `toulbar2` without transforming the GM bounding constraint into linear constraints; `tb2 no pre`: two-phase method using `toulbar2` without extra preprocessing techniques; `cplex`: two-phase method using `cplex`; RL: results from [36]; J: from [24]; C: results from [11]; BS: from [5]. N/A: method not applicable or result not available.

Benchmark	#	toulbar2	tb2 no kp	tb2 no pre	cplex	RL/J/C/BS
VertexCover	400	**400** (3.1 s)	**400 (0.5 s)**	N/A	**400** (1.6 s)	RL: **400** (10.2 s)
SetCover	120	44	N/A	N/A	**120 (47.3 s)**	J: 40
Knapsack	20	0	N/A	N/A	**20 (40.2 s)**	C: **20** (41.7 s)
Warehouse	20	**20 (2.5 s)**	20 (33.6 s)	N/A	**20** (4.8 s)	BS: **20** (186.0 s)
Protein	109	66	**69**	46	0	N/A
UAI2022	26	**26 (23.8 s)**	N/A	18	21	N/A

We compare our two-phase method (TwoPhase using `toulbar2`) with the same method using `cplex` and with other approaches when directly available:

- for VertexCover, a multi-objective depth-first branch and bound (MO-BB) using multi-objective minibucket elimination (MO-MB$_{MOMBE}$) (experiments were made on a Pentium IV at 3 GHz with 2 GB) [36];
- for SetCover, a Max-SAT based hybrid method between core-guided and SAT-UNSAT search methods (MSHybrid) [24] (experiments made on Intel Xeon E5-2670 at 2.6 GHz with 64 GB and 1.5-h CPU-time limit);
- for Knapsack, a two-phase method including dedicated instance preprocessing and where the second phase is a branch and bound with an adaptive branching heuristic (UCB) [11,12] (experiments made on Intel Xeon E5620 at 2.40 GHz with 6 GB);
- for Warehouse, a multi-objective hybrid branch and bound combined with scalarization and using `Bensolve/GLPK` for solving the linear relaxations (M2.1.1.2) (experiments made on an Intel i7-8700 at 3.2 GHz with 32 GB) [5].

Notice that some reported experimental results (VertexCover, Warehouse) were performed on older machines, so the comparison should be taken with caution.

For two of the Operations Research benchmarks, TwoPhase using `toulbar2` has similar performance as TwoPhase using `cplex`, `toulbar2` being twice faster (resp. slower) on Warehouse (resp. VertexCover). However, it is clearly dominated by `cplex` on SetCover and Knapsack. Still, it remains faster than dedicated multi-objective branch and bound approaches, being 3.3 (resp. 74) times faster than MO-MB$_{MOMBE}$ (resp. M2.1.1.2) on VertexCover (resp. Warehouse). It also solves more SetCover instances than Max-SAT MSHybrid in less CPU-time (we set a 1-hour limit instead of 1.5 h for MSHybrid to compensate for the difference in CPU frequencies). On Knapsack, TwoPhase using `cplex` is as efficient as a dedicated multi-objective branch and bound (UCB B&B [11]).

For Graphical Model benchmarks, TwoPhase using toulbar2 got much superior performance than with cplex. It solves 63% of the Protein benchmark whereas none were solved by cplex in less than 300 s per internal Solve call. The largest solved instance by our approach contains 99 amino-acids.[13] The GM bounding constraint allows us to solve all the 26 UAI'2022 selected instances within the local 30-second CPU-time limit. In comparison, TwoPhase using cplex could solve only 21 instances. Using a longer 300-second time limit (detailed results in supplementary materials), TwoPhase using cplex was able to solve all the 26 instances requiring x7 more time (166.1 s) than with toulbar2 (23.8 s) on average. We noticed that cplex can still be faster (up to 2.7 times) than toulbar2 for 6 instances (Grids_26, or_chain_8/22/41/60, pedigree1) showing that both approaches can be worthwhile for this benchmark.

Impact of the GM Bounding Constraint. We tested the impact of removing the transformation of the GM bounding constraint into linear constraints when one of the original objectives is linear.[14] The fourth column in Table 1 shows the effect. Surprisingly, it improves the results on VertexCover (being x6.2 faster with x11.75 fewer search nodes in Phase 2) and Protein (solving optimally 3 more instances) whereas it significantly deteriorates on Warehouse. This behavior shows the complex interactions between the cost functions on the performance of the solver.

Impact of the Preprocessing. We also tested the impact of preprocessing techniques. In the Protein benchmark, applying virtual arc consistency (VAC) [13] in preprocessing was shown to be effective [39]. It solves 66 instances instead of 46 without enforcing VAC.[15] Another stronger preprocessing already used in the UAI'2022 competition is to apply virtual pairwise consistency (VPWC) with additional zero-cost ternary cost functions [29] (options -A -pwc $= -1 -t = 1$). Because this preprocessing can be quite time-consuming (2×0.48 seconds on average for UAI2022, up to 5.3 s on or_chain_41), we applied it on every objective only once before the two-phase method starts.[16] On UAI2022, using VPWC allows to solve 8 more instances than without it. We applied these preprocessing techniques on the GM benchmarks in the following part.

Two-Phase Method Analysis and Anytime Pareto Fronts. In Table 2 we report a more detailed analysis of TwoPhase using toulbar2. We compared the first and second phases in terms of CPU-time spent, number of Pareto assignments found, and number of Solve calls. The ratio of supported efficient states

---

[13] 2pko_0009_multi having 15 states in the Pareto front found in 1,723 s with 37 Solve calls.

[14] We could not test it for SetCover and Knapack due to a limitation in toulbar2.

[15] We also tested running VAC (option -A) in preprocessing at every Solve, instead of only once on the quadratic objective before running the two-phase method, but it solved optimally one less instance (65).

[16] CPU-times reported in Tables 1 and 2 do not include this preprocessing time.

Table 2. For each of the 6 benchmarks, number of tested instances, average CPU-time (seconds) and fraction spent in phase 1, average number of efficient states and fraction found in phase 1, average number of calls to **Solve** and fraction in phase 1, number of problems solved to optimality (full Pareto front) and after phase 1 (supported Pareto front), and final optimality gap.

Benchmark	#	Time (s)	p1%	Sols	p1%	Solve #	p1%	opt	opt1	gap%
VertexCover	400	3.1	5.6	8.0	57.2	16.0	59.8	400	400	0.000
SetCover	120	2411.1	5.6	28.9	38.5	46.3	46.3	44	119	1.462
Knapsack	20	2040.7	0.1	486.1	10.6	669.4	15.4	0	20	0.276
Warehouse	20	2.5	3.0	97.8	17.8	135.5	25.1	20	20	0.000
Protein	109	1334.5	82.5	12.7	84.6	32.1	67.9	66	73	14.766
UAI2022	26	23.8	6.8	14.0	45.5	24.9	47.8	26	26	0.000

was around 50% except for Warehouse (17.8%), Knapsack (10.6%), and Protein (85%). Knapsack had also the largest total number of efficient states (up to 851 for $n = 300$ items, average optimality gap of 0.276).

Protein is the only benchmark class where phase 1 takes more time and Solve than phase 2 (which finds two non-supported efficient states on average). The Pareto front of real protein dpbbss_multi (solved in 244.3 s by our approach) is given in Fig. 2 (Left). Two solutions of dpbbss_multi with $F_2 = 6$ or 5 look very attractive compared to the $F_2 = 7$ identified in [43]. Figure 2 also shows the pareto front boundings of an UAI Promedas instance solved to optimality (Center) and partially (Right), highlighting the anytime behaviour of the method.

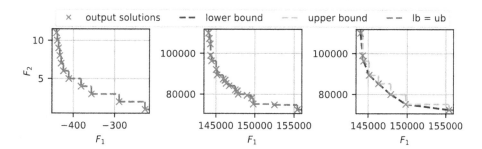

Fig. 2. Three examples of Pareto front bounding: (Left) Protein dpbbss_multi (optimal front), (Center) UAI2022 PROMEDAS OR_CHAIN_60.FG.Q0.5.I3 (optimal), and (Right) UAI2022 PROMEDAS OR_CHAIN_60.FG.Q0.5.I3 (optimality gap of 5.164%).

7   Conclusion

In this paper, motivated by the resolution of bi-objective protein design problems [43], we introduce an original anytime multi-objective two-phase method for solving discrete probabilistic and deterministic Graphical Models optimization problems. Our algorithm relies on a new higher-order GM bounding constraint exploiting Soft Arc Consistencies, a family of convergent message passing algorithms initially introduced for solving the Weighted Constraint Satisfaction Problem [13,37]. Our experimental study, using a variety of real protein design problems as well as Operations Research and UAI'2022 evaluation benchmarks shows that our approach can outperform the state-of-the-art ILP solver CPLEX. At the crossroad of Probabilistic Reasoning and Operations Research, our algorithm offers a first effective answer to bi-objective discrete Graphical Model optimization, a family of problems that could possibly further benefit from additional multi-objective OR technology [20].

Acknowledgments.. This study was funded by the French ANR projects ANR-19-CE09-0032 SpaceHex and ANR-19-PIA3-0004 ANITI. The authors have no competing interests to declare that are relevant to the content of this article.

References

1. Allouche, D., de Givry, S., Katsirelos, G., Schiex, T., Zytnicki, M.: Anytime hybrid best-first search with tree decomposition for weighted CSP. In: Pesant, G. (ed.) CP 2015. LNCS, vol. 9255, pp. 12–29. Springer, Cham (2015). https://doi.org/10.1007/978-3-319-23219-5_2
2. Allouche, D., et al.: Computational protein design as an optimization problem. Artif. Intell. **212**, 59–79 (2014)
3. Allouche, D., et al.: Tractability-preserving transformations of global cost functions. Artif. Intell. **238**, 166–189 (2016)
4. Aneja, Y.P., Nair, K.P.K.: Bicriteria transportation problem. Manage. Sci. **25**(1), 73–78 (1979)
5. Bauß, J., Stiglmayr, M.: Augmenting bi-objective branch and bound by scalarization-based information. arXiv preprint arXiv:2301.11974 (2023)
6. Bessiere, C., Hebrard, E., Hnich, B., Kiziltan, Z., Walsh, T.: Filtering algorithms for the NValue constraint. Constraints **11**, 271–293 (2006)
7. Bistarelli, S., Fargier, H., Montanari, U., Rossi, F., Schiex, T., Verfaillie, G.: Semiring-based CSPs and valued CSPs: frameworks, properties and comparison. Constraints **4**, 199–240 (1999)
8. Boussemart, F., Hemery, F., Lecoutre, C., Sais, L.: Boosting systematic search by weighting constraints. In: ECAI, vol. 16, p. 146 (2004)
9. Braden, B.: The surveyor's area formula. Coll. Math. J. **17**(4), 326–337 (1986)
10. Cabon, B., de Givry, S., Lobjois, L., Schiex, T., Warners, J.: Radio link frequency assignment. Constraints J. **4**, 79–89 (1999)
11. Cerqueus, A.: Bi-objective branch-and-cut algorithms applied to the binary knapsack problem. Ph.D. thesis, université de Nantes (2015). https://hal.science/tel-01242210
12. Cerqueus, A., Gandibleux, X., Przybylski, A., Saubion, F.: On branching heuristics for the bi-objective 0/1 unidimensional knapsack problem. J. Heuristics **23**, 285–319 (2017)
13. Cooper, M., de Givry, S., Sanchez, M., Schiex, T., Zytnicki, M., Werner, T.: Soft arc consistency revisited. Artif. Intell. **174**(7–8), 449–478 (2010)

14. Cooper, M.C., de Givry, S., Schiex, T.: Valued constraint satisfaction problems. In: Marquis, P., Papini, O., Prade, H. (eds.) A Guided Tour of Artificial Intelligence Research, pp. 185–207. Springer, Cham (2020). https://doi.org/10.1007/978-3-030-06167-8_7

15. Defresne, M., Barbe, S., Schiex, T.: Scalable coupling of deep learning with logical reasoning. In: Proceedings of the 32nd IJCAI, Macau, A.S.R., China (2023)

16. Ehrgott, M., Gandibleux, X., Przybylski, A.: Exact methods for multi-objective combinatorial optimisation. In: Greco, S., Ehrgott, M., Figueira, J.R. (eds.) Multiple Criteria Decision Analysis. ISORMS, vol. 233, pp. 817–850. Springer, New York (2016). https://doi.org/10.1007/978-1-4939-3094-4_19

17. Favier, A., de Givry, S., Legarra, A., Schiex, T.: Pairwise decomposition for combinatorial optimization in graphical models. In: Proceedings of IJCAI-11, Barcelona, Spain (2011)

18. de Givry, S., Prestwich, S.D., O'Sullivan, B.: Dead-end elimination for weighted CSP. In: Schulte, C. (ed.) CP 2013. LNCS, vol. 8124, pp. 263–272. Springer, Heidelberg (2013). https://doi.org/10.1007/978-3-642-40627-0_22

19. de Givry, S., Zytnicki, M., Heras, F., Larrosa, J.: Existential arc consistency: getting closer to full arc consistency in weighted CSPs. In: Proceedings of IJCAI-05, Edinburgh, Scotland, pp. 84–89 (2005)

20. Halffmann, P., Schäfer, L.E., Dächert, K., Klamroth, K., Ruzika, S.: Exact algorithms for multiobjective linear optimization problems with integer variables: a state of the art survey. J. Multi-Criteria Decis. Anal. **29**(5–6), 341–363 (2022)

21. Hartert, R., Schaus, P.: A support-based algorithm for the bi-objective pareto constraint. In: Proceedings of the AAAI Conference on Artificial Intelligence, vol. 28 (2014)

22. Hurley, B., et al.: Multi-language evaluation of exact solvers in graphical model discrete optimization (summary). In: Proceedings of CP-AI-OR 2016, Banff, Canada, p. 1 (2016)

23. Jabs, C., Berg, J., Ihalainen, H., Järvisalo, M.: Preprocessing in SAT-based multi-objective combinatorial optimization. In: Proceedings of CP-23, Toronto, Canada (2023)

24. Jabs, C., Berg, J., Niskanen, A., Järvisalo, M.: MaxSAT-based bi-objective Boolean optimization. In: 25th International Conference on Theory and Applications of Satisfiability Testing (SAT 2022) (2022)

25. Kratica, J., Tošic, D., Filipović, V., Ljubić, I.: Solving the simple plant location problem by genetic algorithm. RAIRO-Oper. Res. **35**(1), 127–142 (2001)

26. Larrosa, J.: Boosting search with variable elimination. In: Dechter, R. (ed.) CP 2000. LNCS, vol. 1894, pp. 291–305. Springer, Heidelberg (2000). https://doi.org/10.1007/3-540-45349-0_22

27. Luukkonen, S., van den Maagdenberg, H.W., Emmerich, M.T., van Westen, G.J.: Artificial intelligence in multi-objective drug design. Curr. Opin. Struct. Biol. **79**, 102537 (2023)

28. Malalel, S., Malapert, A., Pelleau, M., Régin, J.C.: MDD archive for boosting the pareto constraint. In: 29th International Conference on Principles and Practice of Constraint Programming (CP 2023). Schloss Dagstuhl-Leibniz-Zentrum für Informatik (2023)

29. Montalbano, P., Allouche, D., de Givry, S., Katsirelos, G., Werner, T.: Virtual pairwise consistency in cost function networks. In: Cire, A.A. (ed.) CPAIOR 2023. LNCS, vol. 13884, pp. 417–426. Springer, Cham (2023). https://doi.org/10.1007/978-3-031-33271-5_27

30. Montalbano, P., de Givry, S., Katsirelos, G.: Multiple-choice knapsack constraint in graphical models. In: Schaus, P. (ed.) CPAIOR 2022. LNCS, vol. 13292, pp. 282–299. Springer, Cham (2022). https://doi.org/10.1007/978-3-031-08011-1_19

31. Nicolaou, C.A., Brown, N.: Multi-objective optimization methods in drug design. Drug Discov. Today Technol. **10**(3), e427–e435 (2013)

32. Noguchi, H., et al.: Computational design of symmetrical eight-bladed β-propeller proteins. IUCrJ **6**(1), 46–55 (2019)

33. o'Rourke, J.: Computational Geometry in C. Cambridge University Press, Cambridge (1998)

34. Rahman, T., Rouhani, S., Gogate, V.: Novel upper bounds for the constrained most probable explanation task. Adv. Neural. Inf. Process. Syst. **34**, 9613–9624 (2021)

35. Rollon, E., Larrosa, J.: Multi-objective propagation in constraint programming. In: Brewka, G., Coradeschi, S., Perini, A., Traverso, P. (eds.) ECAI 2006, 17th European Conference on Artificial Intelligence Proceedings. Frontiers in Artificial Intelligence and Applications, vol. 141, pp. 128–132. IOS Press (2006)

36. Rollon, E., Larrosa, J.: Constraint optimization techniques for multiobjective branch and bound search. In: International conference on logic programming, ICLP (2008)

37. Schiex, T.: Arc consistency for soft constraints. In: Dechter, R. (ed.) CP 2000. LNCS, vol. 1894, pp. 411–425. Springer, Heidelberg (2000). https://doi.org/10.1007/3-540-45349-0_30

38. Simoncini, D., Allouche, D., De Givry, S., Delmas, C., Barbe, S., Schiex, T.: Guaranteed discrete energy optimization on large protein design problems. J. Chem. Theory Comput. **11**(12), 5980–5989 (2015)

39. Traoré, S., et al.: A new framework for computational protein design through cost function network optimization. Bioinformatics **29**(17), 2129–2136 (2013)

40. Ulungu, E.L., Teghem, J.: The two phases method: an efficient procedure to solve bi-objective combinatorial optimization problems. Found. Comput. Decis. Sci. **20**(2), 149–165 (1995)

41. Visée, M., Teghem, J., Pirlot, M., Ulungu, E.L.: Two-phases method and branch and bound procedures to solve the bi-objective knapsack problem. J. Global Optim. **12**(2), 139–155 (1998)

42. Vucinic, J., Simoncini, D., Ruffini, M., Barbe, S., Schiex, T.: Positive multistate protein design. Bioinformatics **36**(1), 122–130 (2020)

43. Yagi, S., et al.: Seven amino acid types suffice to create the core fold of RNA polymerase. J. Am. Chem. Soc. **143**(39), 15998–16006 (2021)

An Exploration of Exact Methods for Effective Network Failure Detection and Diagnosis

Auguste Burlats[(⊠)], Pierre Schaus, and Cristel Pelsser

ICTeam, UCLouvain, Ottignies-Louvain-la-Neuve, Belgium
{auguste.burlats,pierre.schaus,cristel.pelsser}@uclouvain.be

Abstract. In computer networks, swift recovery from failures requires prompt detection and diagnosis. Protocols such as Bidirectional Forwarding Detection (BFD) exists to probe the liveliness of a path and endpoint. These protocols are run on specific nodes that are designated as network monitors. Monitors are responsible for continuously verifying the viability of communication paths. It is important to carefully select monitors as monitoring incurs a cost, necessitating finding a balance between the number of monitor nodes and the monitoring quality. Here, we examine two monitoring challenges from the Boolean network tomography research field: coverage, which involves detecting failures, and 1-identifiability, which additionally requires identifying the failing link or node. We show that minimizing the number of monitors while meeting these requirements constitutes NP-complete problems. We present integer linear programming (ILP), constraint programming (CP) and Maximum Satisfiability (MaxSAT) formulations for these problems and compare their performance. Using 625 network topologies, we demonstrate that employing such exact methods can reduce the number of monitors needed compared to the existing state-of-the-art greedy algorithm.

Keywords: Integer linear programming · Constraint Programming · MaxSAT · Boolean tomography · Network supervision

1 Introduction

Computer networks form the backbone of modern digital communication, and their reliability is crucial for maintaining seamless connectivity across various sectors. Failures within these networks can have significant consequences, leading to service disruption and potential financial loss. As such, it is essential to develop efficient and accurate methods for detecting and diagnosing network failures, enabling swift recovery and minimizing the impact on end users.

Various protocols enable to monitor the liveliness of Internet paths from the well-known `ping` utility present in most operating systems, and used by measurement infrastructures such as RIPE Atlas [23], to more recent protocols such as BFD [15]. In addition to fasten link failure detection when deployed on adjacent routers, BFD enables to quickly detect failures along a path, such

© The Author(s), under exclusive license to Springer Nature Switzerland AG 2024
B. Dilkina (Ed.): CPAIOR 2024, LNCS 14742, pp. 153–169, 2024.
https://doi.org/10.1007/978-3-031-60597-0_11

ability being leveraged in Software Defined Networks (SDN) to quickly detect and report failures to a network controller.

In this study, we focus on Boolean network tomography, a research field that holds great promise for enhancing the resilience of networks. Boolean network tomography combines end-to-end measures, performed with ping or BFD, for example, with inference algorithms to estimate the state of different elements in the network. Its advantage is that it only requires a subset of nodes to be monitors and supervise an entire network. With this approach, monitors send messages to each other through *measurement paths*. When a failure occurs on a node, all paths that cross it fail. Thus, the failure can be detected by observing if some measurement paths are not working. If the set of failed measurement paths forms a unique signature, then it is even possible to identify the failed node.

In the remaining of the paper, we treat the case of node failures. Indeed, we can easily account for edge-failure considerations by transforming the network graph. Adding dummy nodes to represent each link, enables to transform the node failure detection, respectively localization, problem, into detecting and locating link failures, applying the approach in this paper.

Our investigation focuses on minimizing the number of designated monitor nodes while ensuring some level of quality of network monitoring. This is crucial for minimizing monitoring costs without compromising the network's overall health and performance. We explore two critical monitoring challenges: the *cover problem*, which seeks to detect failures, and the *1-identifiability problem*, which requires pinpointing the exact failing node. Figure 1 illustrates these concepts.

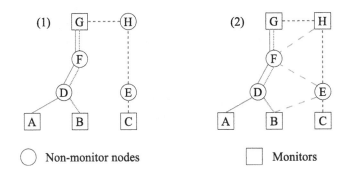

Fig. 1. Illustration of 1-identifiability: Each line style represents a different path. (1) All nodes are covered but not 1-identifiable. (2) All nodes are 1-identifiable. In the situation (1), the path linking nodes B and H is not a measurement path because H is not a monitor. Each node is covered because they are all crossed by a path linking two monitors, but if D or F fails, in each case the paths between A and G and between B and G will both fail. The failure will be detected, but it would be impossible to know which of the two nodes is the origin of the failure. The problem is the same with nodes E and H. In the situation (2), H is a monitor, thus the path linking B and H is a measurement path (in dashed orange). Hence each non-monitor node is now crossed by a unique set of paths: If D fails, paths (A,G) and (B,G) will fail; if E fails it will be paths (C,G) and (B,H); if F is not functional, paths (A,G), (B,G) and (B,H) will not work. By looking at which paths are not working versus the alive paths, it is possible to infer without ambiguity which node is the origin of the breakdown. (Color figure online)

Conceptually, a node failure in a network results in the disruption of all paths traversing it. These affected paths collectively constitute the *symptom* associated with the failing node. A network is covered if there is a non-empty symptom for each node. Additionally, a network is considered 1-identifiable if every node possesses a unique, non-empty symptom, thereby serving as an identifier for the node in the event of a failure. By compiling a comprehensive list of these identifiers, one can efficiently diagnose a failure by simply observing the disrupted paths and cross-referencing a precomputed table that maps the failed paths (symptoms) to the corresponding node.

The *1-identifiability* problem can be generalized as the *k-identifiability* problem for the localization of up to k simultaneous failures. Most precomputed failure protection mechanisms today are tailored for single faults. With BFD and our proposal, locating a single failure enables to use such protections without waiting for the convergence of the routing protocol, speeding up recovery. However, in case of multiple failures, the protections for recovery are usually not in place. Hence, in this situation we are forced to wait for the routing protocol to converge. We have thus limited benefits of quickly identifying k faults. In this paper, we focus on selecting the smallest set of monitors ensuring that the entire network is either covered or 1-identifiable. But our approach can theoretically be extended to k-identifiability.

An important assumption of the considered networks in this study is that the routes between any pairs of nodes are imposed by the routing protocol and known by the planning tool that will select the monitors. A pair of monitors is only able to verify the status of those routes. In practice network operators usually configure link (IGP) weights to influence where the traffic flows in the network assuming they follow shortest paths (see for instance [5] for optimizing IGP weights). Alternatively, other protocols such as segment routing or MPLS [8,9,17] make it possible to introduce deviations or explicit route set-ups between pairs of nodes, deviating from shortest paths. For all these protocols, the monitors are able to determine which data paths between them are affected by a failure.

Our contributions can be summarized as follows.

- We formulate the optimal monitor placement problem for the cover and 1-identifiability (Sect. 3);
- We demonstrate that the node cover and 1-identifiability problems are NP-complete (Sect. 4);
- We introduce an exact model and its formulation for integer linear programming (ILP), constraint programming (CP) and Maximum Satisfiability (MaxSAT). We propose redundant constraints and reductions to shrink the search space (Sect. 5);
- We also introduce a specialized version of the MNMP greedy algorithm [19], tailored for 1-identifiability, to compare our exact methodologies with this approach (Sect. 6).
- We evaluate their performance through a comparative analysis using 625 network topologies (Sect. 7).

Our findings reveal that the introduced models can reduce the amount of monitor nodes compared to a greedy approach while maintaining coverage or 1-identifiability, paving the way for more robust and reliable telecommunication networks.

2   Related Work

Various methods have been investigated to address the monitor placement issue for failure detection, beyond utilizing exact methods. For example, the Maximum Node-identifiability Monitor Placement (MNMP) [19] is a greedy algorithm that progressively adds monitors to achieve the desired k-identifiability and subsequently removes any unnecessary monitors. Unfortunately, it does not guarantee optimality for 1-identifiability problem. Bezerra et al. [3] suggest various improvements to this algorithm to decrease computational cost and enable its use in wireless networks that experience more frequent changes. Our approach here is different, as we use exact methods that allow us to reduce the number of selected monitors, as shown in Sect. 7, and also because we are focusing on coverage and 1-identifiability.

Stanic et al. [25] present an ILP model and a greedy algorithm for the monitor placement for fault localization in transparent all-optical networks, which is an analogous problem to ours. The main difference being that, in their context, each measurement path requires only one monitor. Their model could be applicable here by having monitors probing non-monitor nodes and waiting for their answer. However, it requires the assumption that the routes are symmetrical which is not necessarily the case in practice [12,14,26]. Here we consider routes that can be asymmetrical; for them to be measurement paths, both of their two ends must be monitors.

The dual version of the problem considered in this paper, where the number of monitors is limited and the number of identifiable nodes or links needs to be maximized has also been studied. Ren et al. [22] design a greedy algorithm that chooses monitors such as the number of k-identifiable links is maximized. Bartolini et al. provide in [1] an upper bound for the maximum number of identifiable nodes given a specific measurement path budget. Ma et al. propose the Greedy Maximal identifiability Monitor Placement [18], an algorithm that incrementally adds the monitors that maximize the number of identifiable links, until the maximal budget of monitors is reached. Here, the definition of identifiable is extended to all types of additive metrics (delays, packet delivery ratios, ...) and not only the failure detection.

Related problems, involving tomography for monitoring, have also been examined. He et al. [10] investigate, in the context of Network Function Virtualization (NFV), the challenge of positioning services in a network with a predetermined set of measurement paths to optimize their identifiability while ensuring a high Quality of Service (QoS). Zhang et al. [27] adapt network tomography techniques to supervise traffic in smart cities, which includes determining the optimal placement for monitoring cameras. A possible approach to reduce the load of Boolean network tomography is to partition the network in multiple areas and to locate failures in each area independently. Ogino et al. [20] offer a procedure to divide the network in such areas and a scheme to manage them. This partitioning can be run before the monitor selection, then, instead of dealing with a single large instance of the problem considered in this paper, there would be multiple smaller ones. We refer to chapters 5–6 of the book [11] for a complete review of Boolean-Network Tomography.

3   Problem Formulation

The following two paragraphs formally introduce the problems addressed in this paper, specifically the monitor cover and the 1-identifiability problems.

The Monitor Cover Problem. Assuming that the network topology is known, it can be represented as a connected graph. We further assume that a given route exists between every pair of nodes, but only the ones linking monitor nodes are considered measurement paths. The symptom of a node is defined as the collection of measurement paths passing through it. The objective of this problem is to identify the minimum number of monitors from the network's node set such that every node is traversed by a minimum of one measurement path. The problem can be formalized as follows. We consider an oriented graph $G = (V, E)$ where there is a cycle-free route between each pair of nodes $(i, j) \in V^2$. In this problem the order in which nodes are crossed by a route does not matter, we thus represent routes as unordered sets containing all the nodes crossed by these routes and we denote $P(i, j) = \{i, \ldots, j\}$ the route linking i and j. Routes are not necessarily symmetric; thus $P(i, j)$ and $P(j, i)$ don't necessarily contain the same nodes. The goal is to find the minimal set of monitors $M \subseteq V$ such as $\cup_{(i,j)\in M^2} P(i, j) = V$.

The Monitor 1-Identifiability Problem. This problem adds one constraint over the cover problem. If one node fails, we don't only need the failure to be detectable, but we also aim to be able to locate it without ambiguity. For the failure to be identifiable, the symptom of the failure (set of failed measurement paths) must be unique. The problem can be formalized as follows. We consider the same oriented graph $G = (V, E)$ and set of routes P as for the monitor cover problem. For a set of monitors M, we denote by S_i the symptom of node i: $S_i = \{(i', j') \in M^2 \mid i \in P(i', j')\}$. The goal is to find the minimal set of monitors $M \subseteq V$ such that $\cup_{(i,j)\in M^2} P(i, j) = V$ (cover) and $\forall i \neq j \in V^2, S_i \neq S_j$ (1-identifiability).

4   Complexity of Optimal Monitor Placement

In this section we study the complexity of our two problems and show that they are both NP-complete.

Theorem 1. *The monitor cover problem is NP-complete.*

Proof. We reduce the set cover problem to the monitor cover problem as follows[1].
Consider an instance of the set cover problem defined by $(S = \{S_1, \ldots, S_k\}, U)$,
where S is a set of sets and U is the universe. For the monitor cover problem,
we construct a set of vertices $V = S \cup U \cup \{T\}$, including one vertex for each set
in S, one for each element in U, and a special vertex T, designated as a monitor
in every valid solution.

There is an edge between every pair of nodes (i, j) in V^2. The set of routes is
defined as follows. For each pair of nodes (i, j) in U^2, the route between them is
direct, i.e., $P(i, j) = \{i, j\}$ if $i \neq j$. In addition to these routes, for each $S_i \in S$,
an arbitrary route between S_i and T contains all nodes in S_i. Finally, a route
$P(T, T)$ arbitrarily passes through all nodes in S. This set of routes ensures that
in any optimal solution, only nodes in S plus the node T are monitors.

First, observe that T necessarily needs to be a monitor, as it is only traversed
by routes for which it is the origin or the destination. The simple selection of T
also ensures the coverage of all the nodes in S. Then, choosing a node $S_i \in S$
as a monitor to cover a node $U_j \in U$ is at least as cost-effective as choosing U_j
itself as a monitor. Since we ensured by construction that choosing a set vertex
$S_i \in S$ as a monitor covers all nodes in S_i, the optimal solution to the set cover
problem can be retrieved from the set of nodes in S designated as monitors.   □

Example 1. As an example, Fig. 2 considers a set problem where $U = \{1, 2, 3, 4, 5\}$
and the collection of sets is $S = \{\{1, 2, 3\}, \{2, 4\}, \{3, 4\}, \{4, 5\}\}$. The corresponding vertex set is $V = S \cup U \cup \{T\} = \{\{1, 2, 3\}, \{2, 4\}, \{3, 4\}, \{4, 5\}, 1, 2, 3, 4, 5, T\}$.
In our example, the path $P(\{1, 2, 3\}, T)$ includes nodes 1, 2, and 3 in an arbitrary
order and finishes at T. Similar paths are built from the other nodes in S. The path
$P(T, T)$ includes $\{1, 2, 3\}, \{2, 4\}, \{3, 4\}$, and $\{4, 5\}$ in an arbitrary order. It is easy
to see that selecting $\{1, 2, 3\}$ as a monitor is more cost-effective than selecting 1,
2, and 3 individually.

Theorem 2. *The monitor 1-identifiability problem is NP-complete.*

Proof. We reduce the monitor cover problem (known to be NP-complete from
Theorem 1) to the monitor 1-identifiability problem. Consider a monitor cover
problem instance $(G = (V, E), P)$, where G is the graph and P is the set of
routes. We construct a monitor 1-identifiability problem $(G^* = (V^*, E^*), P^*)$
as follows. The set of vertices is $V^* = V \cup V' \cup \{T\}$ where V is the set of
vertices of the monitor cover problem, V' is a set of companions vertices for
V (each vertex $i \in V$ has its companion $i' \in V'$, thus $|V| = |V'|$) and T is a
special vertex that is linked to each node in V. Nodes in V' will be designated
as monitors in every valid solution. The set of edges E^* is composed of the edges
in $E \cup \{(i, T), \forall i \in V\} \cup \{(i, i') \in V \times V'$ such as i' is the companion of $i\}$.

---

[1] A similar proof can be found here : https://mathoverflow.net/questions/440921/is-
it-np-hard-to-find-the-min-set-of-nodes-in-a-graph-so-that-the-set-of-paths-j.

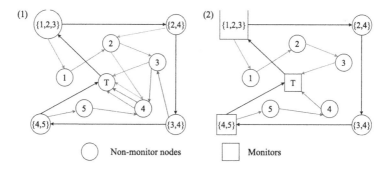

Fig. 2. Illustration of a monitor cover problem. (1) shows the instance with all major paths (2-nodes paths and edges are not shown for readability). (2) shows the optimal solution.

The routes connecting each pair of nodes $(i, j) \in V^2$ are inherited from the monitor cover problem, i.e., $P \subset P^*$. The other routes are represented by the sets in $P' = P^* \backslash P$. For all nodes in V', the route linking them is defined as $P'(i', j') = \{i', i, T, j, j'\} \; \forall (i', j') \in V'^2$. Every other routes in P' are segments of those routes, e.g., the route linking a node $i \in V$ to a companion node $j' \in V'$ of another node $j \in V$ is defined as $P'(i, j') = P'(j', i) = \{i, T, j, j'\}$, or the route linking T to a companion node $i' \in V$ is $P'(i', T) = P'(T, i') = \{i', i, T\}$.

First, note that every node in V' is necessarily a monitor, as they are only present at the extremities of routes. Their selection allows each node in $V*$ to be covered. T is also 1-identifiable as it is the only node crossed by each of these routes. The nodes in V are distinguishable from every other node in the exception of their companion node. Choosing T as a monitor is not effective as the only routes that cross nodes in V without necessarily crossing their companions (and thus make them 1-identifiable) are the routes imported from the cover problem. We ensured by construction that selecting a pair of node $i, j \in V^2$ enables 1-identifiability for every node covered by $P(i, j)$, hence the optimal solution to the monitor cover problem can be retrieved from the nodes in V designated as monitors in the 1-identifiability problem. □

Example 2. As an example, consider a monitor cover problem where $V = \{A, B, C, D, E\}$ and $E = \{(A, B), (B, C), (C, D), (A, D), (C, E)\}$. Figure 3 presents the corresponding topology in the 1-identifiability problem. A', B', C', D' and E' are necessarily monitors, T is thus covered by all the paths connecting them and has a unique symptom. A' and A are both covered by every path starting or ending from A' ($P'(A', B'), P'(B', A'), \ldots$). Their symptoms are equal but different from each other node. Every node in V is in a similar situation, distinguishable from all the other nodes except their companion. If $P(D, B) = \{D, A, B\}$, selecting D and B as monitors allows covering A, B and D in the cover problem. In the 1-identifiability problem, it makes them 1-identifiability because $P(D, B)$ does not contain A', B' and D'.

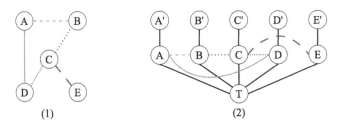

Fig. 3. Illustration of the transformation of a cover problem to an 1-identifiability problem. The routes are not represented. (1) shows the graph for the monitor cover problem. (2) shows the corresponding graph for the monitor 1-identifiability problem, edges from the graph (1) are identifiable by their colors.

5   Model for Optimal Monitor Placement

This section gives the model for the resolution of cover and 1-identifiability problems. We first present the ILP, CP and MaxSAT formulation of the model. Then we introduce some reductions and redundant constraints of the problem to tighten the formulation.

5.1   Model Definition

Integer Linear Programming Formulation. The problem is modeled with two binary variable vectors: x is a vector of size $|V|$ modeling the set of monitors (x_i is true iff node i is a monitor) and y is a vector of size $|P|$ modeling the set of measurement paths ($y_{P(i,j)}$ is true iff the route $P(i,j)$ is a measurement path). We denote by S_i the subset of P containing all the routes crossing node i; we also denote by $D_{i,j}$ the set of routes crossing node i without going by j and vice versa, $D_{i,j} = (S_i \cup S_j) - (S_i \cap S_j)$. The ILP model is composed of the following constraints:

$$\text{minimize} \sum_{i \in V} x_i \tag{1}$$

subject to:

$$y_{P(i,j)} \leq x_i \qquad\qquad \forall P(i,j) \in P \tag{2}$$

$$y_{P(i,j)} \leq x_j \qquad\qquad \forall P(i,j) \in P \tag{3}$$

$$y_{P(i,j)} \geq x_i + x_j - 1 \qquad\qquad \forall P(i,j) \in P \tag{4}$$

$$\sum_{P(i,j) \in S_{i'}} y_{P(i,j)} \geq 1 \qquad\qquad \forall i' \in V \tag{5}$$

$$\sum_{P(i,j) \in D_{i',j'}} y_{P(i,j)} \geq 1 \qquad\qquad \forall i', j' \in V^2 \tag{6}$$

Equation (1) expresses the objective function that is to minimize the number of monitors. Equations (2)–(4) link the monitor selection to the measurement path variables. A route $P(i,j)$ is considered a measurement path ($y_{P(i,j)} = 1$) iff both its starting and ending nodes are monitors ($x_i = 1$ and $x_j = 1$). Equation (5) models the cover constraints. For a node to be covered at least one route among S_i needs to be a measurement path. Thus, for each node i we have the constraint (5). Finally, Eq. (6) models the 1-identifiability constraints. A node is 1-identifiable iff it is distinguishable from each other nodes. Constraining each node to be 1-identifiable is equivalent to constraint each pair of nodes to be distinguishable. For two nodes to be distinguishable, it is sufficient that at least one measurement path crosses one of the nodes without crossing the other. For the 1-identifiability problem, it is important to keep the cover constraints from (5), otherwise a solution with one uncovered node would be a valid solution, as its symptom would be different from each other symptom. Notice that the constraint (6) is defined for pairs of nodes, but the same idea can be applied to all combinations of up to k nodes to model the k-identifiability constraints. Unfortunately, the number of constraints grows exponentially with k.

Constraint Programming Formulation. The CP formulation uses the same set of binary variables x and y than the ILP formulation. The reified constraints (8) ensure that route $P(i,j)$ is considered a measurement path ($y_{P(i,j)} = 1$) iff both its starting and ending nodes are monitors ($x_i = 1$ and $x_j = 1$). The sum constraints ensuring the coverage and the 1-identifiability are replaced by logical or constraints.

$$\text{minimize} \sum_{i \in V} x_i \tag{7}$$

subject to:

$$y_{P(i,j)} \equiv x_i \wedge x_j \qquad \forall P(i,j) \in P \tag{8}$$

$$\bigvee_{P(i,j) \in S_{i'}} y_{P(i,j)} \qquad \forall i' \in V \tag{9}$$

$$\bigvee_{P(i,j) \in D_{i',j'}} y_{P(i,j)} \qquad \forall i', j' \in V^2 \tag{10}$$

Maximum Satisfiability Formulation. The translation to a MaxSAT formulation is straightforward: there are two sets of literals, x and y, that correspond to the binary variables of the CP formulation. The reified constraints in (8) are translated in three sets of clauses defined by Eqs. (12)–(14). The clauses ensuring the coverage and the 1-identifiability, in Eqs. (15) and (16), are inherited from the *or* constraints. All the clauses defined by the Eqs. (12)–(16) are hard constraints that must be satisfied. To translate the objective function we add one weighted clause for each node $i \in V$. Each one of them only contains one literal, x_i, and their cost is fixed to -1. Hence, to maximize the sum of the satisfied constraints' weights, we need to minimize the number of nodes selected to be monitors.

$$\text{maximize} \sum_{i \in V} -1 * x_i \tag{11}$$

subject to:

$$x_i \vee \neg y_{P(i,j)} \qquad \forall P(i,j) \in P \tag{12}$$

$$x_j \vee \neg y_{P(i,j)} \qquad \forall P(i,j) \in P \tag{13}$$

$$\neg x_i \vee \neg x_j \vee y_{P(i,j)} \qquad \forall P(i,j) \in P \tag{14}$$

$$\bigvee_{P(i,j) \in S_{i'}} y_{P(i,j)} \qquad \forall i' \in V \tag{15}$$

$$\bigvee_{P(i,j) \in D_{i',j'}} y_{P(i,j)} \qquad \forall i', j' \in V^2 \tag{16}$$

5.2   Problem Reductions and Redundant Constraints

To reduce the search space and thus the computation costs, one can add nodes that must be monitors and redundant constraints to obtain a tighter model.

Detecting Monitors. The nodes that are only covered by paths originating or ending in them must be monitors to be covered. We refer to these nodes as leaf nodes $= \{i \in V \mid \forall i', j' \neq i, i \notin P(i', j')\}$. Clearly, nodes with degree one are part of the set of leaf nodes. This can be observed in the case of node H in Fig. 4.

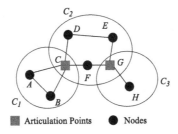

Fig. 4. Illustration of bi-connected components. C_1, C_2 and C_3 represents the 3 bi-connected components. Components C_1 and C_2 are linked together by the articulation node C. Removing this node would result in two connected parts: the first one only composed of C_1 and the other one composed of C_2 and C_3.

Redundant Constraints. A bi-connected component of a graph is a subgraph in which every pair of vertices is connected by at least two disjoint paths, meaning that the subgraph remains connected even if any single vertex or edge is removed. These subgraphs are linked together by articulation points, nodes whose removal would disconnect the total graph. For example, Fig. 4 shows a graph composed of three bi-connected components. The bi-connected components and their artic- ulation points can be identified in linear time with a depth-first search [13]. We denote by $V_C \subseteq V$ the subset of nodes in the bi-connected component C that are not articulation points and $A_C \subseteq V$ the subset of articulation points in C.

Lemma 1. *For a bi-connected component C with exactly one articulation point, ensuring coverage implies that at least one node in V_C must be a monitor.*

Proof. Because routes are cycle-free, a component with only one articulation point can't be crossed by a route connecting two nodes outside of a component. Otherwise, it would require the route to cross the articulation point at least twice, resulting in the presence of a cycle. Thus, it requires at least one monitor in the component to cover its nodes. As an example, in Fig. 4 component C_1 contains only one articulation point (the node C). Because routes are cycle-free, it is clear that if A and B are not monitors, then no measurement paths will go through the component. We can thus add a constraint which enforces that at least one node in such a bi-connected component must be a monitor.   □

We thus add the following constraints to the ILP formulation :

$$\sum_{i \in V_C} x_i \geq 1 \qquad \forall\ \texttt{bi-connected}(C) \in G \text{ with } |A_C| = 1 \qquad (17)$$

Their equivalents in the CP and the MaxSAT formulations are:

$$\bigvee_{i \in V_C} x_i \qquad \forall\ \texttt{bi-connected}(C) \in G \text{ with } |A_C| = 1 \qquad (18)$$

6   A Greedy Algorithm for 1-identifiability

To position our model within the state-of-the-art, we compare them with the greedy algorithm MNMP [19] (more specifically the MNMP-UP version). MNMP is a generic algorithm introduced to solve the broader k-identifiability problem. Its complexity is $O(|P|^2 \cdot |V|^3) = O(|V|^7)$, as $|P| = |V|^2$ in our case. We introduce a version dedicated to the 1-identifiability problem, as outlined in Algorithm 1, that reduces the complexity to $O(|V|^5)$. This is the version we use in our experimental comparisons.

6.1   Description of the Algorithm

First, the set of monitors is initialized with the set of leaf nodes (i.e., nodes that only are at extremities of routes) since it is the only way to cover them. Then the first while loop iteratively adds monitors to the solution until each node in the network is covered. During each iteration, the node selected for inclusion is the one covering the maximum number of currently uncovered nodes, considering the present set of monitors. In the pseudo-code, we denote by U the set of uncovered nodes and $\Psi(M) \subseteq V$ represents the set of nodes covered by the union of all paths between each pair of nodes in M, i.e., $\Psi(M) = \bigcup_{i,j \in M^2} P_{i,j}$. If the objective is solely to cover the nodes, the algorithm halts at this point.

Algorithm 1. MNMP Pseudo-Code for the Cover and 1-identifiability Problems

1: $M \leftarrow$ leaf nodes
2: $U \leftarrow V$ ▷ Uncovered Nodes
3: $U' \leftarrow V^2$ ▷ Indistinguishable pairs of Nodes
4: $U \leftarrow U \backslash \Psi(M)$ ▷ Coverage
5: **while** $U \neq \emptyset$ **do**
6:     $m \leftarrow \arg \max_{w \in V \backslash M} |U \cap \Psi(M \cup \{w\})|$
7:     $U \leftarrow U \backslash \Psi(M \cup \{m\})$
8:     $M \leftarrow M \cup \{m\}$
9: **end while**
10: **return** M if goal is coverage
11: $U' \leftarrow U' \backslash \Omega(M)$
12: **while** $U' \neq \emptyset$ **do**   ▷ 1-identifiability
13:     $m \leftarrow \arg \max_{w \in V \backslash M} |U' \cap \Omega(M \cup \{w\}|$
14:     $U' \leftarrow U' \backslash \Omega(M \cup \{m\})$
15:     $M \leftarrow M \cup \{m\}$
16: **end while**
17: **return** M

Otherwise, the algorithm proceeds with the second 'while' loop, which is designed to make each node 1-identifiable. Once again, the algorithm adds monitors in an iterative manner. However, this time the criterion for adding monitors is the number of pairs of currently indistinguishable nodes that the new monitor would turn distinguishable, taking into account the current set of monitors. We denote by U' the set of pairs of indistinguishable nodes. $\Omega(M) \subseteq V^2$ contains the pair of nodes that are distinguishable under the union of all paths between each pair of nodes in M. The original MNMP algorithm contains a third loop where it iterates on each monitors and test if it is redundant, i.e., if removing the monitor would impact the coverage or the 1-identifiability. If the monitor is redundant, it is then removed from the solution. However, we observed on our instances that no monitors were removed during this loop. Thus, we removed this last loop.

6.2   Time Complexity

Cover Problem. Leaf nodes are detected in $O(|V^3|)$. The set of uncovered nodes U and the routes are represented as bit sets of size $|V|$, where the bit i is set if node i is uncovered or if i is in the route. U and $\Psi(M)$ are computed by logical *or* and *and* operations on these bit sets in $O(|V|)$. Therefore, line 4 requires $O(|V| \cdot |M|^2)$ operation. For the monitor cover problem, the complexity is dominated by the loop in lines 5–9. The most computationally demanding step within the loop is line 6: computing $|U \cap \Psi(M \cup \{w\}|$ for all $w \in V \backslash M$ requires to compute for each route linking a monitor to a non-monitor node $(O(|P|))$ the set of uncovered nodes that it crosses $((O(|V|))$. If we assume that $|P| = |V^2|$, the resulting complexity is $O(|V^3|)$. The worst case is when $O(|V|)$ monitors are required to cover the graph. Hence, the while loop takes $O(|V|^4)$. The overall time complexity of MNMP for the cover problem is $O(|V|^4)$.

1-Identifiability Problem. Computing the set of pairs made distinguishable by a route is done in $O(|V|^2)$. Thereby, computing $\Omega(M)$ requires $O(|M|^2 \cdot |V|^2)$ and the reduction of U' in line 11 takes $O(|V|^2 \cdot |M|^2)$. Line 13 is very similar to line 6. It requires to calculate the set of node pairs $(O(|V|^2))$ that can be distinguished by every route between a monitor and a non-monitor node. The resulting complexity is $O(V^4)$. In the worst case (i.e., the first while loop returns $O(1)$ monitors and $O(|V|)$ monitors are required for 1-identifiability) the while loop in lines 12 to 16 takes $O(|V|^5)$. Hence the time complexity of MNMP for the 1-identifiability problem is $O(|V|^5)$ in the worst case.

7 Experimental Results

In this section we evaluate our different approaches.[2] The ILP model relies on Gurobi 10.0.1 [7]. We run the CP model on OR-Tools 9.7.2996 [21]. For the MaxSAT approach, we use MaxHS 3.0 [2]. Experiences run on a SkyLake CPU @ 2.3 GHz with up to 95 GB of memory. OR-Tools and Gurobi are limited to one thread. For MaxHS, CPLEX has also been limited to one thread. Every test is limited to a runtime of 3 min and is allocated 20 GB of memory.

Dataset. We run our model on topologies from *Rocketfuel* [24], *Internet Topology Zoo* [16] and CAIDA's *ITDK* (IPv4 and from February 2022) [4]. We assume shortest hop-count paths with arbitrary tie breaking when multiple best paths exist. For an even more realistic set of routes, we use topologies from Repetita [6], which contain IGP weights. In non-connected topologies, we keep the largest component. For the largest topologies, the solvers encountered memory issue on the 1-identifiability problem. OR-Tools, Gurobi and MaxHS lacked memory for respectively for 33, 38 and 16 instances. For 2 instances, MNMP reached the timeout and thus returned an invalid solution. To ensure a fair comparison between the solvers, we removed the instances for which at least one solver faced memory issues. As a result, for the cover problem we have 625 topologies with up to 960 nodes (the median number of nodes is 33). For the 1-identifiability problem, there are 587 instances with up to 330 nodes (the median being 31).

Results. Table 1 displays the number of instances where the returned solution is the best among those obtained by different solvers, as well as the number of instances for which the solvers proved the optimality of their solutions. Since MaxHS established the optimality of its solutions for 100% of the instances, the returned solution often corresponds to the optimal one. To compare solvers when they don't return the best solution, Fig. 5 shows the cumulative count of instances for which each solver provides a solution below a given number of extra monitors, compared to the optimal.

Comparison with MNMP. Table 1 shows that the greedy algorithm fails to find the best solution for 5.60% of instances in the monitor cover problem and for 28.45% of instances in the monitor 1-identifiability problem. While all three exact

---

[2] Source code available at https://github.com/BurlatsAuguste/MonitorPlacement.

Table 1. Number of solved instances for each model

Solver	Goal	Best solution found	Optimality proven
Gurobi	Cover	623 (99.68%)	621 (99.36%)
	1-identifiability	583 (99.32%)	583 (99.32%)
OR-Tools	Cover	624 (99.84%)	624 (99.84%)
	1-identifiability	**587 (100.0%)**	585 (99.66%)
MaxHS	Cover	**625 (100.0%)**	**625 (100.0%)**
	1-identifiability	**587 (100.0%)**	**587 (100.0%)**
MNMP	Cover	590 (94.40%)	None
	1-identifiability	420 (71.55%)	None

solvers can find the best solution for a greater number of instances, especially for the 1-identifiability problem (more than 96%). This highlights the effectiveness of our approach, which have the ability to reduce the number of monitors required in numerous scenarios. As we can see in Fig. 5, most of the time the improvement represents 1 to 3 monitors. Each new monitor adds $2*|M-1|$ measurement paths that need to be regularly probed. Thus, in large topologies that require many monitors, having some unnecessary monitors can strongly impact the traffic, especially around the monitors, and congestion can occur. In a context where minimizing the number of monitor is crucial, the exact solvers are a pertinent choice, as they are able to offer better solutions than MNMP.

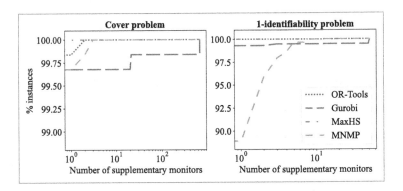

Fig. 5. Proportion of instances solved against the maximal distance on the objective with the best solution found (in number of monitors). Because of the logarithmic scale, the lines start with the proportion of instances for which the solver choose at most one supplementary monitor.

Comparison of the Exact Solvers. The three exact solvers show similar performances. For many instances, the different methods only explore just a few nodes in their search tree. OR-Tools and Gurobi frequently solve the problem at the root node: Gurobi solves at the root node 619 and 577 instances for the cover problem and the 1-identifiability problem respectively. OR-Tools solves 492 instances of the cover problem and 430 instances of the 1-identifiability problem at its root node. MaxHS uses an ILP solver to select the set of soft clause that should be satisfied in order to maximize the sum of weights and use a SAT solver to check if the hard clauses allow satisfying these soft clauses. For 548 instances of the cover problem and 367 instances of the 1-identifiability problem, the solver requires 0 call to the ILP solver. The instances on which the solvers need to perform their search are instances presenting only a few or no leaf nodes in their topologies. As explained in Sect. 5.2, because the routes are cycle-free, the leaf nodes are monitors in every valid solution. These forced monitors simplify the problem enough for the solvers to find the optimal solution with their presolving operations. Finally, for one particular instance of the cover problem, Gurobi returns a solution with more than 600 supplementary monitors. What happens is that the solver reaches the time limit before the end of Gurobi's presolving. For the 1-identifiability problem, for 3 instances Gurobi returns a solution containing 50 more monitors than the best solution. The size of the instance is not the problem. Indeed, they contain fewer than 100 nodes, when the solver is able to find the optimal solution instances three times larger. Instead the diameter of the graph is determinant here.

Scalability to Real-World Instances. The used topologies are frequently employed by the networking community and are deemed realistic, with most derived from probing actual networks. Regarding extending the results further, while feasible, our model is highly demanding in terms of the required constraints. The number of paths rapidly increases, and real-world topologies with several hundred nodes appear to be the limit of this modeling approach. The approach scales to most of today's Internet Service Provider networks. To scale beyond this, a column-generation-like scheme would be necessary, where paths and monitors are generated more lazily.

8   Conclusion

In this paper, we studied the placement of monitors in a network to ensure coverage or 1-identifiability of all nodes. We demonstrated that the problem is NP-complete for both objectives. We proposed an exact model with an Integer Linear Programming (ILP) formulation, a Constraint Programming (CP) formulation and a Maximum Satisfiability (MaxSAT) formulation. Additionally, we specialized and enhanced a state-of-the-art greedy algorithm. For most network topologies, all our exact models were able to find the optimal placement for both problems. Compared to the current state-of-the-art, an exact approach proved to be valuable, often outperforming the MNMP greedy algorithm by finding better solutions, frequently the optimal ones.

168   A. Burlats et al.

References

1. Bartolini, N., He, T., Arrigoni, V., Massini, A., Trombetti, F., Khamfroush, H.: On fundamental bounds on failure identifiability by Boolean network tomography. IEEE/ACM Trans. Netw. **28**(2), 588–601 (2020). https://doi.org/10.1109/TNET.2020.2969523
2. Berg, J., Bacchus, F., Poole, A.: Abstract cores in implicit hitting set MaxSat solving. In: Pulina, L., Seidl, M. (eds.) SAT 2020. LNCS, vol. 12178, pp. 277–294. Springer, Cham (2020). https://doi.org/10.1007/978-3-030-51825-7_20
3. Bezerra, P., Chen, P.Y., McCann, J.A., Yu, W.: Adaptive monitor placement for near real-time node failure localisation in wireless sensor networks. ACM Trans. Sens. Netw. **18**(1), 2:1–2:41 (2021). https://doi.org/10.1145/3466639
4. CAIDA: The CAIDA Macroscopic Internet Topology Data Kit (2022). https://www.caida.org/catalog/datasets/internet-topology-data-kit
5. Fortz, B., Thorup, M.: Internet traffic engineering by optimizing OSPF weights. In: Proceedings IEEE INFOCOM 2000. Conference on Computer Communications. Nineteenth Annual Joint Conference of the IEEE Computer and Communications Societies (Cat. No. 00CH37064), vol. 2, pp. 519–528. IEEE (2000)
6. Gay, S., Schaus, P., Vissicchio, S.: REPETITA: repeatable experiments for performance evaluation of traffic-engineering algorithms (2017)
7. Gurobi Optimization, LLC: Gurobi Optimizer Reference Manual (2023). https://www.gurobi.com
8. Hartert, R., Schaus, P., Vissicchio, S., Bonaventure, O.: Solving segment routing problems with hybrid constraint programming techniques. In: Pesant, G. (ed.) CP 2015. LNCS, vol. 9255, pp. 592–608. Springer, Cham (2015). https://doi.org/10.1007/978-3-319-23219-5_41
9. Hartert, R., et al.: A declarative and expressive approach to control forwarding paths in carrier-grade networks. ACM SIGCOMM Comput. Commun. Rev. **45**(4), 15–28 (2015)
10. He, T., Bartolini, N., Khamfroush, H., Kim, I., Ma, L., La Porta, T.: Service placement for detecting and localizing failures using end-to-end observations. In: 2016 IEEE 36th International Conference on Distributed Computing Systems (ICDCS), pp. 560–569 (2016). https://doi.org/10.1109/ICDCS.2016.21
11. He, T., Ma, L., Swami, A., Towsley, D.: Network Tomography: Identifiability, Measurement Design, and Network State Inference. Cambridge University Press, Cambridge (2021)
12. He, Y., Faloutsos, M., Krishnamurthy, S.: Quantifying routing asymmetry in the internet at the as level. In: 2004 IEEE Global Telecommunications Conference, GLOBECOM 2004, vol. 3, pp. 1474–1479 (2004). https://doi.org/10.1109/GLOCOM.2004.1378227
13. Hopcroft, J., Tarjan, R.: Algorithm 447: efficient algorithms for graph manipulation. Commun. ACM **16**(6), 372–378 (1973). https://doi.org/10.1145/362248.362272
14. John, W., Dusi, M., Claffy, k.: Estimating routing symmetry on single links by passive flow measurements. Technical report, ACM International Workshop on TRaffic Analysis and Classification (TRAC) (2010)
15. Katz, D., Ward, D.: Bidirectional forwarding detection (BFD). RFC 5880 (2010). https://doi.org/10.17487/RFC5880, https://www.rfc-editor.org/info/rfc5880
16. Knight, S., Nguyen, H., Falkner, N., Bowden, R., Roughan, M.: The internet topology zoo. IEEE J. Sel. Areas Commun. **29**(9), 1765–1775 (2011). https://doi.org/10.1109/JSAC.2011.111002

17. Lee, Y., Seok, Y., Choi, Y., Kim, C.: A constrained multipath traffic engineering scheme for MPLS networks. In: 2002 IEEE International Conference on Communications. Conference Proceedings. ICC 2002 (Cat. No. 02CH37333), vol. 4, pp. 2431–2436. IEEE (2002)

18. Ma, L., He, T., Leung, K.K., Swami, A., Towsley, D.: Monitor placement for maximal identifiability in network tomography. In: IEEE INFOCOM 2014 - IEEE Conference on Computer Communications, pp. 1447–1455 (2014). https://doi.org/10.1109/INFOCOM.2014.6848079

19. Ma, L., He, T., Swami, A., Towsley, D., Leung, K.K.: On optimal monitor placement for localizing node failures via network tomography. Perform. Eval. **91**, 16–37 (2015). https://doi.org/10.1016/j.peva.2015.06.003, https://www.sciencedirect.com/science/article/pii/S0166531615000516

20. Ogino, N., Kitahara, T., Arakawa, S., Hasegawa, G., Murata, M.: Decentralized boolean network tomography based on network partitioning. In: NOMS 2016 - 2016 IEEE/IFIP Network Operations and Management Symposium, pp. 162–170 (2016). https://doi.org/10.1109/NOMS.2016.7502809

21. Perron, L., Didier, F.: CP-SAT. https://developers.google.com/optimization/cp/cp_solver/

22. Ren, W., Dong, W.: Robust network tomography: K-identifiability and monitor assignment. In: IEEE INFOCOM 2016 - The 35th Annual IEEE International Conference on Computer Communications, pp. 1–9 (2016). https://doi.org/10.1109/INFOCOM.2016.7524375

23. RIPE NCC: RIPE Atlas. https://atlas.ripe.net. Accessed July 2023

24. Spring, N., Mahajan, R., Wetherall, D.: Measuring ISP topologies with rocketfuel. ACM SIGCOMM Comput. Commun. Rev. **32**(4), 133–145 (2002). https://doi.org/10.1145/964725.633039

25. Stanic, S., Subramaniam, S., Sahin, G., Choi, H., Choi, H.A.: Active monitoring and alarm management for fault localization in transparent all-optical networks. IEEE Trans. Netw. Serv. Manage. **7**(2), 118–131 (2010). https://doi.org/10.1109/TNSM.2010.06.I9P0343

26. de Vries, W., Santanna, J.J., Sperotto, A., Pras, A.: How asymmetric is the internet? In: Latré, S., Charalambides, M., François, J., Schmitt, C., Stiller, B. (eds.) AIMS 2015. LNCS, vol. 9122, pp. 113–125. Springer, Cham (2015). https://doi.org/10.1007/978-3-319-20034-7_12

27. Zhang, R., Newman, S., Ortolani, M., Silvestri, S.: A network tomography approach for traffic monitoring in smart cities. IEEE Trans. Intell. Transp. Syst. **19**(7), 2268–2278 (2018). https://doi.org/10.1109/TITS.2018.2829086

UNSAT Solver Synthesis via Monte Carlo Forest Search

Chris Cameron[1(✉)], Jason Hartford[2], Taylor Lundy[1], Tuan Truong[1], Alan Milligan[1], Rex Chen[3], and Kevin Leyton-Brown[1]

[1] Department of Computer Science, University of British Columbia, Vancouver, BC, Canada
{cchris13,tlundy,kevinlb}@cs.ubc.ca, alanmil@student.ubc.ca
[2] Valence Labs, Montréal, QC, Canada
jason@valencelabs.com
[3] School of Computer Science, Carnegie Mellon University, Pittsburgh, PA, USA
rexc@cmu.edu

Abstract. We introduce Monte Carlo Forest Search (MCFS), a class of reinforcement learning (RL) algorithms for learning policies in tree MDPs, for which policy execution involves traversing an exponential-sized tree. Examples of such problems include proving unsatisfiability of a SAT formula; counting the number of solutions of a satisfiable SAT formula; and finding the optimal solution to a mixed-integer program. MCFS algorithms can be seen as extensions of Monte Carlo Tree Search (MCTS) to cases where, rather than finding a good path (solution) within a tree, the problem is to find a small tree within a forest of candidate trees. We instantiate and evaluate our ideas in an algorithm that we dub Knuth Synthesis, an MCFS algorithm that learns DPLL branching policies for solving the Boolean satisfiability (SAT) problem, with the objective of achieving good average-case performance on a given distribution of unsatisfiable problem instances. Knuth Synthesis is the first RL approach to avoid the prohibitive costs of policy evaluations in an exponentially-sized tree, leveraging two key ideas: first, we estimate tree size by randomly sampling paths and measuring their lengths, drawing on an unbiased approximation due to Knuth (1975); second, we query a strong solver at a user-defined depth rather than learning a policy across the whole tree, to focus our policy search on early decisions that offer the greatest potential for reducing tree size. We matched or exceeded the performance of a strong baseline on three well-known SAT distributions, facing problems that were two orders of magnitude more challenging than those addressed in previous RL studies.

1 Introduction

Silver et al. [61,62] took the world by storm when their AlphaGo system beat world champion Lee Sedol at Go, marking the first time a computer program had achieved superhuman performance on a game with such a large action space. Their key breakthrough was combining Monte Carlo Tree Search (MCTS) rollouts with a neural network-based policy to find increasingly strong paths through the game tree. This breakthrough demonstrated that, with good state-dependent policies, MCTS can asymmetrically explore a game tree to focus on high-reward regions despite massive state spaces. MCTS rollouts avoid the exponential cost of enumerating all subsequent

© The Author(s), under exclusive license to Springer Nature Switzerland AG 2024
B. Dilkina (Ed.): CPAIOR 2024, LNCS 14742, pp. 170–189, 2024.
https://doi.org/10.1007/978-3-031-60597-0_12

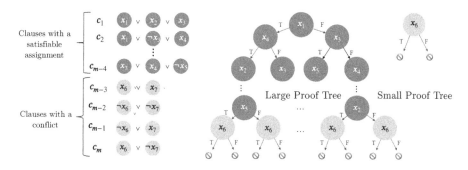

Fig. 1. A proof tree shows that any variable assignment to an unsatisfiable instance leads to a conflict. Its size substantially varies with the choice of branching policy.

sequences of actions [12,30] and can be extremely efficient when leveraging (1) multi-arm bandit policies to trade off exploration and exploitation [37] and (2) function approximation of the policies and values from previous problems to provide priors that further focus rollouts on promising paths [13,46,65]. MCTS is most useful in combinatorial spaces that have a natural hierarchical decomposition into a search tree. As a result, most of the notable applications of MCTS were to search for good paths of actions through game trees [3,5,12,58,71].

MCTS is also useful for searching for solutions to NP-hard combinatorial problems where a solution is a path through a search tree [2,8,33,51]. However, many other combinatorial problems cannot be expressed as searching for a good path. Consider constraint satisfaction problems (CSPs): while solving a satisfiable problem corresponds to finding a path (assigning variables sequentially and checking the solution), existing methods for proving that no solution exists build out a *proof tree* demonstrating that all possible variable assignments lead to a conflict. Rather than finding a high-reward path within a single tree, an algorithm designer's goal is to find a small tree within the forest of possible trees (e.g., see Fig. 1). This difference matters because of the high cost of evaluating each candidate policy: individual trees can be exponentially large.

This paper introduces Monte Carlo Forest Search (MCFS), a class of MCTS-inspired RL algorithms for such settings. We leverage leverage Scavuzzo et al. [56]'s recent concept of *tree MDPs*. When an action is taken in a tree Markov Decision Process (MDP), the environment transitions to two or more new child states; each of these new child states satisfies the Markov property in that each child only depends on its state. Algorithms that recursively decompose problems into two or more simpler sub-problems in a history-independent way can be represented as tree MDP policies. An MCFS rollout is a policy evaluation in the tree MDP and hence builds out a tree rather than a path, with the value of each node corresponding to the sum of rewards over its descendants in the rollout tree. Similar to MCTS, MCFS is a broad class of algorithms that can be adapted for different domains and problem sizes through choices in how policies are evaluated, how actions are selected and how rewards are aggregated. For example, an MCFS algorithm with complete policy evaluation (evaluating every child in a rollout) would be sufficient for tree MDPs giving rise to extremely small policy trees (e.g., log-linear trees in MergeSort). However, we are interested in tree MDP prob-

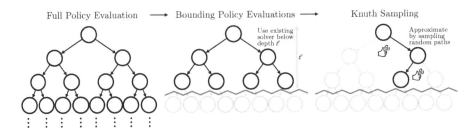

Fig. 2. Two keys ideas of Knuth Synthesis to avoid the prohibitive costs of exact tree-size policy evaluations: (1) bounded-depth search and (2) Knuth samples.

lems where policies produce exponential-sized trees, making exact policy evaluation prohibitive. We are not aware of any method that addresses this issue; existing work using RL [19,39,41,49,56,68] require exact policy evaluation and have only been used to train on very easy instances by industry standards (solvable on the order of 1000 decisions).

To solve this problem, we present Knuth Synthesis, an MCFS algorithm that allows for cheaper policy evaluation and is tailored for *pure-DPLL* [15] algorithms, a popular class of SAT algorithms that can straightforwardly be defined as tree MDPs.[1] We use the idea of *Knuth sampling* [36] to obtain linear and unbiased Monte Carlo approximations of tree sizes, following Lobjois and Lemaître [44] who showed these can be effective for cheaply comparing algorithms despite their high variance. Knuth samples correspond to path-based rollouts, allowing us to integrate them with MCTS. Extensive follow-up work has developed alternatives for approximating tree size (e.g., [10,11,34,52]), but these estimates are not decomposable into path-based root-to-leaf rollouts, and would require new ideas to be integrated into MCFS. Knuth Synthesis is designed to be used offline to synthesize new solvers under the data-driven algorithm design paradigm [4], where we optimize algorithm performance over a training set of instances. As is standard in the literature, we encode our policies using deep neural networks, which are far more expensive to evaluate than standard heuristics. Such policies are too expensive to evaluate at every node of the search tree. We mitigate this cost by limiting our learned policy to the search tree's most important nodes, querying some existing subsolver below a certain depth in the proof tree; this also substantially reduces the policy space. To ensure that we find strong policies for online use, we enforce the same procedure offline for constraining on which nodes to call the policy. Figure 2 illustrates the two keys ideas for making policy evaluation tractable.

Knuth Synthesis is the first MCTS-like method that has been used to learn a branching policy for combinatorial search. We deviated from standard MCTS ideas as appropriate, notably making a novel change to the bandit algorithm to account for our tree size cost function. Unlike AlphaZero, which uses MCTS both offline and online, we use Knuth Synthesis only offline, as we cannot afford the computational overhead online;

---

[1] Other more practically useful algorithms such as CDCL and Branch-and-Bound cannot easily be represented as tree MDPs because they correspond to deeply history-dependent policies; we are not aware of any RL methods that are tractable for learning policies for such algorithms.

we only query our policy online. We evaluated our method on specific unsatisfiable problem distributions where $\approx 100,000$ decisions were required to solve problems during training, a considerable leap beyond other prominent published work. We note that focusing on either satisfiable or unsatisfiable instances in commonplace in SAT. There are many examples of specialized algorithms targeting one of satisfaible or unsatisfiable instances; indeed, there have been specialized tracks of the SAT competition for both cases. For example, local search is a very well-studied class of algorithms that cannot prove unsatisfiability. In practice, one can always use a portfolio strategy running specialized SAT and UNSAT algorithms in parallel and terminating when either returns an answer.

First, we evaluated our method on uniform random 3-SAT at the solubility phase transition, perhaps the best-studied SAT distribution (e.g., featured in the SAT Competition [1] from 2002 to 2018). We matched the performance of kcnfs, which was specifically designed to target this distribution. Second, we evaluated our method on the sgen [64] distribution, which is notoriously difficult for its problem size. We improved running time on sgen by 8% over kcnfs, which in turn is $3.2\times$ faster on this distribution than the hKis solver that solved the most unsatisfiable instances in the 2021 SAT Competition. Lastly, we evaluated our method on the satfc [20] distribution, consisting of radio-spectrum repacking feasibility problems from an FCC spectrum auction, improving running time by 28% over kcnfs. These results show the initial promise of MCFS; through further scaling, we believe these ideas will lead to stronger industrial solvers for specific applications.

2   Related Work

Tree search is a fundamental algorithm for combinatorial optimization that iteratively partitions a search space, e.g., by choosing variables to branch on. At a high level, the efficiency of tree search is measured by the product of (1) the number of branches and (2) the time required to make each branching decision. The order in which variables are assigned—the branching policy—has a dramatic effect on the size of the tree and the corresponding time to solve the problem, so algorithm designers seek branching policies that lead to small trees. In domains such as Mixed Integer Programming (MIP) solving, a "smart and expensive" paradigm has won out, where entire linear programs are solved to determine a branching decision. In other domains such as SAT solving, a "simple and cheap" paradigm dominates: high-performance SAT solvers make very cheap branching decisions that depend only superficially on subproblem state. After 20 years, Variable State Independent Decaying Sum (VSIDS) [47] has remained among the state of the art for Quantified Boolean Formula (QBF) and SAT; it is a simple heuristic that only keeps track of how often a variable occurs in conflict analysis. Some interesting early work developed more expensive heuristics [29], but these generally could not compete with VSIDS in terms of running time. A few notable exceptions are: (1) look-ahead SAT solvers, which have a similar concept of a lookahead as MCFS and choose a branching variable after looking ahead at the resulting state for different possible branches [24,26]; (2) the bsh heuristic used in the kcnfs solver [16] which branches based on a measure of how constraining the assignment of one variable is on the assignment of another; (3)

Bayesian moment matching to initialize VSIDS per-variable scores for literals based on how likely they are to be part of a satisfying assignment [17].

There is now a long history of machine learning being used for automated algorithm design [6,9,27,38]. One might expect that machine learning could learn more informative branching heuristics for tree search that are worth their cost, especially for shallow-depth branches where branching decisions are most consequential. Two approaches have shown promise for learning models to make branching decisions: imitation learning and reinforcement learning. First, imitation learning works by learning a cheap approximation of (1) an expensive existing heuristic [21,28,32,48] or (2) an expensive feature that is a good proxy for a good branching decision [59,69]. Relevant to SAT, Wang et al. [69] predict backdoor variables and integrate these predictions into the VSIDS heuristic while Selsam and Bjørner [59] learned to approximate small unsatisfiable core computation and then branched on variables predicted to belong to a core. Both report practical performance improvements. Second, it is tempting to use reinforcement learning to directly synthesize heuristic policies that are optimized for problem distributions of interest. This idea dates back to Lagoudakis and Littman [40], who used TD-learning to train a policy to select between seven predefined branch heuristics based on simple hand-crafted features. With the advent of modern deep learning architectures, practitioners train models that take the raw representation of a problem instance as the model input. Yolcu and Póczos [70] learned a local search heuristic for SAT; Tonshöff et al. [67] built a generic graph neural network-based method for iteratively changing assignments in CSPs; Lederman et al. [41] used the REINFORCE algorithm to learn an alternative to the VSIDS heuristic for QBF that improved CPU time within a competitive solver; and Vaezipoor et al. [68] used a black-box evolutionary strategy to learn a state-of-the-art branching heuristic for model counting. There are also various applications of RL for learning to branch in MIP [19,49,56]. Finally, Kurin et al. [39]'s work, which used Q-learning for branching in the CDCL solver Minisat, is closest to our own. They trained on random satisfiable problems with 50 variables and generalized online to 250 variables for both satisfiable and unsatisfiable instances and improved Minisat running time. Instances in their training set required on the order of 100 decisions. The authors found that increasing the difficulty of the training set generalized poorly and training was less stable. They resolved the path/tree distinction by treating a traversal through a tree as a path (which could be exponential in length) and allowing backtracking state transitions. This, coupled with not incorporating the stack of backtracking points into their state violates the Markov assumption when backtracking, which is necessary for proving unsatisfiability and is important for satisfiability for all but the optimal policy.

We think that the problem of using RL to train a branching policy deserves study even if the field remains a few steps behind practical relevance. The primary challenge arising in this body of past work is scaling RL methods to more practical problems requiring more decisions. Each of these RL methods do exact policy evaluation and have only been trained with ≤ 1000 decisions and are therefore constrained to training on easy instances. From the perspective of a modern SAT solver, many of these results may appear trivial, but they nevertheless quantify the extent to which RL methods can learn policies that improve on hand-tuned heuristics.

Scavuzzo et al. [56] were the first to cleanly pinpoint how the structure of tree-search algorithm changes credit assignment in RL, introducing the concept of *tree MDPs* and formulate a policy gradient for tree policies. Much earlier, Lagoudakis and Littman [40] recognized that the one-to-two state transition violated the MDP definition and resolved this by cloning the MDP and creating one copy for each transition. A number of other recent papers have also realized the inefficiency of credit assignment when treating an episode as a path (rather than a tree) in a tree-search algorithm [19, 49, 63].

Various researchers have used MCTS as an algorithm to directly search for a satisfying solution to a CSP that lie somewhere between local search and DPLL [31, 45, 51, 57]. In these cases, a rollout can be interpreted as a guess at a satisfying assignment. If an unseen node is reached or a conflict is reached along a rollout path, a reward is assigned based on some measure of how close the path is to being a satisfying assignment (e.g., number of satisfied constraints). We see two main differences between these approaches and ours: (1) they use MCTS online to solve CSPs rather than as an offline procedure for training model-based branching policies and (2) they are not designed for the unsatisfiable case where policies produce trees rather than paths.

3 Preliminaries

We now provide the required technical background for MCTS, the Boolean satisfiability problem (our application area), the DPLL algorithm (the framework that defines our policy space), and tree MDPs (the class of problems to which we apply MCFS).

3.1 Monte Carlo Tree Search

Monte Carlo Tree Search (MCTS) is a general-purpose RL algorithm framework for MDPs. An MDP $M(\mathcal{S}, \mathcal{A}, p, r)$ is defined by a set of states $s \in \mathcal{S}$, actions $a \in \mathcal{A}$, transition distribution $p(s_{i+1}|s_i, a_i)$, and reward function $r : \mathcal{S} \times \mathcal{A} \to \mathbb{R}$. For every state s_i that is visited, MCTS stores a vector of counts $c_i \in \mathbb{Z}^{|\mathcal{A}|}$ and value estimates $v_i \in \mathbb{R}^{|\mathcal{A}|}$. A *rollout* θ is a sequence of state, action pairs $((s_0, a_0), ..., (s_n, a_n))$. MCTS makes a series of rollouts starting from the current state s_0 defined with the below four steps. Our definition is more general than presented in Sutton and Barto [66] to help us later define MCFS, notably the introduction of γ that separates the concepts of (1) choosing an action from (2) terminating a rollout.

1. **Selection.** A *tree policy* $\alpha : s \in \mathcal{S}, c \in \mathbb{R}^{|\mathcal{A}|}, v \in \mathbb{R}^{|\mathcal{A}|} \mapsto \mathcal{A}$ selects an action based on the action values v, counts c, and often a prior that depends on s. Then a *step policy* $\gamma : \theta \in (\mathcal{S} \times \mathcal{A})^n, a \in \mathcal{A}, s \in \mathcal{S} \mapsto \mathcal{A} \cup \emptyset$ takes in the state s_i, action a_i, and the current rollout history $((s_0, a_0), ..., (s_i, a_i))$ and chooses to either play the action selected by α or to play no action ($\emptyset$), terminating the rollout at that state.
2. **Expansion.** Counts and value estimates are tracked for any state reached by our selection step and together form a tree rooted at the current state. For any previously unexplored state s_i along the path, we add a node with vectors c_i and v_i to our tree.
3. **Simulation.** *rollout policy* $\pi : \mathcal{S} \mapsto \mathbb{R}$ then estimates the sum of rewards before reaching terminal state of the MDP from the node where no action was played. π is commonly a value network estimating the value of a path originating from the node's state.
4. **Backup.** Pass back rewards through the path of the MCTS tree, so that v_i at s_i are updated with the sum of rewards from its descendants: $\pi(s_n) + \Sigma_{x=i+1}^{n-1} r(s_x)$.

After a fixed number of rollouts, MCTS commits to the action at the root according to accumulated statistics of the tree (e.g., for AlphaZero, the largest number of samples) and the resulting state becomes the new root node. This procedure repeats until MCTS finds a path to a leaf of the tree. It is then possible to train a policy network to approximate MCTS by using the counts c_i and values v_i of each node along this path as training examples. It is common to use such a policy network within α to further focus rollouts on promising paths.

3.2 Boolean Satisfiability

A Boolean satisfiability (SAT) instance S is defined by a set of clauses $C = \{c_1, \ldots, c_m\}$ over a set of variables $X = \{x_1, \ldots, x_n\}$. Each clause consists of a set of Boolean *literals*, defined as either a variable x_i or its negation $\neg x_i$. Each clause is evaluated as *True* iff at least one of its literals is true (i.e., the literals in a clause are joined by OR operators). S is *True* if there exists an assignment of values to variables for which all the clauses simultaneously evaluate to *True* (i.e., the clauses are joined by AND operators). If such an assignment exists, the instance is called *satisfiable*; it is called *unsatisfiable* otherwise. SAT solvers try to find an assignment of variables to demonstrate that a problem is satisfiable, or to construct proofs showing that no setting of the variables can satisfy the problem. We consider only unsatisfiable problems.

3.3 DPLL and Variable Selection Policies

Many SAT solvers rely on the Davis-Putnam-Logemann-Loveland algorithm (DPLL), which assigns variables in an order given by some (potentially state-dependent) variable selection policy.

Definition 1. *Let S be a SAT instance or any subproblem within a larger instance. A policy ϕ is a mapping $\phi : S \to (v)$ that determines which variable v to assign in DPLL.*

Given a policy, the DPLL algorithm selects a variable to branch on and recursively checks both the *True* and *False* assignments in a tree-like fashion, performing *unit propagation* at each step. Unit propagation assigns variables forced by single-variable (*unit*) clauses; propagates them to other clauses; and repeats until no unit clauses remain. Each recursive DPLL call terminates when a *conflict* (variable assignments form a contradiction) is found, forming a *proof tree* (e.g., Fig. 1). There can be massive gaps in performance between variable selection policies. For example, Fig. 1 shows a formula that leads to a three-node proof tree if x_6 is selected first by the policy (assigning x_6 results in $x_7 \wedge \neg x_7$, implying a contradiction), but a tree that could have as many as $2^{|X|} - 1$ nodes if x_6 is selected last. This also illustrates how top-level decisions are much more powerful; the proof tree doubles in size for every level at which a good branching decision (branch on x_6) is not made. Another way that policies can affect tree size is through DPLL's unit propagation step: policies that cause more unit propagation earlier in the search require fewer decisions overall and therefore yield smaller search trees.

For an instance S solved using variable selection policy ϕ, we denote the size of the resulting proof tree as $T_\phi(S)$. For a given distribution over problems $\mathbb{P}$, our goal is to find a policy ϕ^* that minimizes the average proof tree size $\mathcal{L}(\phi; \mathbb{P}) = \mathbb{E}_{S' \sim \mathbb{P}}[T_\phi(S')]$. Finding policies is computationally challenging; for an n-variable problem, there are $O(n^{3^n})$ possible variable selection policies (3^n states representing each variable as *True*, *False*, or unassigned, and n choices per state); and exact evaluation of $T_\phi(S)$ takes $O(2^n)$ operations; Liberatore [42] showed that identifying the optimal DPLL branching variable at the root decision is both NP-hard and coNP-hard.[2] If we assume that the optimal variable selection policy, $\phi_{\mathbb{P}}^* = \arg\min_{\phi'} \mathcal{L}(\phi'; \mathbb{P})$, is learnable by an appropriate model family, we could in principle learn an approximation to the optimal policy, $\hat{\phi}^*$, to use within our solver. The challenge is to design a procedure that efficiently minimizes $\mathcal{L}(\phi; \mathbb{P})$ so that labeled training examples can be collected through rollouts of learned policies. This is a challenging reinforcement learning problem. For a particular variable selection policy, even evaluating the loss function on a single instance takes time exponential in n, and the number of policies is doubly exponential in n.

3.4   Tree MDPs

Tree MDPs [56] are a generalization of MDPs with 1-to-many transitions. Similar to the Markov property of MDPs, tree MDPs have a *tree Markov property*: each subtree depends only on its preceding state and action. We show how the DPLL algorithm can be represented with tree MDPs in Appendix.

Definition 2 (Scavuzzo et al. [56]). *Tree MDPs are augmented MDPs $tM = (\mathcal{S}, \mathcal{A}, p_{init}, p_{ch}^L, p_{ch}^R, r, l)$, with states $s \in \mathcal{S}$, actions $a \in \mathcal{A}$, initial state distribution $p_{init}(s_0)$, left and right child transition distributions $p_{ch}^L(s_{ch_i}^L|s_i, a_i)$ and $p_{ch}^R(s_{ch_i}^R|s_i, a_i)$, reward function $r : \mathcal{S} \times \mathcal{A} \to \mathbb{R}$ and leaf indicator $l : \mathcal{S} \to \{0, 1\}$. Each non-leaf state s_i (i.e., such that $l(s_i) = 0$), together with an action a_i, produces two new states $s_{ch_i}^L$ (left child) and $s_{ch_i}^L$ (right child). Leaf states (i.e., such that $l(s_i) = 1$) are the leaf nodes of the tree, below which no action can be taken.*

4   Monte Carlo Forest Search

We define Monte Carlo Forest Search (MCFS) as a class of RL algorithms for learning policies in *tree MDPs*. An MCFS rollout θ is a tree of (state, action) pairs structured according to the underlying tree MDP where a parent node s_i shares edges with its children $s_{ch_i}^L$ and $s_{ch_i}^R$. The aggregation of *rollout trees* forms a forest, hence the name. The Expansion step is defined the same as MCTS and so is the Simulation step with the exception of being applied to many states rather than just one. We define the Selection and Backup step for an MCFS rollout below. See Fig. 3 for a side-by-side illustration of MCTS and MCFS in their most general forms and find pseudocode in Appendix.

Selection. When an action selected by α is played by γ at state s_i, the state transitions to new states $s_{ch_i}^L$ and $s_{ch_i}^R$ from sampling $p_{ch}^L(s_{ch_i}^L|s_i, a_i)$ and $p_{ch}^R(s_{ch_i}^R|s_i, a_i)$. These new states are added to a queue q. MCFS iterates over q, calling α and adding the corresponding new states back to q if γ chooses to step and terminates when q is empty.

---

[2] Liberatore [43] later showed the problem is $\Delta_2^p[log(n)]$-hard i.e., at least poly time with $log(n)$ oracle queries to an NP problem.

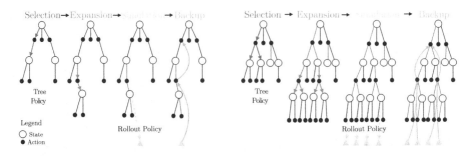

Fig. 3. Side-by-side comparison of MCTS and MCFS highlighted by the Selection, Expansion, Simulation and Backup steps (Adapted from Sutton and Barto [66]).

Backup. The rewards are passed back through θ, so that the value estimate v_i at s_i is updated with the sum of rewards from its descendants in θ.

We can now see the importance of defining the step policy γ; it allows us to evaluate any subset of the full proof tree in a *rollout tree* which is important when exact policy evaluation is intractable. MCFS commits to the action a at the root in the same way as MCTS does and the corresponding child nodes $s_{ch_0}^L$, $s_{ch_0}^R$ become the roots of new search trees and are solved sequentially. This procedure repeats until MCFS finds a tree where every leaf s_i of the tree is a tree MDP leaf ($l(s_i) = 1$).

5   Knuth Synthesis

For tree MDP problems like DPLL for UNSAT or model counting, the tree depth can be linear in the number of actions; therefore the tree size (and thus the cost of evaluating a policy) can grow exponentially with the number of actions. Knuth Synthesis is an MCFS implementation based on two key ideas that avoid the prohibitive costs of exact policy evaluations (See Fig. 2). First, in the step policy we nest a sampling procedure within a given Monte Carlo rollout, using a set of random paths through the rollout tree to approximate tree size. Second, we bound the depth of our forest search; below this point, we call out to a rollout policy that leverages existing SAT solvers. We describe these ideas in the following subsections and then provide our full implementation.

5.1   Nested Monte Carlo Sampling

To obtain a Monte Carlo sample through our tree MDP and determine the associated tree size $T_\alpha(S)$, we may need to visit $O(2^n)$ states. This is computationally expensive, both because of the absolute number of states and because at each node in the tree we need to query α to decide which variable to assign next (which is nontrivial when α is parameterized by a neural network). To address this issue, we nest a sampling procedure within a given Monte Carlo rollout sample, effectively Monte Carlo sampling a Monte Carlo rollout of our tree policy. More precisely, we stochastically approximate $T_\alpha(S)$ by using Knuth samples [36]. We first state the general version of Knuth's theorem for estimating the total weight of a weighted k-ary tree with a random path through the tree.

Theorem 1 (Knuth [36]). *Let N be the set of nodes in a weighted k-ary tree T where each node $n \in N$ has weight $w(n)$. The total weight of this tree is $w_T = \sum_{n \in N} w(n)$. Let $(n_0, \ldots, n_\ell)$ be a path of nodes through the tree from root to leaf that is chosen uniformly at random. Then, $w_T = \mathbb{E}_{(n_0, \ldots n_\ell)} \sum_{i=0}^{\ell} k^i w(n_i)$.*

In the case of binary trees where all nodes have weight 1, a simple corollary holds.

Corollary 1. *Let ℓ_P be the length of a path P $(n_0, \ldots, n_{\ell_P})$ where $\forall_i, w(n_i) = 1$ and sampled uniformly at random from a binary tree T with size s_T. Then, $s_T = \mathbb{E}_P[2^{\ell_P+1} - 1]$.*

Proof. When $w(n_i) = 1$, $s_T = E_P[\frac{k^{\ell_P} - 1}{k - 1}]$. This is true because $\sum_0^{n-1} k^i = \frac{k^n - 1}{k - 1}$. Plugging in $k = 2$ (binary tree in UNSAT), we get $s_T = E_P[2^{\ell_P} - 1]$.

For the tree T produced by α, we can use Theorem 1 to get an unbiased estimate of $T_\alpha(S)$. In the context of a DPLL solver, a Knuth sample amounts to replacing a complete traversal of the binary tree of all *True* / *False* assignments with a path through the tree where assignments are chosen uniformly at random. We can take the length of the resulting path, ℓ, and update the tree size estimate of each node at depth d with $2^{\ell-d} - 1$ for the corresponding decision by α. The average of these estimates across a set of paths yields an unbiased approximation of the tree size.

5.2   Bounding Policy Evaluations

Knuth Synthesis is an offline procedure for training a variable selection policy ϕ. At test time, we want proofs of unsatisfiability rather than estimates of tree size, and so cannot use Knuth samples. We represent ϕ using a deep neural network, which is far more expensive to evaluate than conventional tree-search heuristics. This raises the risk that the computational cost of evaluating ϕ will exceed its benefits via proof tree size reductions, and indeed makes this risk a certainty for small enough subproblems, such as those for which existing solvers are faster than single network calls. Any online policy must therefore have a procedure for constraining the nodes at which ϕ will be queried. As we saw earlier, decisions become more important the closer they are to the root of an UNSAT search tree. We thus apply ϕ at states having depth $\leq \ell$, after which we either (1) at both test and training time, call a "subsolver" (a pre-existing variable selection policy); or (2) at training time, sometimes instead call a value network that more cheaply approximates the tree size of this fixed policy (see "Rollout Policy" for details). Although we could potentially afford to explore larger ℓ values at training time, we do not do so; this ensures that the training phase identifies a policy that will leverage the subsolver appropriately at test time. To determine the appropriate tree size estimation, we add a weight of $T_{\phi_{sub}}(S^*)$ to any node representing state S^* at which we call the subsolver policy ϕ_{sub}. This represents the additional tree size incurred from the subsolver call at this node. For a given Knuth sample of length ℓ that terminates at S^*, we can use the weighted version of Knuth's theorem to update the value of a node at depth d along the path with $2^{\ell-d} T_{\phi_{sub}}(S^*) + 2^{\ell-d} - 1$.

5.3 Implementation

Beyond changes to avoid exponential evaluation costs, Knuth Synthesis has a few other noteworthy components. We alter the Action Selection step to account for the tree size cost function that scales with node depth; we adapt the Simulation step to account for an unreliable value network and a fixed policy; we use a model architecture to respect the structural invariances of SAT; we use a graph-based state-transition data structure to improve sample efficiency; and we share policy priors with child nodes to speed up lookaheads. Find pseudocode in Appendix: https://www.cs.ubc.ca/labs/algorithms/Projects/MCFS/appendix.pdf.

Tree Policy α. At every state s, α returns action $a = \arg\min_{a'}(Q(s,a') - Q_d U(s,a'))$. $Q(s,a)$ is the cost estimate of action a at state s, Q_d is the running average tree size of nodes at depth d, which is used to calibrate confidence intervals across different depths, and $U(s,a) = c_{PUCT} P(s,a) \frac{\sqrt{\sum_{a'} N(s,a')}}{1+N(s,a)}$ is the corresponding confidence interval [53,60]. The confidence interval is parameterized by $N(s,a)$, the number of lookaheads that branch on action a at state s, c_{PUCT}, a constant that controls exploration, and $P(s,a)$, a prior distribution over the actions for a given state, predicted by the policy network. $N(s,a)$ is initialized to 1 and $Q(s,a)$ is initialized with the tree size of the first lookahead at that state. This provides an unbiased measure of the performance of our incumbent policy (i.e., the first sample is exactly the neural network policy), which we seek to improve upon. Because $Q(s,v)$ scales exponentially with the depth of s, we introduce Q_d to calibrate confidence intervals. Since $U(s,v)$ is independent of the scale of costs (it is only a function of counts), we calibrate each depth with Q_d to keep $U(s,v)$ at the same scale as $Q(s,v)$. Otherwise, the choice of the c_{PUCT} would trade off tight confidence intervals at shallow nodes against loose confidence intervals at deeper nodes.

Equivariant Architecture. SAT problems vary in size and are invariant to permutations of their clauses and variables within clauses. We enforce these invariances using a permutation equivariant architecture [23] that can handle input matrices of any size.

Sharing Prior with Child Nodes. Calls to the policy network tend to dominate running time, so we save time by only computing $P(s,a)$ once at the root of the MCFS forest and passing down the prediction to its child nodes. A child state has a subset of the action space of its parent state. For child state s', we set $P(s',a) \leftarrow P(s,a)$, and renormalize for the subset of actions remaining in s'.

Directed Acyclic Graph as Forest Search Data Structure. We improve sample efficiency by changing our forest search data structure from a tree to a graph, which leverages the fact that states are reached independently of the order of previous decisions [14].

Step Policy γ. An MCFS Step policy γ controls how much of the proof tree to evaluate in a rollout. In order for γ to implement Knuth sampling, it should play actions along a single path through the rollout tree, choosing actions uniformly at random. We must therefore ensure that for every pair of newly visited states (s^L, s^R), γ only plays an action at one of them. If γ is called at a state s and its sibling s^{sib} has yet to be processed, then γ chooses between $\alpha(s,)$ and $\emptyset$ with uniform probability. If s^{sib} has already been processed, then γ returns the opposite of the decision made at s^{sib}. γ will also return $\emptyset$ if it encounters a state s at depth ℓ or it encounters a leaf node (s s.t. $l(s) = 0$) i.e., a conflict in the underlying SAT problem.

Rollout Policy π. At each state s, π can be described via two cases: (1) if s's sibling s^{sib} played an action (i.e. $\gamma(s^{sib}) \neq \emptyset$) then $\pi(s)$ returns the cumulative reward at s^{sib}, ensuring that the reward estimates at any given node reflect the correct Knuth estimate ; (2) no action was played at s or its sibling because s is at depth ℓ. In this case our rollout policy π is to call the subsolver at a state S^* at depth ℓ of our Knuth sample. Rather than explicitly solving the subproblem at depth ℓ, we use a value network to access a much cheaper reward signal. This network is trained from an initial batch of subsolver calls and then retrained as we collect new batches of data, so its accuracy varies across time steps of a training run. We address the issue of an unreliable network by randomly deciding between calling the subsolver or the value network, with the probability depending on an online estimate of the value network's accuracy. Specifically, we track the mean multiplicative error ϵ of our value network over time by querying the value network with every subsolver call. For a user-defined accuracy threshold parameter t (we use $t = 0.5$), we sample the value network with probability $1 - \min(1, \epsilon/t)$, so that the probability of calling the subsolver halves as the error halves.

6 Experimental Setup

6.1 Benchmarks

We targeted instance distributions that are well known and difficult for modern SAT solvers. We controlled instance size so that (1) the size of the action space was similar to Go and (2) solving required $\approx$ 100,000 decisions. We did not consider industrial SAT Competition instances, as they often contain millions of variables with state spaces significantly larger than any deployed MCTS application of which we are aware. We evaluated our approach on three distinct distributions: (1) a canonical random distribution: uniform random 3-SAT at the solubility phase transition (R3SAT), (2) a notoriously difficult crafted distribution (sgen [64]), and one real-world application: station repacking problems from the 2016 FCC incentive auction (satfc [20]). For R3SAT, we trained on 300-variable instances, for which calling a subsolver was quick ($\approx$ 1 s solving time), and filtered out satisfiable instances for training and testing. To evaluate upward size generalization, we set aside 100 test instances at both our training size of 300 variables as well as at 350, 400, and 450 variables. For sgen, we trained on 65 variables; to evaluate upward size generalization we set aside 100 test instances at both our training size of 65 variables as well as at 75, 85, and 95 variables. For satfc, we trained on small (<2000-variable) instances from localized regions of the U.S. interference graph, and filtered out satisfiable instances. We set aside 100 test instances of our training distributions. For full details, see Appendix.

6.2 Baselines

Since Knuth Synthesis learns DPLL policies, we sought to evaluate against other purely DPLL solvers rather than to make apples-to-oranges comparisons to clause learning

(CDCL) solvers. We ultimately aim to integrate our MCFS ideas into the much richer design space of CDCL solvers, but anticipate that this will require going beyond the tree MDP formalism. To make our comparisons as fair as possible, we set our baseline and subsolver to be the same: kcnfs [16], which is a pure-DPLL solver and is specifically designed for and is among the strongest solvers for R3SAT. We used its most recent competition submission at the 2007 SAT Competition [7], where it won the silver medal in the Random UNSAT track and had previously won the gold medal in the 2005 Random UNSAT track. The lookahead march solver [25] is $\approx 10\%$ faster but we chose kcnfs because it is a pure-DPLL solver. kcnfs is a much stronger baseline than the minisat baseline which Kurin et al. [39] compare to (two orders of magnitude faster on their benchmarks). We also evaluated a uni+kcnfs baseline on each dataset, where we replaced Knuth Synthesis neural network calls with uniform-at-random decisions and called the kcnfs solver for subproblems at the same user-defined depth. We tried a purely random policy without calling kcnfs as a subsolver on 65-variable sgen, which led to poor performance $40\times$ slower than kcnfs.

6.3 Knuth Synthesis

We integrated our Knuth Synthesis algorithm into the CDCL solver Maple_ LCM_Dist_ChronoBT [55], which won the 2018 SAT Competition and is based on the MiniSAT framework [18]. We removed all clause-learning components so that the solver ran pure DPLL search. We trained our neural networks with PyTorch [50] in Python and ported them to our C++ solver using tracing. Find our code here: https:// github.com/ChrisCameron1/MCFS.

Two important hyperparameters were (1) the constant for the level of exploration c_{PUCT} and (2) the number of lookaheads k. Using a coarse grid search, we selected $c_{PUCT} = 0.5$ and $k = 100,000$, which found the best policies within a 48-h window. We chose $\ell = 5, 6,$ and 8 for the satfc, R3SAT and sgen distributions respectively, based on the average tree size of Knuth Synthesis after 48 h for values between 2 and 10 over a few instances. We made decisions with Knuth Synthesis until depth ℓ; each MCFS decision yielded a training point consisting of a (state, policy vector, Q-value) triple, where the policy vector represents normalized action counts from MCFS.

Before a good policy network is learned, Knuth Synthesis tends to be less efficient. We pretrained our policy and value networks by running Knuth Synthesis with 10,000 lookaheads on 1,000 instances for R3SAT and sgen. We ran one iteration of Knuth Synthesis with 10,000 lookaheads on 1,000 instances to further improve the policy and value network, and then a final iteration with 100,000 lookaheads on 2,000 instances to train our final model. Pretraining runs of Knuth Synthesis took approximately 24 h per instance and were respectively run on 300-variable and 55-variable problems from R3SAT and sgen. The subsequent iterations took approximately 48 h per instance. On R3SAT and sgen, they were respectively run on 300-variable and 65-variable problems. For satfc, we pretrained with 10,000 lookaheads on 441 instances and ran a final iteration using the prior with 100,000 lookaheads. The parameters and architecture described in this paper are only for the last iteration of Knuth Synthesis. We made several minor improvements across iterations.

6.4   Model Training

We used the exchangeable architecture of Hartford et al. [23]. We represented a CNF SAT instance with n clauses and m variables as an $n \times m \times 128$ clause-variable permutation-equivariant tensor, where entry (i, j) is t_v if the true literal for variable i appears in clause j, f_v if the false literal for variable i appears in clause j, and 0 otherwise. t_v and f_v are 128-dimensional trainable embeddings representing the true and false literal. Following Hamilton et al. [22], we also added a node degree feature to every literal embedding. We instantiated the permutation-equivariant portion of the exchangeable architecture as four exchangeable matrix layers with 512 output channels, with leaky RELU as the activation function. We mean pooled the output to a vector, with each index representing a different variable.

We added three feed-forward heads: a policy head, a Q-value head, and a value head. The policy and Q-value heads both had two feed-forward layers with 512 channels and a final layer that mapped to the single output channel. The value head was the same, except there was a final mean pool that output a single scalar. The policy head and the Q-value head were trained to predict the normalized counts and Q-values of Knuth Synthesis, respectively. We optimized cross-entropy loss for both the policy and Q-value head. The purpose of the Q-value head was as an auxiliary task to help learn a better shared representation for the policy head; we observed a 5% reduction in tree size after adding the Q-value head. We trained the value network with mean-squared error against the $\log_2$ tree size of subsolver calls at leaf nodes. The value head was not backpropagated through the exchangeable layers. Using a shared representation was important for training the value head; loss tended to double when using a network trained with only a value head. We used the Adam optimizer [35], with a learning rate of 0.0001 and a batch size of 1. We evaluated mean tree size on a held-out validation to select a model.

Table 1. Mean tree size (1000 s of nodes) and running time (CPU+GPU seconds) comparing Knuth Synthesis (Ours), kcnfs, and uni+kcnfs over 100 test instances from each distribution. Reductions in decisions and running time are relative to kcnfs. In/Out denotes whether the benchmark was in or out of the training distribution.

	Size	In/Out	Tree size (1000 s)			CPU+GPU time (s)		
			uni+kcnfs	kcnfs	Ours	uni+kcnfs	kcnfs	Ours
R3SAT	300	In	44.3	8.9	**8.6 (2%)**	15.7	**1.1**	10.8
	350	Out	215.7	43.3	**42.7 (1%)**	59.5	**6.1**	15.8
	400	Out	1,103.5	226.9	**223.5 (2%)**	289.1	**30.8**	36.5
	450	Out	5,207.9	**989.7**	994.9	1591.1	**168.9**	173.4
sgen	65	In	158.2	162.3	**132.2 (23%)**	8.5	**2.1**	8.0
	75	Out	1,799.3	1,792.3	**1,594.0 (12%)**	26.6	**23.1**	26.1
	85	Out	8,932.9	8,874.8	**8,156.2 (9%)**	115.0	114.4	**105.6(8%)**
	95	Out	98,534.0	97,979.7	**92,407.9 (6%)**	1214.7	1272.2	**1,178.7 (8%)**
satfc		In	631.3	250.1	**67.1 (372%)**	24.8	8.3	**6.5 (128%)**

7   Results

We evaluated (1) search tree size and (2) running time for Knuth Synthesis against the two baselines on each of our instance sets; the full results are presented in Table 1. We measured running time as the cumulative CPU and GPU time with `runsolver` [54]. Overall, we reduced tree size on each of the three training distributions as well as 5/6 upward-size generalization distributions. This led to running time improvements on 3/9 datasets where the neural network overhead was manageable. R3SAT was the most challenging benchmark as `kcnfs` is an extremely strong solver specialized to this distribution; it far surpassed random branching on top-level decisions (4–5× reduction in tree size and walltime) and it was unclear whether it could be improved upon. Despite the strength of `kcnfs`, we squeezed out performance improvements of 1–2% in average tree size over `kcnfs` on up to 400 variables. Given the overhead of our neural network calls, these reductions in tree size did not lead to improvements in running time. For `sgen`, `kcnfs` was also very strong (see Appendix) however there is greater scope for improvement since it was not optimized for `sgen`. We reduced average tree size over `kcnfs` on our training distribution (65 variables) by 1.23×. Our model trained on the 65-variable distribution generalized well to larger problem sizes; it reduced tree size even on 95 variables, which took $\approx$ 20 min to solve (700× more difficult). Our solver incurred a roughly constant overhead that prevented us from improving running time on 65 variables and 75 variables. We were able to improve the running time over `kcnfs` by 8% on 85 variables and 8% ($\approx$1.5 min faster) on 95 variables. Even without using a GPU for model queries, we improved running time by 8% at 95 variables (constant policy-query overhead is swamped by tree-search time at higher running times). For `satfc`, evaluating against `kcnfs` allowed us to evaluate the (more realistic) scenario where there was no strong hand-tailored baseline. Such settings offer the likelihood of large scope for branching policy improvements; we saw the question of whether Knuth Synthesis discovered such policies as an important test. `kcnfs` ran $\approx$ 3× faster than `uni+kcnfs`. We reduced tree size by 3.72× and reduced running time by 1.28× over `kcnfs`. Find expanded results in Appendix including an experiment that shows the infeasibility of an existing RL approach on our datasets.

7.1   Knuth Speed up

We evaluated how much quicker we could find good policies using Knuth samples. Each d-bounded Knuth sample is in expectation 2^d cheaper than evaluating the full proof tree but the variance of the estimates can be high for unbalanced proof trees. We hoped variance would be sufficiently small that we would need much fewer than 2^d samples to distinguish between good and bad policies, allowing us to more quickly move to better parts of the policy space. Figure 4 shows average tree size over time comparing Knuth Synthesis to full-tree evaluation for our three training benchmarks. We normalized tree size across instances by the per-instance minimum and maximum tree size ever observed. In all cases, there was a clear efficiency improvement using Knuth samples. On R3SAT and `sgen`, Knuth sampling reached a given tree size approximately an order of magnitude more quickly. The improvement on `satfc` was substantial but

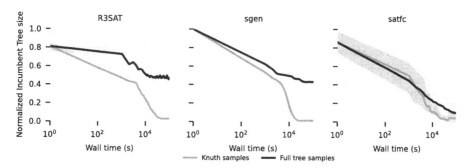

Fig. 4. Convergence rate improvements with Knuth samples. We normalize tree size across instances by the per-instance minimum and maximum tree size ever observed. Shaded regions represent 95% confidence intervals.

not as pronounced; we hypothesized this was because the satfc proof trees tended to be less balanced, which resulted in higher-variance Knuth estimates.

8   Conclusions and Future Work

We presented MCFS, a class of RL algorithms for learning policies in tree MDPs. We introduced an MCFS implementation called Knuth Synthesis that approximates tree size with Knuth samples to avoid the prohibitive costs of exact policy evaluation that existing RL approaches suffer from. We matched or improved performance over a strong baseline in a diverse trio of distributions, tackling problems within these distributions that were two orders of magnitude more challenging than those in previous studies. In future work, we would like to generalize MCFS to CDCL solvers. Most high-performance industrial solvers use the CDCL algorithm, which adds a clause-learning component to DPLL to allow information sharing across the search tree. This information sharing means there is no straightforward way to encode the problem as a tree MDP. Approximating the size of a CDCL search tree with Knuth samples is non-trivial: unlike DPLL, single paths from the root to leaves cannot be evaluated independently to approximate tree size.

Acknowledgments. We thank the reviewers for all the constructive feedback. This work was funded by an NSERC Discovery Grant, a DND/NSERC Discovery Grant Supplement, a CIFAR Canada AI Research Chair (Alberta Machine Intelligence Institute), a Compute Canada RAC Allocation, awards from Facebook Research and Amazon Research, and DARPA award FA8750-19-2-0222, CFDA #12.910 (Air Force Research Laboratory). We thank Greg d'Eon for many useful conversations, especially one infamous AlphaGo tutorial.

References

1. SAT competition. The International SAT Competition Web Page (2002–2024). http://www.satcompetition.org/

2. Abe, K., Xu, Z., Sato, I., Sugiyama, M.: Solving NP-hard problems on graphs by reinforcement learning without domain knowledge. arXiv preprint arXiv:1905.11623, pp. 1–24 (2019)

3. Agostinelli, F., McAleer, S., Shmakov, A., Baldi, P.: Solving the Rubik's cube with deep reinforcement learning and search. Nat. Mach. Intell. **1**(8), 356–363 (2019)

4. Balcan, M.: Data-driven algorithm design. CoRR abs/2011.07177 (2020). https://arxiv.org/abs/2011.07177

5. Baudiš, P., Gailly, J.: PACHI: state of the art open source go program. In: van den Herik, H.J., Plaat, A. (eds.) ACG 2011. LNCS, vol. 7168, pp. 24–38. Springer, Heidelberg (2012). https://doi.org/10.1007/978-3-642-31866-5_3

6. Bengio, Y., Lodi, A., Prouvost, A.: Machine learning for combinatorial optimization: a methodological tour d'horizon. Eur. J. Oper. Res. **290**(2), 405–421 (2021)

7. Berre, D.L., Roussel, O., Simon, L.: SAT 2007 competition. The International SAT Competition Web Page (2007). http://www.satcompetition.org/

8. Browne, C.B., et al.: A survey of Monte Carlo tree search methods. IEEE Trans. Comput. Intell. AI Games **4**(1), 1–43 (2012)

9. Cappart, Q., Chételat, D., Khalil, E.B., Lodi, A., Morris, C., Veličković, P.: Combinatorial optimization and reasoning with graph neural networks. J. Mach. Learn. Res. **24**(130), 1–61 (2023)

10. Chen, P.C.: Heuristic sampling: a method for predicting the performance of tree searching programs. SIAM J. Comput. **21**(2), 295–315 (1992)

11. Cornuéjols, G., Karamanov, M., Li, Y.: Early estimates of the size of branch-and-bound trees. INFORMS J. Comput. **18**(1), 86–96 (2006)

12. Coulom, R.: Efficient selectivity and backup operators in Monte-Carlo tree search. In: van den Herik, H.J., Ciancarini, P., Donkers, H.H.L.M.J. (eds.) CG 2006. LNCS, vol. 4630, pp. 72–83. Springer, Heidelberg (2007). https://doi.org/10.1007/978-3-540-75538-8_7

13. Coulom, R.: Computing "Elo ratings" of move patterns in the game of Go. ICGA J. **30**(4), 198–208 (2007)

14. Czech, J., Korus, P., Kersting, K.: Monte-Carlo graph search for AlphaZero. arXiv preprint arXiv:2012.11045, pp. 1–11 (2020)

15. Davis, M., Logemann, G., Loveland, D.: A machine program for theorem-proving. Commun. ACM **5**(7), 394–397 (1962)

16. Dequen, G., Dubois, O.: *kcnfs*: an efficient solver for random *k*-SAT formulae. In: Giunchiglia, E., Tacchella, A. (eds.) SAT 2003. LNCS, vol. 2919, pp. 486–501. Springer, Heidelberg (2004). https://doi.org/10.1007/978-3-540-24605-3_36

17. Duan, H., Nejati, S., Trimponias, G., Poupart, P., Ganesh, V.: Online Bayesian moment matching based SAT solver heuristics. In: Proceedings of the 37th International Conference on Machine Learning, ICML 2020, pp. 2710–2719 (2020)

18. Eén, N., Sörensson, N.: An extensible SAT-solver. In: Giunchiglia, E., Tacchella, A. (eds.) SAT 2003. LNCS, vol. 2919, pp. 502–518. Springer, Heidelberg (2004). https://doi.org/10.1007/978-3-540-24605-3_37

19. Etheve, M., Alès, Z., Bissuel, C., Juan, O., Kedad-Sidhoum, S.: Reinforcement learning for variable selection in a branch and bound algorithm. In: Hebrard, E., Musliu, N. (eds.) CPAIOR 2020. LNCS, vol. 12296, pp. 176–185. Springer, Cham (2020). https://doi.org/10.1007/978-3-030-58942-4_12

20. Fréchette, A., Newman, N., Leyton-Brown, K.: Solving the station repacking problem. In: Proceedings of the 30th AAAI Conference on Artificial Intelligence, AAAI 2016, pp. 702–709 (2016)

21. Gasse, M., Chételat, D., Ferroni, N., Charlin, L., Lodia, A.: Exact combinatorial optimization with graph convolutional neural networks. In: Proceedings of the 33rd International Conference on Neural Information Processing Systems, NeurIPS 2019, pp. 15580–15592 (2019)

22. Hamilton, W., Ying, Z., Leskovec, J.: Inductive representation learning on large graphs. In: Proceedings of the 31st International Conference on Neural Information Processing Systems, NeurIPS 2017, pp. 1024–1034 (2017)

23. Hartford, J.S., Graham, D.R., Leyton-Brown, K., Ravanbakhsh, S.: Deep models of interactions across sets. In: Proceedings of the 35th International Conference on Machine Learning, ICML 2018, vol. 80, pp. 1914–1923 (2018)

24. Heule, M., van Maaren, H.: Look-ahead based sat solvers. Handb. Satisf. **185**, 155–184 (2009)

25. Heule, M., van Maaren, H.: march_hi. In: SAT Competition 2009: Solver and Benchmark Descriptions, SAT 2009, pp. 27–28 (2009)

26. Heule, M.J.H., Kullmann, O., Wieringa, S., Biere, A.: Cube and conquer: guiding CDCL SAT solvers by lookaheads. In: Eder, K., Lourenço, J., Shehory, O. (eds.) HVC 2011. LNCS, vol. 7261, pp. 50–65. Springer, Heidelberg (2012). https://doi.org/10.1007/978-3-642-34188-5_8

27. Hoos, H.H.: Automated algorithm configuration and parameter tuning. In: Hamadi, Y., Monfroy, E., Saubion, F. (eds.) Autonomous Search, pp. 37–71. Springer, Heidelberg (2011). https://doi.org/10.1007/978-3-642-21434-9_3

28. Hottung, A., Tanaka, S., Tierney, K.: Deep learning assisted heuristic tree search for the container pre-marshalling problem. Comput. Oper. Res. **113**, 104781 (2020)

29. Huang, J., Darwiche, A.: A structure-based variable ordering heuristic for SAT. In: Proceedings of the 18th International Joint Conference on Artificial intelligence, IJCAI 2003, pp. 1167–1172 (2003)

30. Kearns, M., Mansour, Y., Ng, A.Y.: A sparse sampling algorithm for near-optimal planning in large Markov decision processes. Mach. Learn. **49**(2), 193–208 (2002)

31. Keszocze, O., Schmitz, K., Schloeter, J., Drechsler, R.: Improving SAT solving using Monte Carlo tree search-based clause learning. In: Drechsler, R., Soeken, M. (eds.) Advanced Boolean Techniques, pp. 107–133. Springer, Cham (2020). https://doi.org/10.1007/978-3-030-20323-8_5

32. Khalil, E., Le Bodic, P., Song, L., Nemhauser, G., Dilkina, B.: Learning to branch in mixed integer programming. In: Proceedings of the AAAI Conference on Artificial Intelligence, vol. 30 (2016)

33. Khalil, E.B., Vaezipoor, P., Dilkina, B.: Finding backdoors to integer programs: a Monte Carlo tree search framework. In: Proceedings of the 36th AAAI Conference on Artificial Intelligence, AAAI 2022, pp. 1–10 (2022)

34. Kilby, P., Slaney, J., Thiébaux, S., Walsh, T.: Estimating search tree size. In: Proceedings of the 21st National Conference of DArtificial Intelligence, AAAI 2006, pp. 1014–1019 (2006)

35. Kingma, D.P., Ba, J.: Adam: A method for stochastic optimization. In: Proceedings of the 3rd International Conference on Learning Representations, pp. 1–15, ICLR 2014 (2014)

36. Knuth, D.E.: Estimating the efficiency of backtrack programs. Math. Comput. **29**(129), 122–136 (1975)

37. Kocsis, L., Szepesvári, C.: Bandit based Monte-Carlo planning. In: Fürnkranz, J., Scheffer, T., Spiliopoulou, M. (eds.) ECML 2006. LNCS (LNAI), vol. 4212, pp. 282–293. Springer, Heidelberg (2006). https://doi.org/10.1007/11871842_29

38. Kotthoff, L.: Algorithm selection for combinatorial search problems: a survey. In: Bessiere, C., De Raedt, L., Kotthoff, L., Nijssen, S., O'Sullivan, B., Pedreschi, D. (eds.) Data Mining and Constraint Programming. LNCS (LNAI), vol. 10101, pp. 149–190. Springer, Cham (2016). https://doi.org/10.1007/978-3-319-50137-6_7

39. Kurin, V., Godil, S., Whiteson, S., Catanzaro, B.: Can q-learning with graph networks learn a generalizable branching heuristic for a SAT solver? In: Proceedings of the 34th International Conference on Neural Information Processing Systems, NeurIPS 2020, pp. 9608–9621 (2019)

40. Lagoudakis, M.G., Littman, M.L.: Learning to select branching rules in the DPLL procedure for satisfiability. Electron. Notes Discrete Math. **9**, 344–359 (2001)
41. Lederman, G., Rabe, M., Lee, E.A., Seshia, S.A.: Learning heuristics for quantified Boolean formulas through reinforcement learning. In: Proceedings of the 8th International Conference on Learning Representations, ICLR 2019, pp. 1–18 (2019)
42. Liberatore, P.: On the complexity of choosing the branching literal in DPLL. Artif. Intell. **116**(1–2), 315–326 (2000)
43. Liberatore, P.: Complexity results on DPLL and resolution. ACM Trans. Comput. Log. (TOCL) **7**(1), 84–107 (2006)
44. Lobjois, L., Lemaître, M.: Branch and bound algorithm selection by performance prediction. In: Proceedings of the 15th National/10th Conference on Artificial Intelligence, AAAI 1998, pp. 353–358 (1998)
45. Loth, M., Sebag, M., Hamadi, Y., Schoenauer, M.: Bandit-based search for constraint programming. In: Schulte, C. (ed.) CP 2013. LNCS, vol. 8124, pp. 464–480. Springer, Heidelberg (2013). https://doi.org/10.1007/978-3-642-40627-0_36
46. Maddison, C.J., Huang, A., Sutskever, I., Silver, D.: Move evaluation in Go using deep convolutional neural networks. arXiv preprint arXiv:1412.6564, pp. 1–8 (2014)
47. Moskewicz, M.W., Madigan, C.F., Zhao, Y., Zhang, L., Malik, S.: Chaff: Engineering an efficient SAT solver. In: Proceedings of the 38th Annual Design Automation Conference, DAC 2001, pp. 530–535 (2001)
48. Nair, V., et al.: Solving mixed integer programs using neural networks. arXiv preprint arXiv:2012.13349, pp. 1–57 (2020)
49. Parsonson, C.W., Laterre, A., Barrett, T.D.: Reinforcement learning for branch-and-bound optimisation using retrospective trajectories. arXiv preprint arXiv:2205.14345 (2022)
50. Paszke, A., et al.: PyTorch: an imperative style, high-performance deep learning library. In: Proceedings of the 33rd International Conference on Neural Information Processing Systems, NeurIPS 2019, pp. 8024–8035 (2019)
51. Previti, A., Ramanujan, R., Schaerf, M., Selman, B.: Monte-Carlo style UCT search for Boolean satisfiability. In: Pirrone, R., Sorbello, F. (eds.) AI*IA 2011. LNCS (LNAI), vol. 6934, pp. 177–188. Springer, Heidelberg (2011). https://doi.org/10.1007/978-3-642-23954-0_18
52. Purdom, P.W.: Tree size by partial backtracking. SIAM J. Comput. **7**(4), 481–491 (1978)
53. Rosin, C.D.: Multi-armed bandits with episode context. Ann. Math. Artif. Intell. **61**(3), 203–230 (2011)
54. Roussel, O.: Controlling a solver execution with the runsolver tool. J. Satisf. Boolean Model. Comput. **7**(4), 139–144 (2011)
55. Ryvchin, V., Nadel, A.: Maple_LCM_Dist_ChronoBT: featuring chronological backtracking. In: Proceedings of SAT Competition 2018—Solver and Benchmark Descriptions, vol. B-2018-1, p. 29. Department of Computer Science Series of Publications B (2018)
56. Scavuzzo, L., et al.: Learning to branch with tree MDPs. Adv. Neural. Inf. Process. Syst. **35**, 18514–18526 (2022)
57. Schloeter, J.: A Monte Carlo tree search based conflict-driven clause learning SAT solver. In: Proceedings of the 2017 INFORMATIK Conference, INFORMATIK 2017, pp. 2549–2560 (2017)
58. Schrittwieser, J., Antonoglou, I., Hubert, T., et al.: Mastering Atari, Go, chess and shogi by planning with a learned model. Nature **588**(7839), 604–609 (2020)
59. Selsam, D., Bjørner, N.: Guiding high-performance SAT solvers with Unsat-core predictions. In: Janota, M., Lynce, I. (eds.) SAT 2019. LNCS, vol. 11628, pp. 336–353. Springer, Cham (2019). https://doi.org/10.1007/978-3-030-24258-9_24
60. Silver, D., Huang, A., Maddison, C.J., et al.: Mastering the game of Go with deep neural networks and tree search. Nature **529**, 484–489 (2016)

61. Silver, D., Hubert, T., Schrittwieser, J., et al.: A general reinforcement learning algorithm that masters chess, shogi, and Go through self-play. Science **362**(6419), 1140–1144 (2018)
62. Silver, D., Schrittwieser, J., Simonyan, K., et al.: Mastering the game of Go without human knowledge. Nature **550**, 354–359 (2017)
63. Song, W., Cao, Z., Zhang, J., Xu, C., Lim, A.: Learning variable ordering heuristics for solving constraint satisfaction problems. Eng. Appl. Artif. Intell. **109**, 104603 (2022)
64. Spence, I.: sgen1: A generator of small but difficult satisfiability benchmarks. ACM J. Exp. Algorithmics **15**, 1.1–1.15 (2010)
65. Sutskever, I., Nair, V.: Mimicking Go experts with convolutional neural networks. In: Kůrková, V., Neruda, R., Koutník, J. (eds.) ICANN 2008. LNCS, vol. 5164, pp. 101–110. Springer, Heidelberg (2008). https://doi.org/10.1007/978-3-540-87559-8_11
66. Sutton, R.S., Barto, A.G.: Reinforcement Learning: An Introduction. MIT Press, Cambridge (2018)
67. Tönshoff, J., Kisin, B., Lindner, J., Grohe, M.: One model, any CSP: graph neural networks as fast global search heuristics for constraint satisfaction. arXiv preprint arXiv:2208.10227, pp. 1–23 (2022)
68. Vaezipoor, P., et al.: Learning branching heuristics for propositional model counting. In: Proceedings of the 35th AAAI Conference on Artificial Intelligence, AAAI 2021, pp. 12427–12435 (2021)
69. Wang, W., Hu, Y., Tiwari, M., Khurshid, S., McMillan, K., Miikkulainen, R.: NeuroComb: improving sat solving with graph neural networks (2022)
70. Yolcu, E., Póczos, B.: Learning local search heuristics for Boolean satisfiability. In: Proceedings of the 33rd International Conference on Neural Information Processing Systems, NeurIPS 2019, pp. 7992–8003 (2019)
71. Zook, A., Harrison, B., Riedl, M.O.: Monte-Carlo tree search for simulation-based strategy analysis. arXiv preprint arXiv:1908.01423, pp. 1–9 (2019)

A Hybrid Approach Integrating Generalized Arc Consistency and Differential Evolution for Global Optimization

Mariane R. S. Cassenote[(✉)] [ID], Guilherme A. Derenievicz[ID], and Fabiano Silva[ID]

Informatics Department, Federal University of Paraná, Curitiba, Brazil
{mrscassenote,guilherme,fabiano}@inf.ufpr.br

Abstract. Domain filtering techniques such as Generalized Arc Consistency (GAC) are powerful tools to prune partially inconsistent assignments in a branch-and-prune scheme. Although such techniques are usually applied to exact approaches, we show that metaheuristic algorithms can benefit from domain filtering. In this paper, we propose a variation of the well-known AC3 algorithm that iteratively approximates GAC over the structural information of the constraint network. We integrate the algorithm with an interval-based Differential Evolution (DE) to tackle numerical constrained global optimization problems. Our method consists of a hybrid approach: at first, the interval-based DE iteratively explores the search space generating a population of interval boxes; then, these boxes are pruned by the local consistency algorithm; at last, a real-based DE is used as a local search within each box. The experimental results, over a subset of COCONUT Benchmark instances with up to 100 dimensions, show that the use of the proposed domain filtering to prune the boxes can improve the quality of the solutions by 37.15%, at the cost of 7.82% in processing time. The results indicate the benefits of integrating constraint propagation techniques and metaheuristic algorithms to tackle optimization problems.

Keywords: Generalized Arc Consistency · Metaheuristics · Global Optimization

1 Introduction

In recent decades, many algorithms have been reported combining several components from different areas of optimization research [5,28,33]. The primary motivation for hybridization between algorithms is to explore their particular advantages and complementary nature. Among several different approaches, some hybrid methods have been proposed integrating constraint programming techniques, such as branch-and-prune, domain filtering, and constraint propagation, with metaheuristics [1,8,26,36]. Despite the promising results obtained by these hybridizations, there is a lack of studies evaluating the benefit of constraint propagation in Differential Evolution (DE) metaheuristic, one of the most used evolutionary algorithms to tackle global optimization problems.

A *Numerical Constrained Global Optimization Problem* (NCOP) consists of finding an assignment of values to a set of variables $\{x_1, x_2, \ldots, x_D\}$ that minimizes an objective function $f : \mathbb{R}^D \mapsto \mathbb{R}$ subject to a set of numerical constraints $g_j(x_{j_1}, \ldots, x_{j_k}) \leq 0$ and $x_i \in X_i \subseteq \mathbb{R}$, where X_i is the *domain* of the variable x_i. An assignment of values that satisfies all constraints is called a *feasible solution*. A feasible solution that minimizes the objective function is called an *optimum solution*.

Since the late eighties, *interval arithmetic* [25] has been used to tackle NCOPs by combining consistency techniques with *Branch and Bound* (B&B) algorithms [2,19,22]. Generally, such methods are composed of three main phases that are repeatedly applied until a solution is found:

- **Pruning phase**: domain filtering algorithms that guarantee some level of local consistency are applied (e.g., AC3 [23]), excluding partial assignments that surely do not constitute a feasible solution to the problem.
- **Local search phase**: cheap search methods are used to detect if the current domain contains a feasible solution of the instance that can be used as an upper bound for the optimum solution.
- **Exploration phase**: the current domain is partitioned to continue the search recursively.

This paper proposes a hybrid optimizer where the exploration phase is done by a stochastic method of domain segmentation. The search space is explored through an interval-based DE [9] that generates a population of promising interval boxes (a set of subdomains). Furthermore, we propose a variation of the well-known AC3 algorithm [23] that iteratively approximates *Generalized Arc Consistency* (GAC) over the structural information of the instance provided by a decomposition of the constraint network called *epiphytic decomposition* [14]. This algorithm is used in the pruning phase to either detect that a subdomain is inconsistent (i.e., there is no feasible solution within it) or narrow the search space.

Finally, as a local search phase, we apply another DE, here called real-based DE, which uses a population of points to explore the narrowed box of the subdomain. The best point found represents the box quality regarding constraint violation and objective function values. If such point is a feasible solution, it can be used to update the upper bound of the optimal solution that GAC will propagate in the next pruning phase. Moreover, according to the interval-based DE policy, the box's quality is used in the next exploration phase to generate new boxes. With this, we seek an appropriate compromise between solution quality and reasonable computational time.

The experimental evaluation on 157 instances with up to 100 dimensions extracted from the COCONUT Benchmark [29] demonstrates that the advantages of incorporating a pruning phase based on GAC are not limited to exact optimization scenarios. The addition of GAC domain filtering significantly improved the performance of our stochastic optimization method, as it directs computational resources more assertively toward promising regions of the search space.

The remainder of this paper is organized as follows: Sect. 2 lists some related works. Section 3 contains basic definitions of constraint processing and local consistency techniques. Section 4 provides details of the proposed GAC domain filtering algorithm and the implementation of our hybrid approach. The experimental analysis is presented in Sect. 5, and Sect. 6 concludes this work.

2   Related Works

Several optimizers that benefit from integrating rigorous methods and metaheuristics have been recently proposed. In combinatorial optimization, Cotta and Troya [13] integrated a B&B algorithm into the recombination operation of an evolutionary algorithm to select the optimal candidate solution. Similarly, Ozkan et al. [26] proposed hybridizations involving B&B, Simulated Annealing (SA), and Genetic Algorithms (GA) to repair infeasible solutions or improve their quality.

Zhang and Liu [39] introduced a hybrid interval algorithm that combines B&B with a GA. Similarly, Bunnag [8] presented combinations of B&B algorithms and stochastic search for unconstrained numerical optimization problems, including hybridizations with SA and GA. In both cases, the approaches collaborate to update the priority queue of boxes to be explored and exchange information about the search domain and the optimal solution's upper bound.

Handling the economic emissions load dispatch problem, Gupta and Ray [17] introduced a DE method utilizing three-stage interval analysis. In the first stage, interval arithmetic establishes the limits of the objective function. Second, an interval B&B algorithm is applied to the instance domain until its boxes reach a predefined minimum size. Then, the central points of these boxes are defined as the initial population of the DE. In the third stage, a process to shrink the boxes is applied to iteratively update the boundaries of the global optimum. Alliot et al. [1] combined a GA with a B&B to exchange upper bounds and solutions through a shared memory mechanism.

Vanaret et al. [36] proposed an exact hybrid solver in which a B&B cooperates with a DE algorithm. The two independent algorithms run in parallel and exchange bounds, solutions, and search space via message passing. The collaboration process can be broken down into three stages: 1) whenever the DE finds a new feasible solution, it can update the upper bound of the objective function; 2) whenever B&B finds a better point solution (e.g., the center of a box), it is injected into the DE population; 3) an exploration strategy periodically reduces the DE search space, then regenerates the population in the new search space. To the best of our knowledge, this is the closest related work to the hybridization presented in this paper.

In [10], we introduced the Interval Differential Evolution (InDE) solver to integrate the DE metaheuristic with OGRe [14]. This method uses interval-based techniques and exploits structural information encoded in the instance's constraint network. InDE extends typical DE operators to intervals, taking advantage of OGRe's core functionality to identify a critical subset of variables for the search process, with the remaining variables evaluated through constraint propagation. Furthermore, local consistency techniques are employed to prune infeasible solutions.

More recently, we presented I2DE [9], an Improved Interval Differential Evolution, which extends InDE and the operators supported by the OGRe multi-interval core, which allowed us to tackle a greater diversity of NCOP instances. Furthermore, we presented improved versions of three interval evolutionary operators and incorporated several heuristics from recent solvers. An experimental analysis performed on the COCONUT [29] and CEC2018 [37] benchmarks revealed that our I2DE significantly outperforms InDE, OGRe, and a proposed real-based version of I2DE. Such real-based DE obtained better results than several recent state-of-the-art metaheuristic solvers.

3   Background

In this paper, we tackle NCOP instances which objective function and constraints can be rewritten as a network of ternary constraints[1] [16] of the form $x = y \circ z$ or $x = \diamond y$, where $\circ$ ($\diamond$) is a well-defined binary (unary) operation. For example, the NCOP instance $\min (x - 1)^2 + (y - 1)^2$ s.t. $\{x + y \leq 1, x \geq 0\}$ can be rewritten by adding auxiliary variables: $\min z$ s.t. $\{z = a_1 + a_2, a_1 = a_3{}^2, a_2 = a_4{}^2, a_3 = x - 1, a_4 = y - 1, c = x + y, x \in [0, \infty), c \in (-\infty, 1]\}$, where the new variable z encodes the objective function in the constraint network.

An interval X is a continuous set of real numbers. For instance, $X = [\underline{x}, \overline{x}] = \{x \in \mathbb{R} \mid \underline{x} \leq x \leq \overline{x}\}$ is a closed interval, where $\underline{x}, \overline{x} \in \mathbb{R}$ are the *endpoints* of X. Given intervals X, Y and Z the *interval extension* of any binary (unary) operation $\circ$ ($\diamond$) well defined in $\mathbb{R}$ is defined by:

$$X \circ Y = \{x \circ y \mid x \in X, y \in Y \text{ and } x \circ y \text{ is defined in } \mathbb{R}\},$$
$$\diamond Z = \{\diamond z \mid z \in Z \text{ and } \diamond z \text{ is defined in } \mathbb{R}\}.$$

It is possible to compute $X \circ Y$ or $\diamond Z$ for all algebraic and common transcendental functions only by analyzing the endpoints of X and Y or Z [18, 25], e.g., $[\underline{x}, \overline{x}] + [\underline{y}, \overline{y}] = [\underline{x} + \underline{y}, \overline{x} + \overline{y}]$, $[\underline{x}, \overline{x}] \cdot [\underline{y}, \overline{y}] = [\min\{\underline{x}\underline{y}, \underline{x}\overline{y}, \overline{x}\underline{y}, \overline{x}\overline{y}\}, \max\{\underline{x}\underline{y}, \underline{x}\overline{y}, \overline{x}\underline{y}, \overline{x}\overline{y}\}]$, $2^{[\underline{z}, \overline{z}]} = [2^{\underline{z}}, 2^{\overline{z}}]$, etc.

The sets $X \circ Y$ and $\diamond Z$ can be intervals, empty sets or multi-intervals[2]. A multi-interval $\mathcal{X}$ is an ordered set of disjointed intervals. Given multi-intervals $\mathcal{X}$, $\mathcal{Y}$ and $\mathcal{Z}$, the operation $\mathcal{X} \circ \mathcal{Y}$ is the tightest multi-interval that contains $\{X \circ Y \mid X \in \mathcal{X}, Y \in \mathcal{Y}\}$, and $\diamond \mathcal{Z}$ is the tightest multi-interval that contains

---

[1] A constraint is said to be *ternary* if it involves at most three variables.
[2] For example, $[1, 2]/[-1, 1] = (-\infty, -1] \cup [1, \infty)$.

$\{\diamond Z \mid Z \in \mathcal{Z}\}$. The *hull* of a multi-interval $\mathcal{X}$ is the tightest interval that contains $\bigcup_{X \in \mathcal{X}} X$. A *box* is a tuple of (multi-)intervals.

In the constraint programming field, local consistency allows the narrowing of variable domains by excluding values that can not satisfy the constraint network. The most important of these consistencies, called (generalized) arc consistency (GAC), guarantees that any valuation of a single variable can be extended to other variables while satisfying a constraint [24]. In the continuous case, a ternary constraint $x = y \circ_1 z$ is GAC w.r.t. a box (X, Y, Z) iff $X \subseteq Y \circ_1 Z$, $Y \subseteq X \circ_2 Z$ and $Z \subseteq X \circ_3 Y$, where $\circ_2$ and $\circ_3$ are the inverse operations of $\circ_1$ that hold the condition $(x = y \circ_1 z) \iff (y = x \circ_2 z) \iff (z = x \circ_1 y)$.

If a ternary constraint is not GAC, this propriety can be achieved by computing the intersection of each relevant domain with the respective *projection function*:

$$\texttt{GAC\_revise}(x_1 = x_2 \circ_1 x_3) := \begin{cases} X_1' \leftarrow X_1 \cap (X_2 \circ_1 X_3) \\ X_2' \leftarrow X_2 \cap (X_1' \circ_2 X_3) \\ X_3' \leftarrow X_3 \cap (X_1' \circ_3 X_2'). \end{cases} \tag{1}$$

The box (X_1', X_2', X_3') narrowed by (1) may contain multi-intervals. Relaxed consistencies such as Hull Consistency [3] have been proposed to avoid dealing with multi-intervals, since there is no consensus in the literature about the performance of maintaining a multi-interval representation [11,15,30]. Our proposal uses the multi-interval representation to improve effectiveness in pruning inconsistent values from variable domains.

Many algorithms have been proposed to achieve GAC on the whole network. The well-known AC3 algorithm [23] consists of iteratively applying $\texttt{GAC\_revise}$ on the constraints according to a queue Q. Initially, all network constraints are enqueued in Q. While Q is not empty, a constraint C is dequeued from Q and $\texttt{GAC\_revise}(C)$ is called. If the domain of some variable x is narrowed, all the other constraints over x are enqueued in Q. If the empty domain is generated, an inconsistency is detected and AC3 returns.

Generally, this procedure may only terminate if a maximum number of steps is imposed or the network holds specific structural properties [12,15]. For instance, if the constraint network is acyclic[3] then GAC can be achieved by considering an ordering $d = (C_1, C_2, \ldots, C_m)$ of the constraints such that each constraint C_i has at most one variable in predecessor constraints $C_{j<i}$; it suffices apply $\texttt{GAC\_revise}$ backward and forward along d. We call such ordering d a *valuation ordering* of the constraint network. Even if the network is not acyclic, this procedure can be used if the solution is generated by search along d. In this case, the network will be *directionally arc consistent* (DAC) along d.

---

[3] The structure of a constraint network can be represented by a hypergraph in which vertices are the variables, and for each constraint there is a hyperedge connecting its respective vertices. Therefore, a network is acyclic if its hypergraph is Berge-acyclic [4].

4   A Hybridization of Constraint Propagation and Metaheuristics

This work proposes hybridizing constraint propagation techniques and meta-heuristic algorithms to tackle NCOP. Specifically, our objective is to evaluate the ability of GAC domain filtering to improve the quality of solutions found by a stochastic optimization method.

Our approach manipulates subdomains (boxes) of the search space by assigning intervals to variables. Any point inside a box can be considered a candidate solution for the instance. Initially, a predefined number of boxes are generated through a recursive binary segmentation of the search space. As the initial boxes are formed by the leaf nodes of the segmentation tree, full coverage of the search space is guaranteed.

Inspired by the schemes used in the Branch-and-Prune algorithms, our approach is divided into three repeatedly applied phases. The pruning phase involves applying GAC over the ternary representation to prune suboptimal or infeasible regions of the box domain. Next, we apply a real-based DE to find a point whose objective function and constraint violation values represent the quality of the box and may update the upper bound of the instance's optimal solution. Finally, an interval-based DE iteratively generates new boxes to be explored. These last two phases are independent of the ternary representation of the instance. The process stops when a predefined maximum number of evaluated solutions is reached. Some details are discussed in the following subsections.

4.1   Pruning Phase

The GAC domain filtering we propose (Algorithm 1) uses structural information of the constraint network to apply a combination of DAC and AC3. In [14] it was proposed an *epiphytic decomposition* of a constraint network as a tuple $(\mathcal{A}, \Omega, t)$, where $\mathcal{A}$ is an ordered set of acyclic networks obtained by removing from the network a set of constraints Ω, and $t : \Omega \mapsto V_\Omega$ is a function that associates each constraint $C \in \Omega$ with one of its variables $t(C)$ satisfying the following: if $t(C)$ belongs to the network A_i then the remaining variables of C belongs to previous networks $A_{j<i}$ and there is no other constraint $C' \in \Omega$ such that $t(C')$ belongs to A_i.

A constraint network may have many different epiphytic decompositions, but it is desired $|\Omega|$ to be minimum [14]. We use a heuristic approach that tries to maximize the size of the first acyclic network of the decomposition because it is cheaper to enforce GAC in acyclic networks. Figure 1 shows an epiphytic decomposition of the Rosenbrock function $f(x,y) = 100(x - y^2)^2 + (1 - y)^2$ encoded as a ternary network, where $\Omega = \{C_7\}$.

First, the usual DAC pruning is applied in each acyclic network identified by the epiphytic decomposition. Except for numerical errors due to float-pointing arithmetic, GAC is efficiently achieved in all sub-networks since we maintain a multi-interval representation of the box. However, the entire network may not

---

Algorithm 1. GAC domain filtering

Require: Epiphytic decomposition $(\mathcal{A}, \Omega, t)$ of the ternary constraint network, variable z that encodes the objective function, upper bound u, box B and MAX number of steps.
Ensure: In any moment, if the empty interval is obtained returns **conflict**. Otherwise, B is narrowed.

```
1  D_z ← D_z ∩ (−∞, u)                              ▷ narrow the domain of z using the upper bound
2  for all A_i ∈ A do                               ▷ directional GAC in A_i
3     let (C_1, C_2, ..., C_m) be a valuation ordering of A_i
4     GACrevise(C_j, B), for each j = m ... 1
5     GACrevise(C_j, B), for each j = 1 ... m
6  end for
7  Create empty queue Q
8  for all C_j ∈ Ω do                               ▷ process the Ω set of the decomposition
9     Q ← GACrevise(C_j, B)                         ▷ enqueue variables that have been contracted
10 end for
11 k ← 0
12 while Q is not empty and k < MAX do              ▷ AC3-like algorithm
13    x ← dequeue(Q)
14    for all constraint C_j over x do
15       Q ← GACrevise(C_j, B)
16    end for
17    k ← k + 1
18 end while
19 return Hull(B)
```

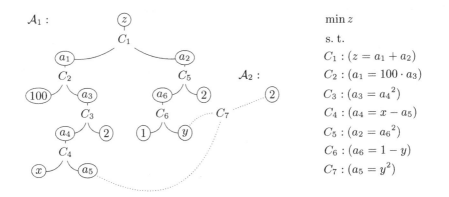

Fig. 1. An epiphytic decomposition of the ternary encoded Rosenbrock function.

be consistent because the constraints in the Ω set of the decomposition were not processed. Therefore, `GAC_revise` is applied to all these constraints, initiating the queue Q of variables to be processed by AC3. Our AC3 implementation uses a counter of processed variables: in each iteration all constraints over a variable are processed. The total number of calls to `GAC_revise` depends on the structure of the constraint network, since dense networks may have, in general, more constraints per variable.

It is worth noting that Algorithm 1 takes an interval box as input, but during the GACrevise process, it may generate multi-intervals[4]. We maintain the multi-

---

[4] Considering the network $x = y^2$ and $y = [+1, +1]/[-1, +1]$, hull consistency narrows the domains to $y = (-\infty, +\infty)$ and $x = [0, +\infty)$, while our approach obtains a tighter result $y = (-\infty, +\infty)$ and $x = [1, +\infty)$. This is because during the algorithm, y is represented by the multi-interval $y = \{(-\infty, -1], [+1, +\infty)\}$.

interval representation over the entire process, but return an interval box where each interval is the hull of the respective multi-interval domain. With this, we simplify the operations of the interval-based DE in the exploration phase.

4.2   Local Search Phase

Using interval boxes to represent subdomains enables local search methods to identify search space regions with feasible solutions. In our previous work, I2DE [9], we initially enforced GAC on the boxes. Then, we tried to instantiate the constraint network in a backtrack-free fashion, starting with the initial valuation $z \leftarrow \underline{z}$, where $Z = [\underline{z}, \overline{z}]$ is the domain interval of the variable that encodes the objective function. Although this instantiation provides suitable reference points, local search methods that widely explore the box could provide more accurate estimates of its quality.

In this way, we propose a local search phase that uses a real-based DE, a populational metaheuristic inspired by the natural selection process. Storn and Price [31] presented it primarily to solve global optimization problems in several areas [27]. Due to its stochastic nature, DE does not guarantee the optimal solution but usually finds good solutions in reasonable computational time.

The best point found by DE within a box is adopted as a representative of the box's quality, and its cost and constraint violation values are calculated considering only the original variables of the instance. From this, it becomes possible to define which boxes are most promising and should be better explored in the search process. Furthermore, the cost values of the feasible points found by DE are used as an upper bound of the optimal solution for the instance, which tends to speed up the domain filtering process.

As an evolutionary metaheuristic, DE runs in a loop of G generations (iterations) over a population of NP candidate solutions (individuals). The initial population is randomly generated according to a uniform distribution over the box domain. At each generation, NP new individuals are generated by applying mutation, crossover, and selection operators. The most popular mutation operator is DE/rand/1 (Eq. 2), where each individual $\mathbf{x}_i$ generates a new candidate solution $\mathbf{v}_i$:

$$\mathbf{v}_i = \mathbf{r}_1 + F \cdot (\mathbf{r}_2 - \mathbf{r}_3), \tag{2}$$

where F is the scaling factor and $\mathbf{r}_1$, $\mathbf{r}_2$ and $\mathbf{r}_3$ are three mutually distinct individuals randomly selected from the population. As proposed in IUDE [35], we use other two mutation operators: DE-current-to-rand/1 [20] (Eq. 3) and DE/current-to-$p$best/1 [38] (Eq. 4).

$$\mathbf{v}_i = \mathbf{x}_i + s \cdot (\mathbf{r}_1 - \mathbf{x}_i) + F \cdot (\mathbf{r}_2 - \mathbf{r}_3), \tag{3}$$

$$\mathbf{v}_i = \mathbf{x}_i + F \cdot (\mathbf{r}_p - \mathbf{x}_i) + F \cdot (\mathbf{r}_1 - \mathbf{r}_2), \tag{4}$$

where s is a uniformly distributed random number between 0 and 1, and $\mathbf{r}_p$ is an individual randomly chosen from the $p = [12.5, 25.0]\%$ best solutions on the current population.

The crossover operator is employed on the solution v_i to generate a novel candidate solution u_i utilizing a probability defined by the crossover rate CR. An integer $d \in [1, D]$ is the starting point, where D is the number of instance dimensions. Another integer value l is also chosen from $[1, D]$ with probability $P(l \geq L) = CR^{L-1}$ for any $L > 0$ which denotes how many consecutive decision variables are selected from v_i starting at d position. Then, the final candidate solution u_i is generated as follows:

$$u_{ij} = \begin{cases} v_{ij}, & \text{if } j \in \{m_D(d), m_D(d+1), \ldots, m_D(d+l-1)\}, \\ x_{ij}, & \text{otherwise,} \end{cases} \tag{5}$$

where $m_D(n) = 1 + ((n-1) \mod D)$ allows to iterate cyclically through the solution dimensions. The values of the parameters F and CR are automatically defined according to the heuristics proposed by [6, 7, 34].

Finally, a selection operator based on ε-constrained method [32] is performed to define whether x_i or u_i will be selected for the next generation. ε-comparisons are a lexicographic order in which constraint violation precedes the objective function value. This precedence is adjusted by the ε-tolerance, which decreases until the execution exceeds a predefined threshold. From this point forward, ε is set to 0 to prioritize solutions with minimal constraint violation.

The population is sorted using the ε-method and subdivided into two subpopulations A and B of size $NP/2$, allowing A to contain the best candidate solutions. The three mutation strategies are systematically applied to prioritize exploring these promising solutions and accelerate the convergence process. The superior performance among the generated candidate solutions is compared with the original solution during the selection operation. In sub-population B, an adaptive scheme [35] is employed to select mutation strategies dynamically.

The mutation, crossover, and selection operators are executed until a maximum number of evaluated individuals is reached or if there is no solution improvement over a predefined number of generations.

4.3   Exploration Phase

To explore the interval boxes that represent the subdomains of the instance, we employ a version of DE that operates on a population of boxes. This interval-based DE replaces the mutation and crossover operators with corresponding interval versions. Since an interval $X = [\underline{x}, \overline{x}]$ can be defined by its *width* $\omega(X) = \overline{x} - \underline{x}$ and *midpoint* $\mu(X) = (\underline{x} + \overline{x})/2$, mutation operators can be applied to the midpoints and extended to the widths of the intervals. The interval version of DE/rand/1 (Eq. 2) combines the boxes r_1, r_2 and r_3 to generate the candidate box v_i. The j-th dimension of v_i is defined by:

$$\mu(v_{ij}) = \mu(r_{1j}) + F \cdot (\mu(r_{2j}) - \mu(r_{3j})), \tag{6}$$

$$\omega(v_{ij}) = \omega(r_{1j}) \cdot (1 + F \cdot ((\omega(r_{2j})/\omega(r_{3j})) - 1)).$$

The other two mutation operators are defined similarly. The interval version of DE/current-to-rand/1 (Eq. 3) combines the box x_i, of the current population, with three other randomly selected boxes r_1, r_2 and r_3. The new box is defined by:

$$\mu(v_{ij}) = \mu(x_{ij}) + s \cdot (\mu(r_{1j}) - \mu(x_{ij})) + F \cdot (\mu(r_{2j}) - \mu(r_{3j})), \qquad (7)$$
$$\omega(v_{ij}) = \omega(x_{ij}) \cdot (1 + s \cdot ((\omega(r_{1j})/\omega(x_{ij})) - 1)) \cdot (1 + F \cdot ((\omega(r_{2j})/\omega(r_{3j})) - 1)),$$

where s is a random number between 0 and 1. This strategy does not use a crossover operator.

Finally, the interval version of DE/current-to-$p$best/1 (Eq. 4) uses the box x_i and the randomly selected boxes r_1 and r_2, while r_p is selected among the p best boxes in the population:

$$\mu(v_{ij}) = \mu(x_{ij}) + F \cdot (\mu(r_{pj}) - \mu(x_{ij})) + F \cdot (\mu(r_{1j}) - \mu(r_{2j})), \qquad (8)$$
$$\omega(v_{ij}) = \omega(x_{ij}) \cdot (1 + F \cdot ((\omega(r_{pj})/\omega(x_{ij})) - 1)) \cdot (1 + F \cdot ((\omega(r_{1j})/\omega(r_{2j})) - 1)).$$

In the same way as in the local search phase, the values of the parameters F and CR are automatically defined according to the heuristics proposed in [6, 7, 34]. The crossover operator is an interval version of Eq. 5. The population of boxes is kept sorted and subdivided into populations A and B of size $NP/2$, which allows the search process to prioritize exploration around the most promising boxes.

After generating each new box, GAC filtering is applied to prune (Sect. 4.1) suboptimal or infeasible regions of the search space. It is important to note that, unlike I2DE, the mutation and crossover operations are only applied to intervals that represent the instance's original variables, ensuring that the ternary representation of the instance is restricted to the GAC domain filtering process. The box's objective function and constraint violation values obtained in the local search phase guide the choice of which boxes should be further explored.

5   Experimental Results

Our experimental evaluation analyzes the impact of GAC domain filtering in the hybridization of constraint propagation with a metaheuristic, focusing on the quality of the solutions and processing time. Therefore, the first objective is to evaluate the inclusion of GAC domain filtering (Algorithm 1) as an intermediate level between manipulating interval boxes and the local search stage. Additionally, an investigation is carried out into the maximum number of GAC steps required to balance finding suitable solutions and the additional processing time required by this task.

To ensure a fair comparison, all versions of our solver employ the same heuristics described in Sect. 4. However, it is worth noting that the algebraic structure of instances and their ternary representation are only explored during GAC domain filtering. Therefore, in the case of experiments without a pruning phase, our approach became a black-box solver.

In this experimental evaluation, we explore 157 instances of COCONUT Benchmark [29] Library 1 with up to 100 dimensions, different amounts of equality and inequality constraints, and containing operators supported by our multi-interval core [9]. This benchmark suite was chosen due to its wide use in numerical optimization research and because it contains the AMPL (A Mathematical Programming Language) description necessary to explore the structure of instances in a white-box approach.

Due to the stochastic nature of the mutation and crossover operations, all instances were tested in 25 independent runs with different random seeds. A budget of function evaluations is defined for each run, corresponding to the maximum number of candidate solutions to be evaluated. This budget is consumed with each call to Algorithm 1. Furthermore, the number of candidate solutions explored by the local search DE is deducted from this budget. The execution of our approach ends when the budget is completely consumed.

The tests include function evaluation budgets ranging from $1 \times 10^6 \times D$ to $2 \times 10^7 \times D$, where D is the number of variables (dimensions) of the NCOP instance. The population size is $5 \times D$ interval boxes. A maximum of $2 \times 10^3 \times D$ candidate solutions are evaluated for each local search call on a box, and a population of $5 \times D$ candidate solutions is maintained. The local search prematurely terminates if there is no improvement in the candidate solutions over 50 generations. The parameters used to adjust the ε-tolerance were the same as those for I2DE [9]. The experiments ran on a computer with an Intel Xeon E5-4627v2 3.30 GHz processor, 256 GB of RAM, and Debian Linux 11.7.

The first aspect to be considered is the quality of the solutions obtained by our approach. All analyses of this experimental evaluation were based on the average values obtained in each of the 25 runs. Solutions whose sum of constraint violations was less than 10^{-8} were considered feasible. Similarly, instances with an absolute error to the optimal objective function value lower than 10^{-4} were considered optimal.

Figure 2 and Table 1 show the number of feasible and optimal solutions in relation to different maximum numbers of GAC steps (parameter MAX on Algorithm 1) and function evaluation budgets. The line "no pruning" means that GAC domain filtering was not used, and the line "DAC + Ω" means that only directional GAC was applied in the acyclic networks of the epiphytic decomposition and GACrevise was applied in constraints of the Ω set (i.e., $MAX = 0$). Observing the number of instances with feasible (92.99%) and optimal (32.48%) solutions, it is evident that GAC domain filtering can significantly improve the quality of solutions obtained by the version with no pruning (73.24% and 26.11%, respectively). There is an increasing gain concerning the feasibility of the solutions for the maximum number of GAC steps (10^3). Regarding the quantity of optimal solutions, GAC steps of 10^2 and 10^3 exhibit superior performance.

We seek to estimate the improvement obtained by each maximum value of the GAC steps given each function evaluation budget. Thus, we calculated the percentage variation of the average absolute error of feasible solutions from runs with and without GAC pruning. To this end, we have considered three cases:

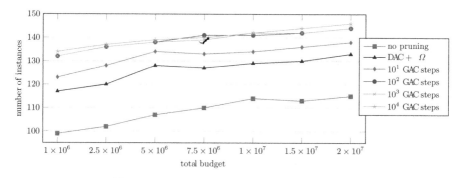

(a) Number of instances with feasible solutions.

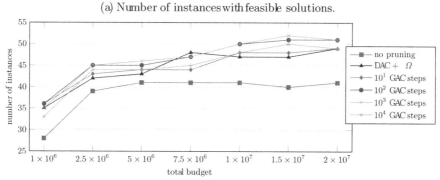

(b) Number of instances with optimal solutions.

Fig. 2. Average number of feasible and optimal instances considering 25 runs.

1. given that both runs obtain feasible solutions, the variation is given by $(E - E_{gac})/\text{Max}\{E, E_{gac}\}$, where E_{gac} and E represent the average absolute error[5] of the version with and without GAC pruning, respectively;
2. if the solution obtained by the version with no pruning is feasible and the solution with GAC pruning is infeasible, the variation is considered as -100%;
3. for cases with a solution that is infeasible when executed with no pruning and feasible when executed with GAC pruning, the variation is considered as $+100\%$.

Figure 3 shows the improvement in the quality of solutions as the number of GAC steps increases. Our approach benefits significantly from the domain filtering process, even with small function evaluation budgets. However, this gain appears more discreet from 10^2 GAC steps onwards, corroborating the information presented in Fig. 2, and may indicate stagnation. It is possible that, after a certain point, the pruning phase has less impact on the search for better solutions or that a smaller number of steps is sufficient to prune suboptimal and unfeasible values from the box.

---

[5] The difference between the value of the objective function for the solution found and the value of the optimal solution provided by COCONUT Benchmark.

Table 1. For each budget and pruning configuration, the table shows the average number of instances with feasible and optimal solutions, the improvement in absolute error, the sum of average processing times, the percentages of: additional time demanded by GAC, inconsistent boxes discarded, average box contraction, and average consistent box contraction, i.e., excluding the discarded ones.

Budget	GAC	Feas	Opt	Error improv. (%)	Time (s)	Time (%)	Inconsist. (%)	Contract. (%)	Contract. consist. (%)
1×10^6	no pruning	99	28	–	6786.54	–	–	–	–
	DAC + Ω	117	35	30.67	6966.16	4.97	14.63	71.74	12.98
	10^1 steps	123	36	32.66	6985.22	5.71	16.11	67.88	14.38
	10^2 steps	132	36	38.92	7234.04	8.53	16.55	68.36	14.78
	10^3 steps	132	35	36.82	7370.23	12.80	17.06	69.26	14.13
	10^4 steps	134	33	38.11	7459.68	16.80	16.78	69.58	13.84
2.5×10^6	no pruning	102	39	–	13030.96	–	–	–	–
	DAC + Ω	120	42	29.21	13926.84	5.69	15.51	71.25	12.53
	10^1 steps	128	43	33.43	13946.99	6.58	16.80	67.58	14.08
	10^2 steps	136	45	38.97	14436.99	8.44	17.03	67.89	14.77
	10^3 steps	136	45	38.76	14658.43	12.39	17.52	68.86	14.02
	10^4 steps	137	44	39.49	15278.27	17.09	17.45	69.03	13.71
5×10^6	no pruning	107	41	–	24406.10	–	–	–	–
	DAC + Ω	128	43	25.48	25152.30	4.37	15.85	71.17	12.50
	10^1 steps	134	44	30.52	24951.21	5.20	17.38	67.47	13.91
	10^2 steps	138	45	35.06	25491.43	6.48	17.31	67.64	14.93
	10^3 steps	138	46	35.72	25896.27	11.09	17.51	68.67	14.31
	10^4 steps	139	44	35.80	27881.36	16.01	17.45	68.83	13.91
7.5×10^6	no pruning	110	41	–	37487.15	–	–	–	–
	DAC + Ω	127	48	26.56	38288.78	4.98	16.07	71.08	12.41
	10^1 steps	133	44	31.62	38095.46	5.78	17.75	67.36	13.87
	10^2 steps	141	47	38.07	38904.27	7.78	17.43	67.55	15.04
	10^3 steps	139	47	36.25	39684.68	12.09	18.07	68.60	14.44
	10^4 steps	140	45	37.25	41792.15	16.68	17.90	68.73	14.07
1×10^7	no pruning	114	41	–	47996.11	–	–	–	–
	DAC + Ω	129	47	25.21	49560.77	4.97	16.25	71.15	12.48
	10^1 steps	134	48	28.19	49385.20	5.63	17.95	67.35	13.92
	10^2 steps	141	50	34.48	50569.65	7.69	17.63	67.52	15.15
	10^3 steps	142	50	36.90	51272.04	12.10	18.18	68.62	14.59
	10^4 steps	142	48	35.48	54263.11	17.04	18.11	68.65	14.21
1.5×10^7	no pruning	113	40	–	71028.06	–	–	–	–
	DAC + Ω	130	47	23.88	72348.05	4.97	16.62	70.65	12.83
	10^1 steps	136	48	27.94	72795.75	5.90	18.46	67.04	13.96
	10^2 steps	142	51	35.74	73624.10	8.01	18.03	67.28	15.29
	10^3 steps	142	52	36.67	74609.19	12.11	18.52	68.46	14.64
	10^4 steps	144	50	37.44	78963.27	17.35	18.42	68.51	14.28
2×10^7	no pruning	115	41	–	94381.01	–	–	–	–
	DAC + Ω	133	49	26.90	96732.07	5.27	16.68	70.37	12.90
	10^1 steps	138	49	33.19	96781.30	6.38	18.80	66.81	14.07
	10^2 steps	144	51	38.80	97982.04	7.81	18.26	67.11	15.63
	10^3 steps	144	51	38.78	99162.37	12.27	18.81	68.29	14.82
	10^4 steps	146	49	39.68	104035.78	17.17	18.60	68.35	14.43

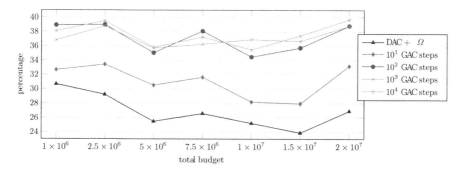

Fig. 3. Percentage of improvement in the absolute error of the objective function obtained by applying GAC domain filtering.

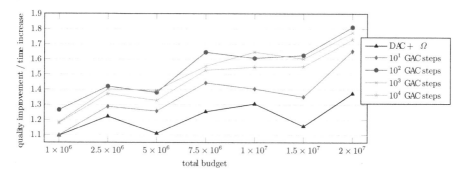

Fig. 4. Relationship between quality gain in the solution and processing time required by GAC-based pruning.

Another critical aspect to be considered is the additional processing time required by the domain pruning phase. Table 1 shows the sum of the average time of 25 runs of each instance over each budget of function evaluations (Time(s)). For a more precise observation, we also show the percentage of additional time required by each value of GAC steps defined versus the implementation without the pruning phase (Time(%)). The best results obtained by no pruning were 115 feasible and 41 optimal solutions, with a budget of 2×10^7 and a time of 94381.01 s. For the 1×10^7 budget, all variants with pruning have better values for feasible and optimal, but with a time of 54263.11 s. Even almost doubling the processing time (47,996.11 s to 94,381.01 s) with no pruning, the solutions found are inferior. The same relation can be observed in other budget pairs.

Finally, to estimate the cost-benefit ratio among different maximum GAC steps, we divided the improvement in the absolute error of the objective function described in Fig. 3 by the increase in processing time required by the application of GAC domain filtering, as depicted in Table 1 and Fig. 4. In general, all configurations of GAC steps exhibit a gradual performance improvement as the budget of function evaluations increases. However, the 10^2 GAC steps configuration was the best cost-benefit ratio for most budgets.

Given the results obtained, we show that applying a GAC-based pruning phase can significantly improve the results obtained by the proposed hybrid approach. As depicted in Table 1, the domain filtering could discard, on average, 16.3% of the boxes by detecting inconsistency and contract 14.3% the size of the remaining boxes on average. Considering that discarded boxes were contracted 100% (narrowed to the empty set), the total average of box contraction was about 69.5%. By pruning suboptimal and infeasible regions of the boxes, the local search phase focuses on the most promising regions of the search space, which provides better estimates of the quality of the boxes and the upper bound of the optimal solution for the instance. Furthermore, whenever the GAC pruning finds a box that surely does not contain the optimal solution for the instance, it is discarded from the evolutionary process before the local search is applied.

6   Conclusion

This study introduced a hybrid optimization approach combining an AC3-like domain filtering algorithm with an interval-based DE and a real-based DE designed to address numerical constrained global optimization problems. An essential contribution of this research involves a modified version of the well-established AC3 algorithm, which iteratively approximates GAC on the structural information derived from an epiphytic decomposition of the constraint network of the instance. In our implementation, this algorithm eliminates suboptimal and infeasible values from the search space. We also employ a DE-based local search to estimate the solution boxes' quality and the optimal solution's upper bound for the given instance.

The experimental analysis on 157 instances with up to 100 dimensions selected from the COCONUT Benchmark reveals that the benefits of using a GAC-based pruning step are not restricted to exact optimization scenarios and can significantly improve the performance of stochastic optimization methods. The results showed that investing processing time in GAC filtering is more effective than increasing the stochastic search evaluation budget.

By pruning inconsistent and suboptimal regions, it becomes possible to direct computational efforts to more promising regions of the search space. We obtained 92.99% and 32.48% of feasible and optimal solutions, respectively, and 39.68% of maximum quality gain in the solution if compared to the solver version without the pruning phase, with a processing time increase of 17.17%. Assuming that a limit of 100 GAC steps is a good cost-benefit configuration, we have an average quality gain of 37.15% and an increase in time of just 7.82%.

Some future works include exploring methods such as DIRECT [21] and B&B for branching boxes to achieve complete and exact search space coverage. Additionally, it is expected to add other constraint programming techniques to prune the domain of instances efficiently without losing solutions.

Acknowledgments. Coordination for the Improvement of Higher Education Personnel (CAPES) - Program of Academic Excellence (PROEX).

Disclosure of Interests. The authors have no competing interests to declare that are relevant to the content of this article.

References

1. Alliot, J.M., Durand, N., Gianazza, D., Gotteland, J.B.: Finding and proving the optimum: cooperative stochastic and deterministic search. In: ECAI 2012, 20th European Conference on Artificial Intelligence, pp. 55–60 (2012)
2. Araya, I., Reyes, V.: Interval branch-and-bound algorithms for optimization and constraint satisfaction: a survey and prospects. J. Glob. Optim. **65**(4), 837–866 (2016)
3. Benhamou, F., Older, W.J.: Applying interval arithmetic to real, integer and boolean constraints. Technical report, BNR, Bell Northern Research (1992)
4. Berge, C.: Graphs and Hypergraphs. Elsevier Science Ltd., Oxford (1985)
5. Blum, C., Raidl, G.R.: Hybrid Metaheuristics: Powerful Tools for Optimization. Springer, Heidelberg (2016). https://doi.org/10.1007/978-3-319-30883-8
6. Brest, J., Maučec, M.S., Bošković, B.: iL-SHADE: improved L-SHADE algorithm for single objective real-parameter optimization. In: 2016 IEEE Congress on Evolutionary Computation (CEC), pp. 1188–1195. IEEE (2016)
7. Brest, J., Maučec, M.S., Bošković, B.: Single objective real-parameter optimization: algorithm jSO. In: 2017 IEEE Congress on Evolutionary Computation (CEC), pp. 1311–1318. IEEE (2017)
8. Bunnag, D.: Combining interval branch and bound and stochastic search. In: Abstract and Applied Analysis, vol. 2014. Hindawi (2014)
9. Cassenote, M.R.S., Derenievicz, G.A., Silva, F.: I2de: Improved interval differential evolution for numerical constrained global optimization. In: Britto, A., Valdivia Delgado, K. (eds.) BRACIS 2021, pp. 186–201. Springer, Heidelberg (2021). https://doi.org/10.1007/978-3-030-91702-9_13
10. Cassenote, M.R.S., Derenievicz, G.A., Silva, F.: Interval differential evolution using structural information of global optimization problems. In: Moura Oliveira, P., Novais, P., Reis, L. (eds.) EPIA Portuguese Conference on Artificial Intelligence, pp. 724–736. Springer (2019). https://doi.org/10.1007/978-3-030-30241-2_60
11. Chabert, G., Trombettoni, G., Neveu, B.: New light on arc consistency over continuous domains. Technical Report. RR-5365, INRIA (2004)
12. Cohen, D.A., Jeavons, P.G.: The power of propagation: when GAC is enough. Constraints **22**(1), 3–23 (2017). https://doi.org/10.1007/s10601-016-9251-0
13. Cotta, C., Troya, J.M.: Embedding branch and bound within evolutionary algorithms. Appl. Intell. **18**(2), 137–153 (2003)
14. Derenievicz, G.A., Silva, F.: Epiphytic trees: relational consistency applied to global optimization problems. In: van Hoeve, W.J. (ed.) CPAIOR 2018, vol. 10848, pp. 153–169. Springer, Cham (2018). https://doi.org/10.1007/978-3-319-93031-2_11
15. Faltings, B.: Arc consistency for continuous variables. Artif. Intell. **65**, 363–376 (1998)
16. Faltings, B., Gelle, E.M.: Local consistency for ternary numeric constraints. In: 15th International Joint Conference on Artificial Intelligence, pp. 392–397 (1997)

17. Gupta, A., Ray, S.: Economic emission load dispatch using interval differential evolution algorithm. In: 4th International Workshop on reliable Engineering Computing (REC 2010). Citeseer (2010)

18. Hansen, E., Walster, G.W.: Global Optimization Using Interval Analysis: Revised and Expanded, vol. 264. CRC Press, Boca Raton (2003)

19. Hansen, E., Walster, G.W.: Global optimization using interval analysis. In: Monographs and Textbooks in Pure and Applied Mathematics. Marcel Dekker, New York (2004)

20. Iorio, A.W., Li, X.: Solving rotated multi-objective optimization problems using Differential Evolution. In: Webb, G.I., Yu, X. (eds.) AI 2004, pp. 861–872. Springer, Heidelberg (2004). https://doi.org/10.1007/978-3-540-30549-1_74

21. Jones, D.R., Martins, J.R.: The direct algorithm: 25 years later. J. Glob. Optim. **79**(3), 521–566 (2021)

22. Kearfott, R.B.: An interval branch and bound algorithm for bound constrained optimization problems. J. Glob. Optim. **2**(3), 259–280 (1992)

23. Mackworth, A.K.: Consistency in networks of relations. Artif. Intell. **8**(1), 99–118 (1977)

24. Mackworth, A.K.: On reading sketch maps. In: Proceedings of the Fifth International Joint Conference on Artificial Intelligence, IJCAI 1977, pp. 598–606. MIT, Cambridge (1977)

25. Moore, R.E.: Interval Analysis. Prentice-Hall, Englewood Cliffs (1966)

26. Ozkan, O., Ermis, M., Bekmezci, I.: Reliable communication network design: the hybridisation of metaheuristics with the branch and bound method. J. Oper. Res. Soc. **71**(5), 784–799 (2019)

27. Pant, M., Zaheer, H., Garcia-Hernandez, L., Abraham, A., et al.: Differential evolution: a review of more than two decades of research. Eng. Appl. Artif. Intell. **90**, 103479 (2020)

28. Raidl, G.R., Puchinger, J., Blum, C.: Metaheuristic hybrids. In: Gendreau, M., Potvin, J.Y. (eds.) Handbook of Metaheuristics, pp. 385–417. Springer, Heidelberg (2019). https://doi.org/10.1007/978-3-319-91086-4_12

29. Shcherbina, O., Neumaier, A., Sam-Haroud, D., Vu, X.H., Nguyen, T.V.: Benchmarking global optimization and constraint satisfaction codes. In: Bliek, C., Jermann, C., Neumaier, A. (eds.) COCOS 2002, pp. 211–222. Springer, Heidelberg (2003). https://doi.org/10.1007/978-3-540-39901-8_16

30. Sidebottom, G., Havens, W.S.: Hierarchical arc consistency for disjoint real intervals in constraint logic programming. Comput. Intell. **8**, 601–623 (1992)

31. Storn, R., Price, K.: Differential Evolution – a simple and efficient heuristic for global optimization over continuous spaces. J. Glob. Optim. **11**(4), 341–359 (1997)

32. Takahama, T., Sakai, S.: Constrained optimization by the ε constrained Differential Evolution with an archive and gradient-based mutation. In: 2010 IEEE Congress on Evolutionary Computation, pp. 1–9. IEEE (2010)

33. Talbi, E.G.: Combining metaheuristics with mathematical programming, constraint programming and machine learning. Ann. Oper. Res. **240**(1), 171–215 (2016)

34. Tanabe, R., Fukunaga, A.S.: Improving the search performance of SHADE using linear population size reduction. In: 2014 IEEE Congress on Evolutionary Computation (CEC), pp. 1658–1665. IEEE (2014)

35. Trivedi, A., Srinivasan, D., Biswas, N.: An improved unified differential evolution algorithm for constrained optimization problems. In: Proceedings of 2018 IEEE Congress on Evolutionary Computation, pp. 1–10. IEEE (2018)

36. Vanaret, C., Gotteland, J.B., Durand, N., Alliot, J.M.: Hybridization of interval cp and evolutionary algorithms for optimizing difficult problems. In: Pesant, G. (ed.) International Conference on Principles and Practice of Constraint Programming, pp. 446–462. Springer, Heidelberg (2015). https://doi.org/10.1007/978-3-319-23219-5_32

37. Wu, G., Mallipeddi, R., Suganthan, P.: Problem definitions and evaluation criteria for the CEC 2017 competition on constrained real-parameter optimization. Technical report (2017)

38. Zhang, J., Sanderson, A.C.: JADE: adaptive Differential Evolution with optional external archive. IEEE Trans. Evol. Comput. **13**(5), 945–958 (2009)

39. Zhang, X., Liu, S.: A new interval-genetic algorithm. In: Third International Conference on Natural Computation (ICNC 2007), vol. 4, pp. 193–197. IEEE (2007)

Assessing Group Fairness with Social Welfare Optimization

Violet Chen[1], J. N. Hooker[2(✉)], and Derek Leben[2]

[1] Stevens Institute of Technology, Hoboken, USA
vchen3@stevens.edu
[2] Carnegie Mellon University, Pittsburgh, USA
{jh38,dleben}@andrew.cmu.edu

Abstract. Statistical parity metrics have been widely studied and endorsed in the AI community as a means of achieving fairness, but they suffer from at least two weaknesses. They disregard the actual welfare consequences of decisions and may therefore fail to achieve the kind of fairness that is desired for disadvantaged groups. In addition, they are often incompatible with each other, and there is no convincing justification for selecting one rather than another. This paper explores whether a broader conception of social justice, based on optimizing a social welfare function (SWF), can be useful for assessing various definitions of parity. We focus on the well-known alpha fairness SWF, which has been defended by axiomatic and bargaining arguments over a period of 70 years. We analyze the optimal solution and show that it can justify demographic parity or equalized odds under certain conditions, but frequently requires a departure from these types of parity. In addition, we find that predictive rate parity is of limited usefulness. These results suggest that optimization theory can shed light on the intensely discussed question of how to achieve group fairness in AI.

Keywords: Social welfare optimization · group parity in AI

1 Introduction

There is growing demand within industry and government for assurance that machine learning (ML) models respect and promote equality of impact across protected groups [18,25] and comply with legal requirements [17,42]. This concern arises in contexts that range from hiring and parole decisions to mortgage lending and credit ratings. One prominent method of satisfying these ethical and legal goals is the use of statistical parity metrics [3]. For example, one might assess two groups have equal approval rates (*demographic parity*), whether the approval and rejection rates of qualified candidates are equal (*equalized odds*), or whether the fraction of qualified candidates among those approved is the same (*predictive rate parity*).

There are at least two problems, however, with reliance on statistical parity as a measure of fairness. One is that parity metrics take no account of the actual utility consequences of being selected or rejected. Presumably, group disparities are viewed as unjust because different groups derive unequal benefits from the selection process. Yet an assessment of these benefits requires consideration of the actual welfare outcomes of selecting or rejecting individuals. For example, rejecting a member of a disadvantaged group may have greater negative consequences than rejecting a member of an advantaged group. The standard parity metrics take account only of the number of individuals selected or rejected, not the impacts of these decisions.

A second problem is that parity metrics are frequently incompatible with each other [14,19,29] and, in particular, imply different trade-offs between fairness and accuracy [5,29]. As a result, there is often no consensus on which metric is appropriate in a given context. This is illustrated by the famous debate over parole decisions between ProPublica and Northpointe (now Equivant) regarding whether the latter's COMPAS product is fair, with one side claiming that the model is unfair because it fails to achieve equalize odds, and the other side claiming it is fair because it achieves predictive rate parity [1,16]. Lacking any further grounds for settling this dispute, the debate has (for now) reached a stalemate. Ideally, one would justify (or reject) a parity metric by appealing to a broader principle of justice.

In this paper, we explore an approach for evaluating group parity metrics via their effects on the welfare of individuals in each group. The aim is to connect the debate about group parity with the rich tradition of welfare economics, where policies are evaluated by their effects on social welfare, as measured by a social welfare function (SWF). Such a function can take into account the distribution of utilities as well as overall welfare. We ask whether a selection policy that optimizes social welfare, as measured by a SWF, results in some particular form of group parity or requires departure from the standard parity measures. Our underlying hypothesis is that insights obtained from optimization theory can shed light on the vexing problem of fairness in AI.

As a first step in this research program, we investigate the parity implications of *alpha fairness* [36,43], a well-known family of SWFs parameterized by a nonnegative real number α. Larger values of α indicate a stronger emphasis on fairness as opposed to maximizing total utility, the latter corresponding to $\alpha = 0$. Alpha fairness can therefore evaluate the trade-off of fairness and accuracy, a perennial issue in machine learning. Other special cases include the maximin (Rawlsian) criterion ($\alpha = \infty$) and *proportional fairness*, also known as the Nash bargaining solution ($\alpha = 1$). We ask what are the parity implications of a given level of fairness as indicated by α.

Our purpose here is not to defend alpha fairness as a fairness criterion, but to explore the implications of a criterion that has *already* been extensively defended. Alpha fairness in its various forms has been studied for over 70 years by investigators that include two Nobel laureates (John Nash and J. C. Harsanyi). Nash [38] gave an axiomatic argument for his bargaining solution in 1950, while Rubinstein, Harsanyi and Binmore [6,21,41] supplied bargaining arguments. Lan et. al [30,31] provided an axiomatic derivation for general alpha fairness and proposed an interpretation of the α parameter. Bertsimas et al. [5] studied resulting equity/efficiency trade-offs. Alpha fairness has also seen a number of practical applications, particularly in telecommunications and other engineering fields [28,34,36,39,43].

After a brief survey of related work, we first establish a general solution to the problem of maximizing alpha fairness subject to a constraint on the number of individuals selected. We then present a utility model that allows us to relate group characteristics to the implications of alpha fairness. Following this, we describe specific implications for demographic parity, equalized odds, and predictive rate parity, and draw conclusions from these results.

2   Related Work

Statistical group parity metrics are the most widely studied approach to fairness in AI and machine learning. Much of this research is surveyed in [10,35]. However, a welfarist approach is beginning to receive recognition in AI fairness literature, e.g. [4,8,11,13,15,23,24,33]. One motivation is pragmatic: social welfare can provide a "common currency" with which one can justify the choice of parity metric, when the typical justifications are incommensurate [20]. For example, arguments for individual fairness appeal to procedural justice concerns, while arguments for group fairness appeal to distributive justice [7,32]. When a model cannot satisfy both of these values, it is necessary to justify one's choice. Another motivation for a welfarist approach is ethical: one may wish to strive for group parity to make disadvantaged groups better off, rather than to achieve equality for its own sake [9,37].

Social welfare functions have been used in optimization models for some time, as surveyed in [13,27,40]. Aside from alpha fairness, SWFs that balance equity and efficiency include Kalai-Smorodinsky bargaining [26] and threshold functions [12,22,44].

Despite the large literature on SWFs and group parity metrics, we describe here what is, to our knowledge, the first explicit connection between them.

3   The Basic Model

We address the task of selecting individuals from a population to receive a benefit or resource, such as a mortgage loan or a job interview. Some individuals belong to a protected group that is disadvantaged with respect to qualification status. We define binary variables D, Y, Z to indicate whether an individual is selected ($D = 1$), qualified ($Y = 1$), or protected ($Z = 1$). To simplify notation, we use D to represent $D = 1$ and $\neg D$ to represent $D = 0$, and similarly for Y and Z.

We have demographic parity when $P(D|Z) = P(D|\neg Z)$, equalized odds (in the positive sense of equality of opportunity) when $P(D|Y, Z) = P(D|Y, \neg Z)$, and predictive rate parity when $P(Y|D, Z) = P(Y|D, \neg Z)$. We interpret the conditional probability $P(D|Z)$ as the fraction of protected individuals who are selected, and similarly for the other probabilities. The latter two types of parity are typically defined in terms of qualifications that are determined after the fact, such as whether a mortgage recipient repaid the loan, a job interviewee was hired, or a parolee committed no further crimes. In addition, calculation of the odds ratio requires knowledge of how many rejected candidates are qualified.

To assess utilitarian outcomes, we suppose that an individual i experiences expected utility $u_i = a_i + b_i$ if selected, and a baseline utility $u_i = b_i$ if rejected. We refer to a_i as the *selection benefit*. It can be negative (indicating that selection is harmful), but we assume that $b_i > 0$ and $a_i + b_i > 0$ because alpha fairness is not defined for nonpositive utilities. This assumption can be met by a positive translation of the utility scale if necesssary.

We assess the desirability of a utility distribution $\boldsymbol{u} = (u_1, \dots, u_n)$ with the alpha fairness social welfare function, given by

$$W_\alpha(\boldsymbol{u}) = \begin{cases} \dfrac{1}{1-\alpha} \sum_i u_i^{1-\alpha}, & \text{if } \alpha \geq 0 \text{ and } \alpha \neq 1 \\ \sum_i \log(u_i), & \text{if } \alpha = 1 \end{cases} \tag{1}$$

Alpha fairness is achieved by maximizing $W_\alpha(\boldsymbol{u})$ subject to a limit on the number of individuals that can be selected.

We let binary variable $x_i = 1$ when individual i is selected. The expected utility gained by individual i is therefore $a_i x_i + b_i$. The social welfare resulting from a given vector $\boldsymbol{x} = (x_1, \dots, x_n)$ of selection decisions, as measured by the alpha fairness SWF, is

$$W_\alpha(\boldsymbol{x}) = \begin{cases} \dfrac{1}{1-\alpha} \sum_{i=1}^n (a_i x_i + b_i)^{1-\alpha}, & \text{if } \alpha \geq 0 \text{ and } \alpha \neq 1 \\ \sum_{i=1}^n \log(a_i x + b_i), & \text{if } \alpha = 1 \end{cases} \tag{2}$$

If m $(< n)$ individuals are to be selected, one achieves alpha fairness for a given α by maximizing $W_\alpha(\boldsymbol{x})$ subject to $\sum_{i=1}^n x_i = m$. A maximizing vector $\boldsymbol{x}$ can be deduced using a simple greedy algorithm. We first consider the case $\alpha \neq 1$. The top expression in (2) can be written as

$$\frac{1}{1-\alpha} \sum_{i=1}^n b_i^{1-\alpha} + \frac{1}{1-\alpha} \sum_{i=1}^n \left((a_i x_i + b_i)^{1-\alpha} - b_i^{1-\alpha} \right) \tag{3}$$

Since the first term is a constant, we can maximize (3) by maximizing its second term, which can be written as

$$\frac{1}{1-\alpha} \sum_{i|x_i=1} \left((a_i + b_i)^{1-\alpha} - b_i^{1-\alpha} \right) = \sum_{i|x_i=1} \Delta_i(\alpha) \tag{4}$$

where we define

$$\Delta_i(\alpha) = \begin{cases} \frac{1}{1-\alpha}\left((a_i + b_i)^{1-\alpha} - b_i^{1-\alpha} \right), & \text{if } \alpha \geq 0, \ \alpha \neq 1 \\ \log(a_i + b_i) - \log(b_i), & \text{if } \alpha = 1 \end{cases}$$

The term $\Delta_i(\alpha)$ is the increase in welfare that results from selecting individual i (for a given $\alpha \neq 1$). We can maximize (4) subject to $\sum_{i=1}^{n} x_i = m$ by selecting the m individuals with the largest *welfare differential* $\Delta_i(\alpha)$. A similar argument applies for $\alpha = 1$. Thus we have

Theorem 1. *If $\Delta_{\pi_1}(\alpha) \geq \cdots \geq \Delta_{\pi_n}(\alpha)$, where $\pi_1, \ldots, \pi_n$ is a permutation of $1, \ldots, n$, then one can maximize $W_\alpha(\boldsymbol{x})$ subject to $\sum_{i=1}^{n} x_i = m$ by setting $x_i = 1$ for $i = \pi_1, \ldots, \pi_m$, and $x_i = 0$ for $i = \pi_{m+1}, \ldots, \pi_n$.*

At this point we can easily check whether achieving alpha fairness results in the various forms of group parity by observing whether their definitions are satisfied when individuals $\pi_1, \ldots, \pi_m$ are selected.

4   Modeling Protected and Nonprotected Groups

While Theorem 1 specifies an alpha fair selection policy for any given set of individual utility parameters $(a_1, b_1), \ldots (a_n, b_n)$, it yields limited insight into how the utility characteristics of protected and nonprotected groups affect alpha fair selections. In addition, the large number of parameters makes relationships difficult to analyze in a comprehensible fashion.

We address these issues by supposing that the expected utilities in the two groups occur on sliding scale. Specifically, we suppose that the selection benefits a_i in the nonprotected group are distributed uniformly on a scale from a maximum $A_{\max}$ down to a minimum $A_{\min}(< A_{\max})$, and selection benefits in the protected group vary uniformly from $a_{\max}$ down to $a_{\min}(< a_{\max})$. A nonuniform distribution is more realistic, but it requires a complicated analysis that is harder to interpret, while yielding basically the same qualitative results. To further simplify analysis, we suppose that the base utility has the same value B for all nonprotected individuals, and the same value b for all protected individuals. We assume that $B > b$ and, consistent with the previous section, that $A_{\min} + B > 0$ and $a_{\min} + b > 0$. Finally, we suppose that the protected group comprises a fraction β of the population, with $0 < \beta < 1$.

We further assume that individuals within a given group are selected in decreasing order of their selection benefit. Thus if a fraction S of nonprotected individuals are selected, the last individual selected in that group has the selection benefit $A(S) = (1 - S)A_{\max} + SA_{\min}$ and a social welfare differential of

$$\Delta_S(\alpha) = \begin{cases} \frac{1}{1-\alpha}\left((A(S) + B)^{1-\alpha} - B^{1-\alpha}\right), & \text{if } \alpha \geq 0,\ \alpha \neq 1 \\ \log\left(A(S) + B\right) - \log(B), & \text{if } \alpha = 1 \end{cases}$$

Similarly, if a fraction s of individuals are selected in the protected group, the last individual selected has the selection benefit $a(s) = (1 - s)a_{\max} + sa_{\min}$ and the social welfare differential

$$\Delta'_s(\alpha) = \begin{cases} \frac{1}{1-\alpha}\left((a(s) + b)^{1-\alpha} - b^{1-\alpha}\right), & \text{if } \alpha \geq 0,\ \alpha \neq 1 \\ \log\left(a(s) + b\right) - \log(b), & \text{if } \alpha = 1 \end{cases}$$

We will suppose that the population is large enough that S and s can be treated as continuous variables. This simplifies the analysis considerably without materially affecting the conclusions.

Since the social welfare differential is a monotone increasing function of the selection benefit, selecting individuals in order of decreasing welfare differential is, within each group, the same as selecting in order of decreasing selection benefit. By Theorem 1, selection in order of decreasing welfare differential maximizes the alpha fairness SWF subject to $\sum_i x_i = m$ if we select individuals until the desired fraction $\sigma = m/n$ of the population is selected. This occurs when

$$(1 - \beta)S + \beta s = \sigma, \quad \text{or} \quad s = s(S) = \frac{\sigma - (1 - \beta)S}{\beta} \tag{5}$$

We first take note of the ranges within which S and s can vary, subject to (5). Since we must have $0 \leq S \leq 1$ and $0 \leq s \leq 1$, S can vary in the range from $S_{\min}$ to $S_{\max}$, where

$$S_{\min} = \max\left\{0, \frac{\sigma - \beta}{1 - \beta}\right\}, \quad S_{\max} = \min\left\{1, \frac{\sigma}{1 - \beta}\right\}$$

and s can vary from $s(S_{\max})$ to $s(S_{\min})$. Now since $A_{\max} > A_{\min}$, $\Delta_S(\alpha)$ is monotone decreasing in S. Similarly, $\Delta'_s(\alpha)$ is monotone decreasing in s, so that $\Delta'_{s(S)}(\alpha)$ is monotone increasing in S. This means that we can consider three cases, illustrated by Fig. 1:

(a) $\Delta_{S_{\min}}(\alpha) > \Delta'_{s(S_{\min})}(\alpha)$ and $\Delta_{S_{\max}}(\alpha) \geq \Delta'_{s(S_{\max})}(\alpha)$.

(b) $\Delta_{S_{\min}}(\alpha) \leq \Delta'_{s(S_{\min})}(\alpha)$ and $\Delta_{S_{\max}}(\alpha) < \Delta'_{s(S_{\max})}(\alpha)$.

(c) $\Delta_{S_{\min}}(\alpha) > \Delta'_{s(S_{\min})}(\alpha)$ and $\Delta_{S_{\max}}(\alpha) < \Delta'_{s(S_{\max})}(\alpha)$

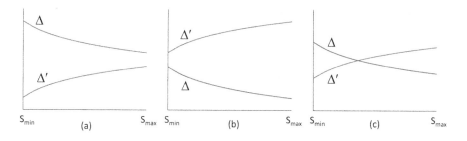

Fig. 1. Cases (a), (b), and (c) in the proof of Theorem 2

Theorem 2. *Suppose that individuals are selected in decreasing order of their selection benefit, and let S^* and $s^* = s(S^*)$, respectively, be the fraction of the nonprotected and protected groups selected at the end of the selection process.*

Then for a sufficiently large population, S^* and s^* achieve alpha fairness if and only if

$$\begin{cases} (S^*, s^*) = \left(\min\left\{1, \dfrac{\sigma}{1-\beta}\right\}, \ \dfrac{\sigma}{\beta}\left[1 - \min\left\{1, \dfrac{1-\beta}{\sigma}\right\}\right]\right), & \textit{in case (a)} \\[2ex] (S^*, s^*) = \left(\dfrac{\sigma}{1-\beta}\left[1 - \min\left\{1, \dfrac{\beta}{\sigma}\right\}\right], \ \min\left\{1, \dfrac{\sigma}{\beta}\right\}\right), & \textit{in case (b)} \\[2ex] \Delta_{S^*}(\alpha) = \Delta'_{s(S^*)}(\alpha), & \textit{in case (c)} \end{cases} \quad (6)$$

Proof. Recall that by Theorem 1, alpha fairness is achieved by selecting individuals in decreasing order of their welfare differential until $S = s(S)$. We consider the three cases separately. (a) Because $\Delta_S(\alpha) \geq \Delta'_{s(S)}(\alpha)$ for all $S \in [S_{\min}, S_{\max}]$, we select entirely from the nonprotected group until it is exhausted, and then move to the protected group if necessary to select a fraction σ of the population. Thus we can set

$$S^* = \min\left\{S_{\max}, \frac{\sigma}{1-\beta}\right\} = \min\left\{1, \frac{\sigma}{1-\beta}\right\}$$

where the first equality is due to the fact that we must have $S^* = \sigma/(1-\beta)$ in order to select a fraction σ if the population if $\sigma \leq 1 - \beta$, and the second equality is due to the definition of $S_{\max}$. The expression given in (6) for $s^* = s(S^*)$ follows directly from the definition of $s(S^*)$, and it is easily checked that $s_{\min} \leq s^* \leq s_{\max}$ using the definitions of $s_{\min}$ and $s_{\max}$. (b) The argument is very similar to that of the previous case. (c) In this case, some but not all individuals are selected in both groups. Let (S, s) be the fraction of the nonprotected and protected individuals selected at any given point in the selection process. We first show that $\Delta_S(\alpha) = \Delta'_s(\alpha)$ for a sufficiently large population. Let Δ_0 and Δ_1 be the welfare differentials of the last two nonprotected individuals selected, and Δ'_0 and Δ'_1 the differentials of the last two protected individuals selected. Their selection order is necessarily one of the following: $(\Delta_0, \Delta'_0, \Delta_1, \Delta'_1)$, $(\Delta_0, \Delta'_0, \Delta'_1, \Delta_1)$, $(\Delta'_0, \Delta_0, \Delta_1, \Delta'_1)$, $(\Delta'_0, \Delta_0, \Delta'_1, \Delta_1)$. In each case, $|\Delta_1 - \Delta'_1|$ is at most $\max\{\Delta_0 - \Delta_1, \Delta'_0 - \Delta'_1\}$. For a sufficiently large population, $\Delta_0 - \Delta_1$ and $\Delta'_0 - \Delta'_1$ are arbitrarily small, and so $|\Delta_1 - \Delta'_1|$ is arbitrarily small. Thus we have $\Delta_S(\alpha) = \Delta'_s(\alpha)$ throughout the selection process, and in particular at the end of the process, when $(S, s) = (S^*, s^*)$. The theorem follows. $\square$

To explore how alpha fair selection policies depend on the utility characteristics of protected and nonprotected groups, we define three scenarios that represent qualitatively different practical situations.

Scenario 1. Protected individuals are somewhat less likely to benefit from being selected, as when those selected for job interviews are less likely to be hired due to less obvious qualifications. Here, $[A_{\min}, A_{\max}] = [0.5, 1.5]$ and $[a_{\min}, a_{\max}] = [0.2, 1]$.

Scenario 2. Some protected individuals can benefit more than anyone else from selection, as when talented but economically disadvantaged individuals are admitted to a university. Here, $[A_{\min}, A_{\max}] = [0.5, 0.8]$ and $[a_{\min}, a_{\max}] = [0.2, 1]$.

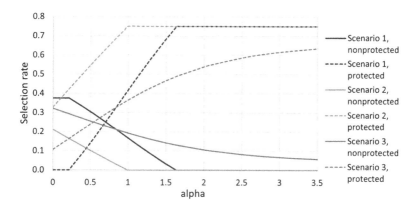

Fig. 2. Alpha fair selection rates, assuming overall selection rate of 0.25

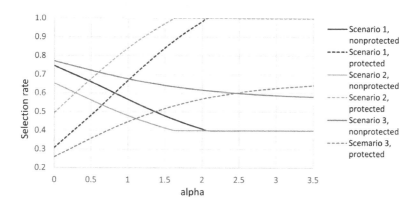

Fig. 3. Alpha fair selection rates, assuming overall selection rate of 0.6

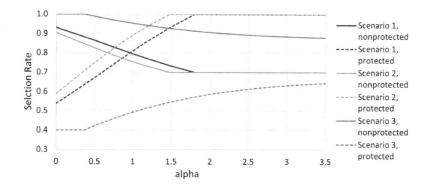

Fig. 4. Alpha fair selection rates, assuming overall selection rate of 0.8

Scenario 3. Significantly many protected individuals are likely to be harmed by selection, as when failure to repay a mortgage results in eviction. Here, $[A_{min}, A_{max}] = [0.5, 1]$ and $[a_{min}, a_{max}] = [-0.5, 1]$.

Plots of alpha fair selection policies in these scenarios appear in each of Figs. 2, 3 and 4. The three figures respectively assume overall selection rates of $\sigma = 0.25, 0.6, 0.8$. These selection rates are chosen to be less than, equal to, and greater than a qualification rate of 0.6, which will be assumed for subsequent plots of alpha fair odds ratios and predictive rates.

The plots show the relationship between alpha fair selection rates (S, s) and the chosen value of α. As expected, larger values of α (indicating a greater emphasis on fairness) result in higher selection rates in the protected group (dashed curves) and lower rates in the nonprotected group (solid curves). Scenario 1 calls for lower section rates in the protected group than Scenario 2 because of the greater utility cost of achieving fairness in Scenario 1; recall that alpha fairness consider total utility as well as Rawlsian fairness. Both scenarios require selecting the entire protected group for sufficiently large α, except when $\sigma = 0.25$, in which case the small number of selections does not exhaust the protected group. In Scenario 3, by contrast, the protected group's selection rate approaches $2/3$ asymptotically, because only $2/3$ of the group benefits from being selected in this scenario.

5   Demographic Parity

Demographic parity is achieved when $P(D|\neg Z) = P(D|Z)$. In the above model, this occurs when $s = S = \sigma$. As it turns out, cases (a) and (b) of Theorem 2 do not apply, and we can achieve demographic parity only by choosing a value of α (if one exists) dictated by case (c).

Theorem 3. *An alpha fair selection policy for a given α results in demographic parity if and only if there exists a selection rate S^* that satisfies the equation $\Delta_{S^*}(\alpha) = \Delta'_{S^*}(\alpha)$, in which case (S^*, S^*) is such a policy.*

Proof. We first note as follows that neither case (a) nor (b) in Theorem 2 applies. In case (a), demographic parity requires that

$$\min\left\{1, \frac{\sigma}{1-\beta}\right\} = \sigma, \text{ or } \min\left\{\frac{1}{\sigma}, \frac{1}{1-\beta}\right\} = 1$$

This cannot hold, because $\beta > 0$ and $\sigma < 1$. Case (b) is similarly ruled out. We are therefore left with case (c), wherein Theorem 2 implies that $S^* = s(S^*)$ if and only if $\Delta_{S^*}(\alpha) = \Delta'_{S^*}(\alpha)$, as claimed. $\square$

In Figs. 2, 3 and 4, demographic parity is achieved at the value of α where the rising and falling curves for a given scenario intersect. For example, if the overall section rate is $\sigma = 0.6$, parity is achieved in Scenario 1 when $\alpha = 0.833$ (Fig. 3). An important lesson in these plots is that a relatively small value of α frequently results in parity. That is, parity achieves a rather modest degree of fairness when utilities are taken into account. Indeed, proportional fairness ($\alpha = 1$), which is something of an industrial benchmark, typically calls for selecting a significantly greater fraction of the protected group than the nonprotected group. This is not the case in Scenario 3, however, where parity requires selecting protected individuals who receive minimal benefit and even harm from being selected. For example, no value of α corresponds to parity when $\sigma = 0.8$ (Fig. 4) because alpha fairness never endorses harmful choices.

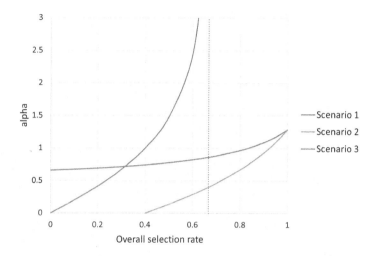

Fig. 5. Values of alpha that achieve demographic parity

Figure 5 provides a fuller picture of the relation between the selection rate σ and parity-achieving values of α. As σ increases, parity corresponds to larger values of α because it becomes necessary to select protected individuals who benefit little from selection. The curves for Scenarios 1 and 2 happen to meet at $\sigma = 1$ in this example because A_{min}, a_{min}, B, and b are the same in the two scenarios. We also note that $\alpha \to \infty$ as $\sigma \to 2/3$ in Scenario 3 because $\sigma > 2/3$ requires selecting individuals who are harmed by selection.

Interestingly, smaller values of α correspond to parity in Scenario 2 than in Scenario 1, despite the fact that rejection can be quite costly to some members of the protected group in Scenario 2 (due to their higher selection benefits). This occurs because a purely utilitarian assessment already takes this cost into account.

6   Equalized Odds

Equalized odds are achieved when $P(D|Y, Z) = P(D|Y, \neg Z)$. To define equalized odds in the above model, we suppose that a fraction Q of nonprotected individuals are qualified, and a fraction q of protected individuals are qualified. The a fraction $(1 - \beta)Q + \beta q$ of the population is qualified. We also make the reasonable assumption that the selection benefit is greater for qualified individuals than unqualified individuals within a given group. Thus since $\Delta_S(\alpha)$ and $\Delta'_s(\alpha)$ are monotone decreasing as S and s increase, the qualified individuals in the nonprotected group consist of the fraction Q with the largest welfare differentials. The odds ratio for the nonprotected group is S/Q when $S \leq Q$ and 1 when $S > Q$, since in the latter case all the qualified individuals are selected. Thus the odds ratio is $\min\{1, S/Q\}$ for the nonprotected group, and similarly for the protected group. This means that we have equalized odds when

$$\min\left\{\frac{S}{Q}, 1\right\} = \min\left\{\frac{s}{q}, 1\right\}$$

This leads to the following theorem. It is convenient to define ρ to be the ratio of the fraction selected to the fraction of the population that is qualified, so that

$$\rho = \frac{\sigma}{(1 - \beta)Q + \beta q}$$

Theorem 4. *An alpha fair selection policy $(S^*, s(S^*))$ for a given α results in equalized odds if and only if one of the following holds:*

$$S^* = Q\rho \leq Q \ \ and \ \ s(S^*) = q\rho \leq q \tag{7}$$
$$S^* \geq Q \ \ and \ \ s(S^*) \geq q \tag{8}$$

Proof. We consider four mutually exclusive and exhaustive cases:
(a) $S^* \leq Q$ and $s(S^*) \leq q$ (c) $S^* > Q$ and $s(S^*) \leq q$
(b) $S^* \leq Q$ and $s(S^*) > q$ (d) $S^* > Q$ and $s(S^*) > q$
     In case (a), equalized odds is equivalent to $S^*(\alpha)/Q = s^*(\alpha)/q$, which implies $S^* = Q\rho$ and $s(S^*) = q\rho$ in (7) due to (5). Conversely, we can see as follows that either of the conditions (7) and (8) implies equalized odds. Under condition (7), the values for S^* and $s(S^*)$ in (7) imply $S^*/Q = s(S^*)/q$, and we have equalized odds. Under condition (8), both odds ratios are 1, and we again have equalized odds. In case (b), equalized odds implies $S^* = Q$, in which case condition (8) is satisfied. Conversely, the case hypothesis is consistent with only condition (8), in which case both odds ratios are 1 and we have equalized odds. Case (c) is similar. In case (d), one of the conditions (7)–(8) is necessarily satisfied (because the latter is satisfied), and we necessarily have equalized odds, because both odds ratios are 1. $\square$

     To continue the example of the previous section, we suppose that the qualification rates are $(Q, q) = (0.65, 0.5)$, so that a fraction 0.6 of the population is qualified. Figures 6 and 7, corresponding to $\sigma = 0.25$ and $\sigma = 0.6$, show alpha fair odds ratios for various α. No plot is given for $\sigma = 0.8$ because nearly all of the odds ratios are 1 due to the fact that considerably more individuals are selected than are qualified. In the important special case where the number selected is equal to the number qualified (Fig. 7), equalized odds is achieved only by an

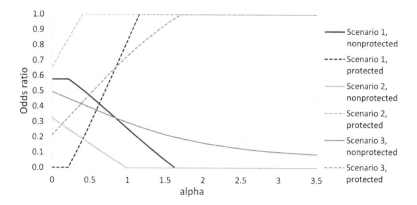

Fig. 6. Alpha fair odds ratios, assuming overall selection rate of 0.25.

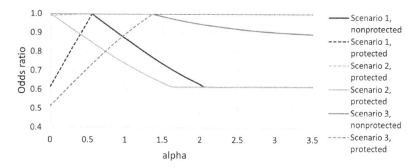

Fig. 7. Alpha fair odds ratios, assuming overall selection rate of 0.6.

accuracy-maximizing solution: precisely the qualified individuals are selected in both groups. This rules out any adjustment for fairness. The odds ratio is perhaps more useful when limited resources compel one to reject significantly many qualified individuals. In this event, somewhat smaller values of α are typically necessary to achieve equalized odds than demographic parity (Fig. 6). In Scenario 2, a purely utilitarian solution already achieves a higher odds ratio for the protected group, since some of its qualified members derive more utility from selection than anyone in the nonprotected group.

7 Predictive Rate Parity

Predictive rate parity is achieved when $P(Y|D, Z) = P(Y|D, \neg Z)$. The predictive rate for the nonprotected group is Q/S when $S \geq Q$ and 1 when $S < Q$, since in the latter case all the selected individuals are qualified. Thus the predictive rate is $\min\{Q/S, 1\}$ for the nonprotected group, and similarly for the protected group. This means that we have predictive rate parity when

$$\min\left\{\frac{Q}{S},1\right\} = \min\left\{\frac{q}{s},1\right\}$$

This leads to the following theorem, whose proof is very similar to the proof of Theorem 4.

Theorem 5. *An alpha fair selection policy* $(S^*, s(S^*))$ *for a given* α *results in predictive rate parity if and only if one of the following holds:*

$$S^* = Q\rho \geq Q \quad and \quad s(S^*) = q\rho \geq q \tag{9}$$
$$S^* \leq Q \quad and \quad s(S^*) \leq q \tag{10}$$

Note that the expressions for S^* *and* $s(S^*)$ *in* (9) *are the same as in* (7).

Figures 8 and 9, corresponding to $\sigma = 0.6$ and $\sigma = 0.8$, show alpha fair predictive rates for various α. There is no plot for $\sigma = 0.25$, because nearly all of the predictive rates are 1. We also note that larger predictive rates correspond to *smaller* values of α.

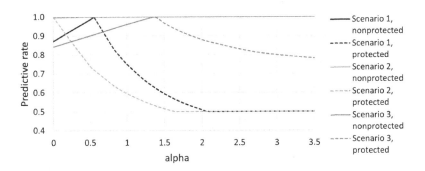

Fig. 8. Alpha fair predictive rates, assuming overall selection rate of 0.6.

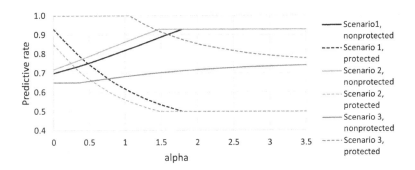

Fig. 9. Alpha fair predictive rates, assuming overall selection rate of 0.8.

8   Conclusion

Our aim in this paper has been to explore the extent to which social welfare optimization can assess well-known statistical parity metrics as criteria for group fairness in AI. Our focus on alpha fairness allows us to address parity questions by appealing to a well-studied concept of just distribution with theoretical underpinning. We conclude in this section by recalling the two problems associated with parity metrics and summarizing how they might be addressed from an optimization perspective.

1. *Accounting for welfare.* The alpha fairness criterion allows us to take explicit account of welfare implications, for various levels of fairness as indicated by the α parameter. We find that for certain values of α and certain group characteristics, an alpha fair selection policy can result in group parity of any of the three types. Yet it can also call for significant statistical *disparity* in order to achieve an acceptable distribution of utilities.

In particular, the alpha values that result in parity typically lie significantly below that corresponding to proportional fairness ($\alpha = 1$)—except when some individuals in the protected group are actually harmed by being selected, in which case larger values of α correspond to parity. Since proportional fairness is the most widely defended and applied variety of alpha fairness, it is noteworthy that it often requires, not parity, but *higher* selection rates for the protected group than for the rest of the population. In addition, a lower level of fairness (i.e., a smaller α) is necessary to achieve parity when rejection is more costly to members of the protected group than the rest of the population, other things being equal. This is because even a purely utilitarian accounting already takes this cost into account.

2. *Selecting and justifying parity metrics.* We derive a number of conclusions regarding the choice of parity metric. In general, we find that the implications of alpha fairness depend heavily on how many individuals are selected relative to the total number qualified, at least where equalized odds and predictive rate parity are concerned.

To elaborate on this, we first suppose that the total number selected is the same (or approximately the same) as the total number who are qualified in the population as a whole. In this case, demographic parity follows the pattern described above, in which relatively small values of α result in parity, except when some protected individuals are harmed by selection. Yet equalized odds, as well as predictive rate parity, are achieved if and only if the odds ratios and the predictive rates are 1 in both groups. This corresponds to an accuracy maximizing policy of selecting all and only qualified individuals. As a result, neither equalized odds nor predictive rate parity reflects any consideration of fairness beyond mere accuracy, and consequently neither is suitable as a fairness criterion in this context.

We next suppose that the total number selected is significantly less than the number qualified, presumably a common situation due to limited resources. In this case, equalized odds is generally achieved for smaller values α than are required for demographic parity, considerably smaller when some protected individuals are harmed by selection. This indicates that demographic parity demands a greater emphasis on fairness than equalized odds. This is consonant with the fact that equalized odds is sometimes seen as more easily defended, perhaps on grounds of equality of opportunity, than is demographic parity, which may reflect a desire to compensate for historically unjust discrimination. As for the predictive rate, it is almost always 1 when a significant number of qualified individuals are rejected, since those who make it through the sieve are almost always qualified. This means that predictive rate parity is likely to be achieved simply due to the high rejection rate and is therefore of little value as a fairness criterion.

Finally, we suppose that the number selected is significantly greater than the number qualified. Here, the odds ratio loses interest because it is almost always 1. While predictive rate parity becomes meaningful in this case, decision makers may be reluctant to select more individuals than are qualified. To the extent this is true, predictive rate parity has limited usefulness. A possible exception arises in the controversy over parole mentioned earlier. Predictive rate parity might be defended on the ground that a lower recidivism rate in the protected group (and therefore a higher predictive rate) may reflect stricter parole criteria than for other inmates [2]. Greater fairness may therefore require a *reduction* in the predictive rate of the protected group, which we have seen can be achieved by choosing a larger value of α. If this is taken as justifying a practice of paroling more individuals than are qualified (perhaps in order to reduce the predictive rate of protected individuals without tightening the criteria for others), then predictive rate parity could be a suitable criterion.

In summary, demographic parity can under certain conditions correspond to an alpha fair policy, but it may result in less fairness than desired for the protected group. Equalized odds can be a useful criterion when fewer individuals are selected than are qualified to be selected, but it corresponds to an even lesser degree of fairness. Predictive parity is a meaningful fairness measure only in the perhaps rather uncommon situation when decision makers select significantly more individuals than are qualified.

The foregoing conclusions regarding equalized odds and predictive rate parity rest on the assumption that, *within a given group*, qualified individuals are selected before unqualified individuals. This assumption might be defended on the ground that (a) qualified individuals are likely to benefit more from being selected, and (b) individuals who benefit more from being selected are selected first in the group. Assumption (a) might be based on observations that less qualified individuals pose a greater risk of defaulting on a mortgage, failing to secure a job, committing a crime while on parole, and so forth, and therefore have less expected benefit. As for (b), there is no apparent rationale, based on either expected utility or fairness, for selecting individuals within a group in any other order. It therefore seems reasonable to suppose (b) is true before assessing fairness.

We believe these results suggest that there is potential in an optimization perspective to inform fairness debates in AI. Further research could explore the parity implications of alternative social welfare functions, such as the Kalai-Smorodinsky and threshold criteria cited earlier. A particularly interesting research issue is the extent to which achieving fairness in the population as a whole can result in a reasonable degree of parity across all groups. This would obviate the necessity of selecting which groups to regard as protected, and how to balance their interests.

References

1. Angwin, J., Larson, J., Mattu, S., Kirchner, L.: Machine bias: There's software used across the country to predict future criminals. And it's biased against blacks. ProPublica (2016). Accessed 23 May 2016
2. Anwar, S., Fang, H.: Testing for racial prejudice in the parole board release process: theory and evidence. J. Legal Stud. **44**, 1–37 (2015)
3. Barocas, S., Hardt, M., Narayanan, A.: Fairness and Machine Learning: Limitations and Opportunities. MIT Press, Cambridge (2023)
4. Baumann, J., Hannó, A., Heitz, C.: Enforcing group fairness in algorithmic decision making: utility maximization under sufficiency. In: Proceedings of FAccT 2022 (2022)
5. Bertsimas, D., Farias, V., Trichakis, N.: On the fairness-efficiency trade-off. Manag. Sci. **58**, 2234–2250 (2012)
6. Binmore, K., Rubinstein, A., Wolinsky, A.: The Nash bargaining solution in economic modelling. RAND J. Econ. 176–188 (1986)
7. Binns, R.: Fairness in machine learning: lessons from political philosophy. Proc. Mach. Learn. Res. **8**, 1–11 (2018)
8. Card, D., Smith, N.: On consequentialism and fairness. Front. Artif. Intell. **3**, 34 (2020)
9. Carter, I., Page, O.: When is equality basic. Aust. J. Philos. **101**, 983–997 (2022)
10. Castelnovo, A., Crupi, R., Greco, G., Regoli, D., Penco, I.G., Cosentini, A.C.: A clarification of the nuances in the fairness metrics landscape. Sci. Rep. **12**, 4209 (2022)
11. Chen, V., Hooker, J.N.: A just approach balancing Rawlsian leximax fairness and utilitarianism. In: Proceedings of the AAAI/ACM Conference on AI, Ethics, and Society, pp. 221–227 (2020)
12. Chen, V., Hooker, J.N.: Combining leximax fairness and efficiency in an optimization model. Eur. J. Oper. Res. **299**, 235–248 (2022)
13. Chen, V., Hooker, J.N.: A guide to formulating fairness in an optimization model. Ann. Oper. Res. **326**, 581–619 (2023)
14. Chouldechova, A.: Fair prediction with disparate impact: a study of bias in recidivism prediction instruments. Big Data **5**(2), 153–163 (2017)
15. Corbett-Davies, S., Gaebler, J.D., Nilforoshan, H., Shroff, R., Goel, S.: The measure and mismeasure of fairness: a critical review of fair machine learning. J. Mach. Learn. Res. **24**, 1–117 (2023)
16. Dieterich, W., Mendoza, C., Brennan, T.: COMPAS risk scales: Demonstrating accuracy equity and predictive parity. Report , Northpointe Inc., Research Department (2016)
17. Feldman, M., Friedler, S.A., Moeller, J., Scheidegger, C., Venkatasubramanian, S.: Certifying and removing disparate impact. In: Proceedings of 21st SIGKDD. ACM (2017)

18. Fjeld, J., Achten, N., Hilligoss, H., Nagy, A., Srikumar, M.: Principled artificial intelligence: mapping consensus in ethical and rights-based approaches to principles for AI. Berkman Klein Center Research Publication No. 2020-1 (2020)
19. Friedler, S.A., Scheidegger, C., Venkatasubramanian, S.: On the (im)possibility of fairness. Commun. ACM **64**, 136–143 (2021)
20. Greene, J.: Moral Tribes: Emotion, Reason, and the Gap between Us and Them. Penguin Press, London (2013)
21. Harsanyi, J.C.: Rational Behaviour and Bargaining Equilibrium in Games and Social Situations. Cambridge University Press, Cambridge (1977)
22. Hooker, J.N., Williams, H.P.: Combining equity and utilitarianism in a mathematical programming model. Manag. Sci. **58**, 1682–1693 (2012)
23. Hu, L., Chen, Y.: Welfare and distributional impacts of fair classification (2018). arXiv preprint arXiv:1807.01134
24. Hu, L., Chen, Y.: Fair classification and social welfare. In: Proceedings of the 2020 Conference on Fairness, Accountability, and Transparency, pp. 535–545 (2020)
25. Jobin, A., Ienca, M., Vayena, E.: The global landscape of AI ethics guidelines. Nat. Mach. Intell. **1**, 389–399 (2019)
26. Kalai, E., Smorodinsky, M.: Other solutions to Nash's bargaining problem. Econometrica **43**, 513–518 (1975)
27. Karsu, O., Morton, A.: Inequality averse optimization in operational research. Eur. J. Oper. Res. **245**, 343–359 (2015)
28. Kelly, F.P., Maulloo, A.K., Tan, D.K.H.: Rate control for communication networks: shadow prices, proportional fairness and stability. J. Oper. Res. Soc. **49**(3), 237–252 (1998)
29. Kleinberg, J., Mullainathan, S., Raghavan, M.: Inherent trade-offs in the fair determination of risk scores. In: Proceedings, Innovations in Theoretical Computer Science (ITCS). Dagstuhl Publishing, Germany (2017)
30. Lan, T., Chiang, M.: An axiomatic theory of fairness in resource allocation. Technical report. Princeton University (2011)
31. Lan, T., Kao, D., Chiang, M., Sabharwal, A.: An axiomatic theory of fairness in network resource allocation. In: Proceedings of the 29th Conference on Information communications (INFOCOM), pp. 1343–1351 (2010)
32. Leben, D.: Normative principles for evaluating fairness in machine learning. In: Proceedings, AAAI/ACM Conference on AI, Ethics, and Society, pp. 86–92 (2020)
33. Loi, M., Herlitz, A., Heidari, H.: A philosophical theory of fairness for prediction-based decisions. SSRN Electron. J. (2019)
34. Mazumdar, R., Mason, L., Douligeris, C.: Fairness in network optimal flow control: optimality of product forms. IEEE Trans. Commun. **39**(5), 775–782 (1991)
35. Mehrabi, N., Morstatter, F., Saxena, N., Lerman, K., Galstyan, A.: A survey on bias and fairness in machine learning. ACM Comput. Surv. (CSUR) **54**(6), 1–35 (2021)
36. Mo, J., Walrand, J.: Fair end-to-end window-based congestion control. IEEE/ACM Trans. Network. **8**, 556–567 (2000)
37. Moss, J.: How to value equality. Philos. Compass **10**, 187–196 (2015)
38. Nash, J.F.: The bargaining problem. Econometrica **18**, 155–162 (1950)
39. Ogryczak, W., Luss, H., Pióro, M., Nace, D., Tomaszewski, A.: Fair optimization and networks: a survey. J. Appl. Math. **2014**, 1–25 (2014)
40. Ogryczak, W., Wierzbicki, A., Milewski, M.: A multi-criteria approach to fair and efficient bandwidth allocation. Omega **36**(3), 451–463 (2008)
41. Rubinstein, A.: Perfect equilibrium in a bargaining model. In: Econometrica, pp. 97–109 (1982)

42. Selbst, A., Barocas, S.: Big data's disparate impact. Calif. Law Rev. **671**, 671–732 (2016)
43. Verloop, I.M., Ayesta, U., Borst, S.: Monotonicity properties for multi-class queue-ing systems. Disc. Event Dyn. Syst. **20**, 473–509 (2010)
44. Williams, A., Cookson, R.: Equity in Health. In: Culyer, A.J., Newhouse, J.P. (eds.) Handbook of Health Economics (2000)

Modeling and Exploiting Dominance Rules for Discrete Optimization with Decision Diagrams

Vianney Coppé[(✉)] [iD], Xavier Gillard[iD], and Pierre Schaus[iD]

UCLouvain, Louvain-la-Neuve, Belgium
{vianney.coppe,xavier.gillard,pierre.schaus}@uclouvain.be

Abstract. Discrete optimization with decision diagrams is a recent approach to solve combinatorial problems that can be formulated with dynamic programming. It consists in a branch-and-bound algorithm that iteratively explores the search space by compiling bounded-width decision diagrams. Those decision diagrams are used both to subdivide a given problem into smaller subproblems – in a divide-and-conquer fashion – and to compute primal and dual bounds for those. It has been previously shown that pruning performed during the compilation of those decision diagrams can greatly impact the quality of the bounds, and consequently the performance of the branch-and-bound algorithm. In this paper, we study the integration of dominance rules inside the decision diagram-based optimization framework. We propose a modeling language for consistently formulating dominance rules for dynamic programming models, and describe how they can be exploited to systematically detect and prune dominated nodes during the search. Furthermore, we explain how to combine this additional filtering mechanism with caching techniques to further improve the performance of the algorithm. Dominance rules are shown to significantly reduce the number of nodes expanded and the running time of the algorithm on four optimization problems.

Keywords: Decision diagrams · Branch-and-bound · Dominance rules

1 Introduction

Discrete optimization with *decision diagrams* (DDs) [3] is a recent framework for solving *dynamic programming* formulations of discrete optimization problems through *branch-and-bound* (B&B). It relies on bounded-width DDs to subdivide the problem into smaller subproblems and compute bounds for those. In particular, the compilation of restricted DDs generates feasible solutions in a *beam search* fashion. Inversely, relaxed DDs automatically compute dual bounds by means of a problem-specific state merging operator. As shown in [9,14], filtering techniques that prune nodes *a priori* during the top-down compilation of approximate DDs can greatly impact the performance of the B&B algorithm. On the one hand, restricted DDs produce better solutions because they are guided towards

© The Author(s), under exclusive license to Springer Nature Switzerland AG 2024
B. Dilkina (Ed.): CPAIOR 2024, LNCS 14742, pp. 226–242, 2024.
https://doi.org/10.1007/978-3-031-60597-0_15

promising parts of the search space. On the other hand, the pruning performed inside relaxed DDs further shrinks the areas of the search space that effectively needs to be explored, which facilitates the work of the B&B algorithm.

Dominance rules are another well-known ingredient that can reduce the size of the search tree by filtering subproblems leading to redundant solutions. They were first formalized in [19,20] in the general case of a B&B framework. Several optimization paradigms successfully applied them, including MIP [12], CP [8,23,25] and DP [4,7,16,31]. In any of those technologies, dominance rules play a crucial role in facilitating the solving process when applicable. Therefore, it is a very natural step to incorporate this ingredient inside DD-based B&B solvers. This paper is, to the best of our knowledge, the first to fill this gap for this particular field of research, although similar work has already been done for the neighboring line of research on state space search for optimization [21]. After a brief summary of the DD-based B&B algorithm in Sect. 2, it starts by providing general definitions about dominance rules within the context of DD-based optimization in Sect. 3, and describe how dominance rules can be formulated for DP models. Section 4 then explains how they can be exploited to systematically detect and prune dominated nodes during the search. Next, a brief explanation on how to combine this additional filtering mechanism with the caching techniques proposed in [9] is given in Sect. 5. Finally, we present in Sect. 6 the experimental evaluation of the integration of dominance rules for four different optimization problems, and discuss the results before concluding.

2   Preliminaries

Dynamic Programming. The DD-based optimization framework introduced in [3] manipulates a discrete optimization problem $\mathcal{P}$ through a DP model composed of the following elements:

- a vector of *control variables* $x = (x_0, \ldots, x_{n-1})$ with $x \in \mathcal{D} = \mathcal{D}_0 \times \cdots \times \mathcal{D}_{n-1}$ and $x_j \in \mathcal{D}_j$ for each $j \in \{0, \ldots, n-1\}$.
- a *state space* $\mathcal{S}$ partitioned into $n+1$ sets $\mathcal{S}_0, \ldots, \mathcal{S}_n$ corresponding to the successive stages of the DP model. In particular, we define the *root – or initial – state* $\hat{r}$, the *terminal* state $\hat{t}$ and the *infeasible* state $\hat{0}$.
- a set of *transition functions* $t_j : \mathcal{S}_j \times \mathcal{D}_j \to \mathcal{S}_{j+1}$ with $j = 0, \ldots, n-1$ encoding the transition from one state s^j to another s^{j+1}, according to the decision d made about variable x_j.
- a set of *transition value functions* $h_j : \mathcal{S}_j \times \mathcal{D}_j \to \mathbb{R}$ that specify the reward of assigning some value $d \in \mathcal{D}_j$ to the variable x_j for each $j = 0, \ldots, n-1$.
- a *root value* v_r to model constant terms in the objective.

Given such a DP model, the optimal solution can be obtained by solving:

$$\text{maximize } f(x) = v_r + \sum_{j=0}^{n-1} h_j(s^j, x_j)$$

$$\text{subject to } s^{j+1} = t_j(s^j, x_j), \text{ for all } j = 0, \ldots, n-1, \text{ with } x_j \in \mathcal{D}_j$$

$$s^j \in \mathcal{S}_j, s^j \neq \hat{0}, j = 0, \ldots, n.$$

Decision Diagrams. DDs are used in a variety of domains to compactly encode a set of solutions, and that also applies to those induced by a DP model. With this specific application in mind, a DD $\mathcal{B} = (U, A, \sigma, l, v)$ is defined as a layered directed acyclic graph with U the set of nodes and A the set of arcs. The state function σ maps each node $u \in U$ to a DP state $\sigma(u) \in \mathcal{S}$. The set of nodes U is partitioned into a set of *layers* $L_0, \ldots, L_n$ that correspond to stages of the DP model. Each transition between pairs of states $s^j \in \mathcal{S}_j$ and $s^{j+1} \in \mathcal{S}_{j+1}$ is materialized by an arc $a = (u_j \xrightarrow{d} u_{j+1})$ that connects the corresponding nodes $u_j \in L_j, u_{j+1} \in L_{j+1}$, with $\sigma(u_j) = s^j$ and $\sigma(u_{j+1}) = s^{j+1}$. The *label* $l(a) = d$ of each arc represents the assignment of decision $d \in \mathcal{D}_j$ to variable x_j, and its value $v(a)$ captures the transition value. Both the first and last layer – L_0 and L_n – contain a single node, respectively the *root* r and the *terminal* node t. Consequently, each $r \rightsquigarrow t$ path $p = (a_0, \ldots, a_{n-1})$ that connects the root and the terminal node through the arcs $a_0, \ldots, a_{n-1}$ encodes a solution $x(p) = (l(a_0), \ldots, l(a_{n-1}))$ with value $v(p) = v_r + \sum_{j=0}^{n-1} v(a_j)$. DD $\mathcal{B}$ is said *exact* if it exactly represents the set of solutions of the corresponding problem, i.e. $Sol(\mathcal{B}) = Sol(\mathcal{P})$ and $v(p) = f(x(p)), \forall p \in \mathcal{B}$, with $Sol(\mathcal{B}) = \{x(p) \mid \exists p : r \rightsquigarrow t, p \in \mathcal{B}\}$. We denote by $v^*(u \mid \mathcal{B})$ the value of the longest path that reaches node u within a DD $\mathcal{B}$, and define $v^*(\mathcal{B}) = v^*(t \mid \mathcal{B})$ for conciseness.

Example 1. Given a set of items $N = \{0, \ldots, n-1\}$, along with their weights $W = \langle w_0, \ldots, w_{n-1} \rangle$ and values $V = \langle v_0, \ldots, v_{n-1} \rangle$, the goal of the *0–1 Knapsack Problem* (KP) is to select a subset of items that maximizes the total value while keeping the total weight under a given capacity C. In its well-known DP formulation, each item $j \in N$ is associated with a binary variable x_j that decides whether to include it in the knapsack. States simply contain the remaining capacity of the knapsack. The state space is thus defined as $S = [0, C]$, with the root state $\hat{r} = C$ starting at maximum capacity, and with root value $v_r = 0$. The transition functions are given by:

$$t_j(s^j, x_j) = \begin{cases} s^j - x_j w_j, & \text{if } x_j w_j \leq s^j, \\ \hat{0}, & \text{otherwise}, \end{cases}$$

meaning that the weight of item j is subtracted from the remaining capacity when it is added to the knapsack. If the capacity constraint is violated, the transition is redirected to the infeasible state. Likewise, the transition value functions $h_j(s^j, x_j) = x_j v_j$ add the value of item j if it is included in the knapsack.

Let us consider an instance of the KP with $n = 4$, $C = 12$, $W = \langle 6, 5, 6, 6 \rangle$ and $V = \langle 5, 6, 1, 6 \rangle$. Figure 1(a) shows the exact DD for that problem. The longest path corresponds to the optimal solution $x^*(\mathcal{B}) = (0, 1, 0, 1)$ for a value of $v^*(\mathcal{B}) = 12$ and a total weight of 11.

DD Compilation. Algorithm 1 describes the top-down compilation of a DD $\mathcal{B}$ for a given DP model. It takes a root node u_r and a maximum width W as input and recursively builds the DD by applying all valid transitions to the last completed layer. When the number of nodes in the last completed layer exceeds

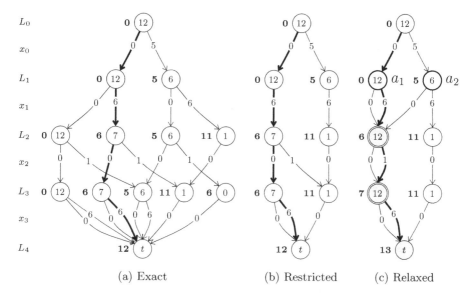

Fig. 1. Exact, restricted and relaxed DDs for the KP instance of Example 1. The value inside each node u corresponds to its state $\sigma(u)$ – the remaining capacity – and the annotation on the left gives the value of the longest path that reaches it $v^*(u \mid \mathcal{B})$. For clarity, only arc values are present. The longest path is highlighted in bold.

the parameter W at line 7, the algorithm compiles an *approximate* DD, as will be detailed next. To create the next layer, the algorithm iterates over all nodes of the last completed layer and applies all valid DP transitions to them at lines 10 to 14 before encoding them as arcs and nodes. Note that a single node is created for each *distinct* state reached by the transitions. The last step of the algorithm is to merge all nodes of the terminal layer into a single terminal node t at line 15.

Approximate DDs. When the size of a layer exceeds the parameter W at line 7 of Algorithm 1, two procedures exist to reduce the number of nodes in the layer. Restricted DDs adopt a simple strategy that consists in heuristically removing the least promising nodes of the layer, as described by Algorithm 2. They thus produce a subset of the solutions of the problem, which provide lower bounds on the optimal solution. For a restricted DD $\mathcal{B}$, we thus have that $Sol(\mathcal{B}) \subseteq Sol(\mathcal{P})$ and $v(p) = f(x(p)), \forall p \in \mathcal{B}$. On the other hand, relaxed DDs over-approximate the set of solutions of the problem by locally relaxing the problem by *merging* surplus nodes together. To this end, state merging operators ensuring that no feasible solutions are removed must be defined for each DP model. If $\mathcal{M}$ is the set of nodes to merge and $\sigma(\mathcal{M}) = \{\sigma(u) \mid u \in \mathcal{M}\}$ the corresponding set of states, the operator $\oplus(\sigma(\mathcal{M}))$ gives the state of the merged node. In Algorithm 2, this operator is used at line 4 to create a single *meta*-node and at lines 5 to 6, the arcs pointing to the merged nodes are redirected to it. A second operator denoted

---

Algorithm 1. Compilation of DD $\mathcal{B}$ rooted at node u_r with max. width W.

1: $i \leftarrow$ index of the layer containing u_r
2: $L_i \leftarrow \{u_r\}$
3: **for** $j = i$ **to** $n - 1$ **do**
4:     $pruned \leftarrow \emptyset$
5:     perform dominance pruning using Algorithm 3
6:     $L'_j \leftarrow L_j \setminus pruned$
7:     **if** $|L'_j| > W$ **then**
8:         restrict or relax the layer to get W nodes with Algorithm 2
9:     $L_{j+1} \leftarrow \emptyset$
10:    **for all** $u \in L'_j$ **do**
11:        **for all** $d \in \mathcal{D}_j$ **do**
12:            create node u' with state $\sigma(u') = t_j(\sigma(u), d)$ or retrieve it from L_{j+1}
13:            create arc $a = (u \xrightarrow{d} u')$ with $v(a) = h_j(\sigma(u), d)$ and $l(a) = d$
14:            add u' to L_{j+1} and add a to A
15: merge nodes in L_n into terminal node t

Algorithm 2. Restriction or relaxation of layer L'_j with maximum width W.

1: **while** $|L_j| > W$ **do**
2:     $\mathcal{M} \leftarrow$ select nodes from L'_j
3:     $L_j \leftarrow L_j \setminus \mathcal{M}$
4:     create node μ with state $\sigma(\mu) = \oplus(\sigma(\mathcal{M}))$ and add it to L_j // relaxation only
5:     **for all** $u \in \mathcal{M}$ and arc $a = (u' \xrightarrow{d} u)$ incident to u **do**
6:         replace a by $a' = (u' \xrightarrow{d} \mu)$ and set $v(a') = \Gamma_{\mathcal{M}}(v(a), u)$

$\Gamma_{\mathcal{M}}$ can be defined to adjust the value of the arcs incident to the merged node at line 6. For valid relaxation operators, a relaxed DD $\overline{\mathcal{B}}$ verifies $Sol(\overline{\mathcal{B}}) \supseteq Sol(\mathcal{P})$ and $v(p) \geq f(x(p)), \forall p \in \overline{\mathcal{B}}$. Whereas restricted DDs only contain *exact* nodes, relaxed DDs also contain *relaxed* nodes that are either merged nodes, or nodes that are reached by at least one path that traverses a merged node.

Branch-and-Bound. The B&B algorithm introduced in [3] builds upon those two types of approximate DDs to solve problems to optimality. Restricted DDs are used to generate feasible solutions from any B&B node, while relaxed DDs decompose the problem further and provide dual bounds for the subproblems thus created. Indeed, in a relaxed DD, it is possible to identify a set of exact nodes whose associated subproblems collectively represent the root problem, and therefore solving them is equivalent to solving the root problem. The B&B algorithm maintains a queue of such nodes to process, and uses restricted DDs to try to improve the best solution found so far, and relaxed DDs to further decompose the problem and prune unpromising subproblems. When the algorithm terminates, all solutions have been either enumerated or pruned. For the sake of conciseness, we do not detail here the additional filtering techniques that have been proposed in [9,14] to speed up this process.

Example 2. Figure 1(b) shows the result of compiling a restricted DD with maximum width $W = 2$, for the KP instance of Example 1. By applying the greedy heuristic that deletes nodes with the lowest prefix values, the best solution that the restricted DD obtains is $x^*(\underline{\mathcal{B}}) = (0, 1, 0, 1)$ with a value of $v^*(\underline{\mathcal{B}}) = 12$. This lower bound is actually the optimal solution to the problem.

To compile a relaxed DD for this problem, we first define a state merging operator: $\oplus(\sigma(\mathcal{M})) = \max_{s \in \sigma(\mathcal{M})} s$, which keeps the maximum remaining capacity among the states to merge. The operator $\Gamma_\mathcal{M}$ is the identity function here since there is no need to modify the arc values. Given those operators, Fig. 1(c) shows a relaxed DD compiled with $W = 2$. The longest path in this diagram corresponds to the solution $x^*(\overline{\mathcal{B}}) = (0, 1, 1, 1)$ for a value of $v^*(\overline{\mathcal{B}}) = 13$ and a total weight of 17. This solution violates the capacity constraint, which can happen because the state merging operator relaxes this constraint. Nevertheless, it provides an upper bound for the problem. If we were to solve the problem to optimality, a set of exact nodes to explore next would be extracted from the relaxed DD. For instance, nodes a_1 and a_2 could be added to the B&B queue.

3 Dominance Rules for Decision Diagrams

Let us now define the concept of node dominance in the DD-based optimization context. It only concerns exact nodes because relaxed nodes have both a relaxed value and state representation, and thus do not produce valid dominance relations. In the following, the operator $\cdot$ denotes the concatenation of two vectors.

Definition 1 (Node Dominance). *Let $u_1 \in \mathcal{B}_1$ and $u_2 \in \mathcal{B}_2$ be two exact nodes respectively obtained in DDs $\mathcal{B}_1$ and $\mathcal{B}_2$ compiled for a problem $\mathcal{P}$, and whose states belong to the j-th stage of the corresponding DP model, meaning that $\sigma(u_1), \sigma(u_2) \in \mathcal{S}_j$. We say that u_1 dominates u_2 – written as $u_1 \succ u_2$ – if for any partial assignment $(x_j, \ldots, x_{n-1}) \in \mathcal{D}_j \times \cdots \times \mathcal{D}_{n-1}$ such that $x_2 = x^*(u_2 \mid \mathcal{B}_2) \cdot (x_j, \ldots, x_{n-1}) \in Sol(\mathcal{P})$, we also have that $x_1 = x^*(u_1 \mid \mathcal{B}_1) \cdot (x_j, \ldots, x_{n-1}) \in Sol(\mathcal{P})$ and either:*

- *$\sigma(u_1) \neq \sigma(u_2)$ and $f(x_1) \geq f(x_2)$,*
- *or, $\sigma(u_1) = \sigma(u_2)$ and $f(x_1) > f(x_2)$.*

If we are interested in finding a single optimal solution to the problem and that such dominance relation exists between nodes u_1 and u_2, then clearly the exploration of node u_2 can be avoided. For some DP models, *dominance rules* that systematically identify scenarios where this kind of node dominance relation exists can be derived. That is, they provide a simple criterion to detect dominated nodes without needing to expand them in the first place and determine algorithmically whether such dominance relation arises. We define such dominance rules through two components:

- The *dominance key* operator $\kappa : \mathcal{S} \to \mathcal{S}'$ that maps each state of the state space $\mathcal{S}$ of a DP model to a *reduced* state in a *reduced* state space $\mathcal{S}'$. This operator partitions the state space $\mathcal{S}$ in equivalence classes $\mathcal{S}^0, \ldots, \mathcal{S}^M$ such

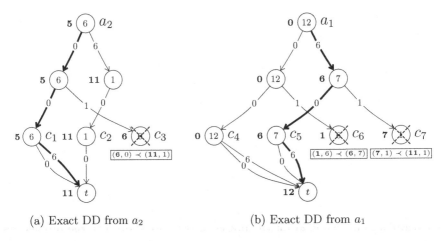

(a) Exact DD from a_2 (b) Exact DD from a_1

Fig. 2. Exact DD compiled from nodes a_1 and a_2 of Fig. 1(b) and exploiting the dominance rule for this problem.

that $\forall s_1, s_2 \in \mathcal{S}^m : \kappa(s_1) = \kappa(s_2)$ for all $m = 0, \ldots, M$. The dominance key typically contains a subset of the original state definition and the equivalence classes group states that are eligible for a dominance relation.

– Furthermore, the *partial dominance utility* operator $\psi : \mathcal{S} \to \mathbb{R}^k$ transforms each state into a vector of k coordinates. Given a node $u \in \mathcal{B}$, we also define the *dominance utility* operator $\Psi(u) = (v^*(u \mid \mathcal{B})) \cdot \psi(\sigma(u))$ that concatenates the node value with the partial utility vector, producing a vector in $\mathbb{R}^{k+1}$ that must characterize the utility of the corresponding node.

These modeling ingredients are similar, yet slightly more flexible than the resource-based approach adopted in [21], since they allow reasoning over quantities other than state variables. The following definition formalizes the connection between those operators and Definition 1, and the necessary condition for those modeling components to constitute a valid dominance rule. It assumes that, given two vectors $x, y \in \mathbb{R}^{k+1}$, we write $x \geq y$ if $x_i \geq y_i$ for $i = 0, \ldots, k$ and $x \neq y$.

Definition 2 (Dominance Rule). *The operators κ and ψ define a valid dominance rule for a given DP model if, for any two exact nodes $u_1 \in \mathcal{B}_1$ and $u_2 \in \mathcal{B}_2$ obtained in the j-th layer of DDs $\mathcal{B}_1$ and $\mathcal{B}_2$, having $\kappa(\sigma(u_1)) = \kappa(\sigma(u_2))$ and $\Psi(u_1) \geq \Psi(u_2)$ implies that $u_1 \succ u_2$ holds.*

Example 3. In the case of the KP, a node u_1 having both higher value and remaining capacity than another node u_2 will always produce better solutions. This dominance rule can be formulated through the following dominance key: $\kappa(s) = \mathbf{0}$ for each state s, which is the zero-dimensional vector, since all states of the same stage can be compared. The partial dominance utility is simply $\psi(s) = s$ so that for a node $u \in \mathcal{B}$, the dominance utility operator compares the value and the remaining capacity $\Psi(u) = (v^*(u \mid \mathcal{B})) \cdot \psi(\sigma(u)) = (v^*(u \mid \mathcal{B}), \sigma(u))$.

Figure 2(a) shows a DD compiled from node a_2 of Fig. 1(c), obtained by performing dominance checks using the rule defined above during the compilation. Node c_3 is dominated by c_2 since $\kappa(c_3) = \kappa(c_2) = \mathbf{0}$ and $\Psi(c_3) = (6,0) \leq \Psi(c_2) = (11,1)$. It can thus be pruned, resulting in an exact DD even with $W = 2$.

4   Filtering the Search Using Dominance Rules

In this section, we explain how to systematically detect and prune dominated nodes within the DD-based B&B algorithm, based on the modeling ingredients previously defined. In this regard, we propose two strategies that both fall in the category of *memory-based dominance relations* [26].

- The first is to perform dominance checks exclusively for nodes belonging to the same layer of the same DD. This way, no extra memory is required and the number of nodes for which dominance relations are checked is kept small.
- On the other hand, the second strategy maintains a persistent collection of non-dominated nodes during the whole search algorithm, and exploits it to also detect dominance relations across DD compilations.

Preliminary experiments convinced us to pursue the second strategy because of its much stronger pruning capacities and relatively small – or even positive – impact on the memory consumption of the algorithm, as will be discussed in Sect. 6. This way of enforcing the dominance rules involves storing all non-dominated nodes found at any stage of the B&B algorithm. We propose to use a hash table denoted by $Fronts_j$ for each DP stage j. Each of these hash tables stores *key-value* pairs of the form $\langle \kappa, Front \rangle$ that associate each dominance key κ with a Pareto front denoted $Front$ containing the set of non-dominated nodes. The only addition to the DD compilation procedure given by Algorithm 1 is that each layer is filtered through the dominance checks before expanding each of its nodes. The *pruned* set collects the pruned nodes of the layer and is used to define L'_j at line 6, a clone of the j-th layer from which the pruned nodes have been removed. In the rest of the algorithm, the pruned layer L'_j is employed instead of L_j to prevent generating any outgoing transition from the pruned nodes.

Algorithm 3 describes the actual dominance detection procedure, which also takes care of updating the *Fronts*. It begins by sorting the nodes of layer in *reverse* lexicographic order of dominance utilities Ψ at line 1. This ensures that if there exist two exact nodes $u_1, u_2 \in L_j$ such that $u_1 \succ u_2$, then u_1 will be processed before u_2 since $\Psi(u_1) \geq \Psi(u_2)$. Then, the algorithm loops over all nodes of the layer and first determines whether a front already exists for the dominance key of the current node. If not, it is simply initialized at lines 18 and 20 as a front containing the utility of the current node only. Otherwise, the existing front is retrieved as *Front* at lines 5 and 6 and the dominance check with respect to this front is initiated. By comparing the utility of the current node against those of the non-dominated nodes found so far, the node is declared dominated or not. In the dominated case, it is added to the *pruned* set at line 15, and otherwise to the *Front* at line 17. Along the way, every entry that the current

---

Algorithm 3. Dominance-based filtering of layer L_j of a DD $\mathcal{B}$.

1: sort nodes u in L_j in *reverse* lexicographic order of $\Psi(u)$
2: **for all** $u \in L_j$ **do**
3:   **if** u is relaxed **then**
4:     **continue**
5:   **if** $Fronts_j.contains(\kappa(\sigma(u)))$ **then**
6:     $Front \leftarrow Fronts_j.get(\kappa(\sigma(u)))$
7:     $dominated \leftarrow False$
8:     **for all** $\Psi' \in Front$ **do**
9:       **if** $\Psi(u) \leq \Psi'$ **then**   // exit if $\Psi(u)$ is dominated
10:        $dominated \leftarrow True$
11:        **break**
12:       **if** $\Psi(u) \geq \Psi'$ **then**   // remove entries that $\Psi(u)$ dominates
13:        $Front \leftarrow Front \setminus \{\Psi'\}$
14:     **if** $dominated$ **then**
15:       $pruned \leftarrow pruned \cup \{u\}$
16:     **else**   // add to front if non-dominated
17:       $Front \leftarrow Front \cup \{\Psi(u)\}$
18:   **else**   // initialize if first with given key
19:     $Front \leftarrow \{\Psi(u)\}$
20:     $Fronts_j.insert(\langle \kappa(\sigma(u)), Front \rangle)$

---

node dominates is removed from the *Front* with line 13 to keep its size as small as possible. Note that this process is performed both for restricted and relaxed DDs, meaning that both types of DD benefit from this filtering mechanism. In addition, the exploratory nature of restricted DDs can help quickly find strong non-dominated nodes, and thus generate a lot of pruning early in the search.

Example 4. Using this procedure, we can derive dominance relations between nodes belonging to the DDs shown on Fig. 2. Let us assume that the *Fronts* have already been filled with the utilities of the nodes reached by the exact DD of Fig. 2(a). We thus have that $Fronts_3 = \{\langle \mathbf{0}, \{(5,6),(11,1)\}\rangle\}$. Now if we consider layer L_3 of the exact DD given by Fig. 2(b), we can first compute the utility – given by $\Psi(u) = (v^*(u \mid \mathcal{B}), \sigma(u))$ – of each node: $\Psi(c_4) = (0,12), \Psi(c_5) = (6,7), \Psi(c_6) = (1,6)$ and $\Psi(c_7) = (7,1)$, and then order the nodes by reverse lexicographic order of those, which produces: $\langle c_7, c_5, c_6, c_4 \rangle$.

- Node c_7 with $\Psi(c_7) = (7,1)$ is dominated by utility $(11,1)$ in the front and is thus added to the *pruned* set.
- Node c_5 with $\Psi(c_5) = (6,7)$ is not dominated by any utility in the front and is thus added to the front. Moreover, it dominates the utility $(5,6)$, which is therefore removed from the front. This gives: $Fronts_3 = \{\langle \mathbf{0}, \{(11,1),(6,7)\}\rangle\}$.
- Node c_6 with $\Psi(c_6) = (1,6)$ is dominated by the utility that was just added to the front and is inserted in the *pruned* set.
- Finally, node c_4 with $\Psi(c_4) = (0,12)$ is added to the front because it has the largest remaining capacity and is therefore non-dominated. The final front is given by $Fronts_3 = \{\langle \mathbf{0}, \{(11,1),(6,7),(0,12)\}\rangle\}$.

5   Synergy with Caching

In [9], a caching mechanism was proposed to mitigate the number of repeated expansions of DP states, which are reached by multiple approximate DDs during the search. It includes a bottom-up procedure that computes an *expansion threshold* for each exact state reached by a relaxed DD, and that exploits the pruning inequalities of each filtering technique involved. Nodes are discarded whenever their value is lower or equal to the threshold. To combine dominance rules with this caching and pruning rule, we specify how expansion thresholds are computed in case of dominance pruning. Given a relaxed DD $\overline{\mathcal{B}}$, an exact node $u \in \overline{\mathcal{B}}$ with $\sigma(u) \in \mathcal{S}_j$ and a utility $\Psi' \in Fronts_j[\kappa(u)]$ such that $\Psi' \geq \Psi(u)$, the *dominance pruning threshold* of u is defined by, with $\Psi' = (v') \cdot \psi'$:

$$
\theta_p(u \mid \overline{\mathcal{B}}) = \begin{cases} v' - 1, & \text{if } \psi' = \psi(u), \\ v', & \text{otherwise.} \end{cases}
$$

Indeed, if u is dominated by a utility with the same partial utility, nodes with the same DP state will always be pruned unless they have a value of v' or higher. On the other hand, if u is dominated by a utility with a better partial utility, then nodes with the same DP state will always be pruned unless their value strictly exceeds v'. As these thresholds are propagated bottom-up in the relaxed DDs, dominance rules will also strengthen expansion thresholds for states of earlier DP stages, and can thus also reinforce the cache-based pruning strategy.

6   Computational Experiments

In this section, we evaluate experimentally the impact of dominance rules within the DD-based solver DDO [15]. To this end, four DP formulations were implemented and applied to the associated benchmark instances. For all problems, 600 s were given to solve each instance to optimality on a single thread, with the techniques described in [9,14] enabled by default. We first give a high-level description of the DP models and dominance rules of each problem, and of the benchmark instances and settings used before discussing the results of the experiments. Whereas the given definition of the dominance utility assumes that *greater is better*, the opposite rule is applied for minimization problems.

6.1   Experimental Setting

TSPTW. The *Traveling Salesman Problem with Time Windows* is a variant of the well-known *Traveling Salesman Problem* where the cities are replaced by a set of customers $N = \{0, \ldots, n - 1\}$ that must each be visited during a given time window. The objective is to find a tour starting and ending at customer 0 – the *depot* – and that visits all customers during their time window in the shortest possible time. The DP model used in the experiments is the one presented in [13], which extends the model introduced in [17] for the TSP. However, the present description omits state components that are not relevant for the dominance rule, but are useful to tighten the relaxation of the problem.

- **DP model:** the state representation contains a tuple (c, t, M), where $c \in N$ and t respectively represent the customer and time of the last visit made by the salesman. The set $M \subseteq N$ contains the customers that still must be visited. Starting from the root state $\hat{r} = (0, 0, N)$, the transitions then model the possible next visits of the salesman and the associated cost.
- **Dominance rule:** if two states represent the salesman at the same location and having visited the same set of customers, then the one arriving earlier is always preferred. This dominance rule can be expressed by specifying the following dominance key: $\kappa(s) = (s.c, s.M)$. Then, the utility of a state is given by the elapsed time: $\psi(s) = s.t$. We could also simply have $\psi(s) = \mathbf{0}$ since the elapsed time is also captured by the node value. However, the definition given is also valid for the *travel time* version of the TSPTW.

All configurations of the DD-based solver were tested on a classical set of benchmark instances introduced in the following papers [1, 11, 22, 28, 30]. A dynamic width was used, where the maximum width for layers at depth j is given by $n \times (j + 1) \times \alpha$ with n the number of variables in the instance.

ALP. The *Aircraft Landing Problem* requires to schedule the landing of a set of aircraft $N = \{0, \ldots, n - 1\}$ on a set of runways $R = \{0, \ldots, r - 1\}$. The aircraft have an earliest and latest landing time. Moreover, the set of aircraft is partitioned in disjoint sets $A_0, \ldots, A_{c-1}$ corresponding to c aircraft classes. For each pair of aircraft classes, a minimum separation time between the landings is given. The goal is to find a feasible schedule for all the aircraft, which minimizes the total waiting time – the delay between the earliest landing times and scheduled landing times – while respecting the latest landing times. The DP model presented in [24] was implemented, with a slightly different dominance rule.

- **DP model:** states are pairs (Q, ROP), with Q a vector that gives the remaining number of aircraft of each class to schedule and ROP a *runway occupation profile*: a vector containing pairs (l, c) that respectively give the time and aircraft class of the latest landing scheduled on each runway. The root state $\hat{r} = ((|A_0|, \ldots, |A_{c-1}|), ((0, \perp), \ldots, (0, \perp)))$ corresponds to the total number of aircraft to schedule for each class and an empty runway occupation profile, and the transitions model the next possible landings for each aircraft class and runway.
- **Dominance rule:** for a fixed remaining number of aircraft to schedule for each class and a same aircraft class previously scheduled on each runway, it is always better to have an earlier previous landing time if it comes with a better or equal objective function. This is expressed by the following dominance key and dominance utility vector: $\kappa(s) = (s.Q, (s.ROP_0.c, \ldots, s.ROP_{r-1}.c))$ and $\psi(s) = (s.ROP_0.l, \ldots, s.ROP_{r-1}.l)$.

A set of 720 random instances was generated, with $n \in \{25, 50, 75, 100\}$ aircraft, $r \in \{1, 2, 3, 4\}$ runways, and $c = 4$ aircraft classes. The target landing times

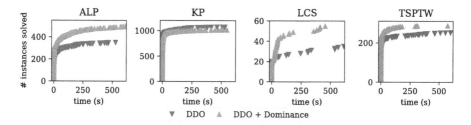

Fig. 3. Cumulative number of instances solved over time by DDO and DDO+D.

were generated by a Poisson arrival process with a mean inter-arrival time of $40/r$, instances with more runways thus require producing denser schedules. An arbitrary width of $W = 100$ is used for all experiments concerning the ALP.

LCS. The *Longest Common Subsequence Problem* considers a set of m strings $S = \{S_0, \ldots, S_{m-1}\}$ and asks for the longest *subsequence* that appears in all of them. We reproduced the formulation presented in [18] almost identically.

– **DP model:** states are defined as tuples $\langle p_0, \ldots, p_{m-1} \rangle$ that give the current position in each string. The root state $\hat{r} = \langle 0, \ldots, 0 \rangle$ corresponds to the beginning of each string, and the transitions model the insertion of one character at the end of the subsequence and adapt the current positions.
– **Dominance rule:** Given states with the same current position in string S_0, it is always better to have both lower other positions and a greater objective value. This is expressed by the following dominance key and dominance utility vector: $\kappa(s) = \mathbf{0}$ and $\psi(s) = s$.

We used the following classical benchmark instances: BB [6], BL [5], RAT, VIRUS and RANDOM [32], POLY and ABSTRACT [27], but limited to instances with $m < 10$. A fixed width of 100 is used for all experiments concerning the LCS.

KP. We solve the KP with all the ingredients presented in Examples 1 to 3. In addition, variables are ordered so that the items are considered in decreasing *profit-to-weight* ratios and the LP bound of [10] is used as an additional dual bound. A set of benchmark instances consisting of a random selection of 2% of the instances from [29] (636 instances) and 10% of the instances from [33] (530 instances). Again, a fixed width of 100 is used for all instances and configurations.

6.2   Results

Number of Instances Solved. Figure 3 shows the number of instances solved by each solver and configuration with respect to the solving time. For all problems except the KP, DDO with dominance rules enabled – referred to as DDO+D from now on – solves more instances than DDO, and by a large margin. As a

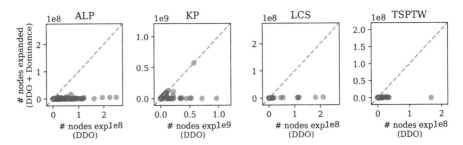

Fig. 4. Comparison of the number of node expansions performed by DDO and DDO+D for each instance solved by both configurations.

result, we can confidently say that the integration of dominance rules has a very positive impact on the performance of the B&B algorithm. In the case of the KP, this low performance gain can be attributed to the fact that, when using the *profit-to-weight* ratio variable ordering and the LP bound, many unpromising partial solutions are quickly discarded and thus fewer dominance relations arise.

Number of Node Expansions. The impact of the dominance rules can also be measured in terms of the number of nodes expanded during the successive approximate DD compilations. Figure 4 shows a pairwise comparison of this measure for each instance solved by both configurations. It appears that solving additional instances with DDO+D is made possible by a reduction in the number of node expansions needed to close them. Moreover, it clearly shows the magnitude of the filtering brought by the dominance rules, since for many instances that are unsolved by DDO, DDO+D requires only a negligible amount of node expansions to close them. For the KP, however, it seems that the decrease in node expansions is more significant than the decrease in time. This is probably due to the formulation of the dominance rule that needs to perform dominance checks for all pairs of nodes belonging to the same layer. For this kind of dominance checks, it might be worth considering a more specialized data structure than a simple list for the *Fronts*, such as k-d trees [2].

Quality of the First Solution. Another important dimension for a solver is the quality of the first solution found, which captures its *anytime* behavior. Figure 5 compares the value of the first solution found by DDO and DDO+D for each instance, as well as the iteration – in terms of B&B nodes – at which this solution is found for the TSPTW and the ALP. Indeed, those problems both have time window constraints that can make it difficult to find feasible solutions. This comparison allows us to make several observations for each problem:

- ALP: the quality of the first solution found by DDO+D is in general slightly better than the one obtained by DDO – in 314 cases, while the opposite occurs in 118 cases. Furthermore, there are 29 instances for which DDO does not

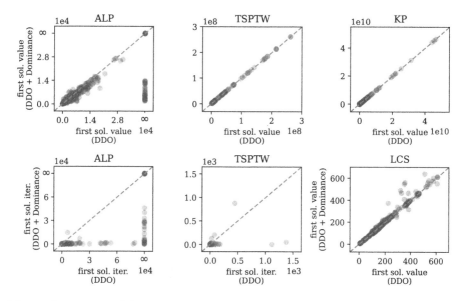

Fig. 5. Comparison of the value of the first solution obtained for each instance by DDO and DDO+D, and of the iteration at which it is found for ALP and TSPTW.

manage to find a single feasible solution to the problem, unlike DDO+D. In addition, when comparing the iteration at which the first solution is found, it appears that DDO+D finds a solution earlier than DDO for 235 instances, whereas the opposite is true for only 26 instances.

- TSPTW: the quality of the first solution found is only moderately impacted by the addition of dominance rules. Still, DDO+D finds a better first solution than DDO in 32 cases, against 4 cases in the other direction. Moreover, when looking at the iteration at which this first solution is found, we can see that DDO+D obtains it slightly earlier in the search for 34 instances, whereas the opposite is true for only 8 instances.
- LCS and KP: they are both maximization problems, so this time DDO+D compares better for data points located above the diagonal line. Although it is difficult to distinguish the solution values for the KP on Fig. 5, DDO+D actually finds a slightly better solution than DDO in 229 cases, compared to only 20 cases in the opposite direction. For LCS, this occurs for 201 instances, whereas DDO finds a better first solution in 93 cases.

Integrating dominance rules therefore also contributes to quickly producing quality solutions by moving the compilation of restricted DDs away from parts of the search space that are not worth spending time on.

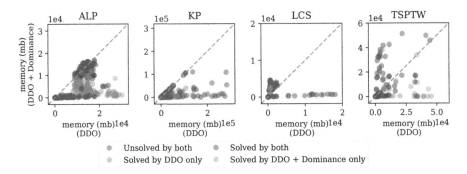

Fig. 6. Comparison of the peak amount of memory used for each instance by DDO and DDO+D, shown in different colors depending on which configurations solved it.

Memory Consumption. Finally, we discuss the memory footprint of maintaining the *Fronts* used to continually derive dominance relations with respect to non-dominated exact nodes previously found. Figure 6 compares the peak amount of memory used by both configurations for each instance. For all problems, we observe that a lower amount of memory is used by DDO+D for the large majority of instances solved by both configurations or only by DDO+D. Even for instances that both configurations fail to solve, DDO+D does not necessarily require more memory than DDO – the only exception is the TSPTW, although the maximum amount of memory used by DDO+D in those cases is not much larger than that reached by DDO for some instances.

7   Conclusion

In this paper, we proposed a formalism for specifying dominance rules of DP models. We then explained how they can be exploited within the DD compilation algorithm, as well as for the B&B algorithm as a whole by introducing a persistent data structure used to detect dominance relations across DD compilations. In addition, we showed how to combine this filtering mechanism with the caching procedure introduced in [9]. The modeling of the dominance rules was illustrated on four optimization problems and its impact was evaluated through extensive computational experiments. The results clearly highlight the interest of this additional ingredient, which significantly reduces the number of node expansions required by the algorithm to close the instances. This is directly reflected by the corresponding solving times, and leads to the resolution of many instances previously unsolved by DDO. Moreover, the experiments demonstrate the beneficial effect that dominance rules have in the ability of the algorithm to quickly find quality solutions, especially when the problem is highly constrained. Finally, the proposed memory-based dominance filtering procedure was shown to reduce the memory consumption of the solver in most cases.

References

1. Ascheuer, N.: Hamiltonian path problems in the on-line optimization of flexible manufacturing systems. Ph.D. thesis, University of Technology Berlin (1996)
2. Bentley, J.L.: Multidimensional binary search trees used for associative searching. Commun. ACM **18**(9), 509–517 (1975)
3. Bergman, D., Cire, A.A., van Hoeve, W.J., Hooker, J.N.: Discrete optimization with decision diagrams. INFORMS J. Comput. **28**(1), 47–66 (2016)
4. Bianco, L., Mingozzi, A., Ricciardelli, S.: The traveling salesman problem with cumulative costs. Networks **23**(2), 81–91 (1993)
5. Blum, C., Festa, P.: Longest common subsequence problems. In: Metaheuristics for String Problems in Bioinformatics, pp. 45–60 (2016)
6. Blum, C., Blesa, M.J.: Probabilistic beam search for the longest common subsequence problem. In: Stutzle, T., Birattari, M., H. Hoos, H. (eds.) International Workshop on Engineering Stochastic Local Search Algorithms, pp. 150–161. Springer, Heidelberg (2007). https://doi.org/10.1007/978-3-540-74446-7_11
7. Chambers, R.J., Carraway, R.L., Lowe, T.J., Morin, T.L.: Dominance and decomposition heuristics for single machine scheduling. Oper. Res. **39**(4), 639–647 (1991)
8. Chu, G., Stuckey, P.J.: Dominance breaking constraints. Constraints **20**, 155–182 (2015)
9. Coppé, V., Gillard, X., Schaus, P.: Decision diagram-based branch-and-bound with caching for dominance and suboptimality detection. INFORMS J. Comput. (2024)
10. Dantzig, G.B.: Discrete-variable extremum problems. Oper. Res. **5**(2), 266–277 (1957)
11. Dumas, Y., Desrosiers, J., Gelinas, E., Solomon, M.M.: An optimal algorithm for the traveling salesman problem with time windows. Oper. Res. **43**(2), 367–371 (1995)
12. Fischetti, M., Salvagnin, D.: Pruning moves. INFORMS J. Comput. **22**(1), 108–119 (2010)
13. Gillard, X.: Discrete optimization with decision diagrams: design of a generic solver, improved bounding techniques, and discovery of good feasible solutions with large neighborhood search. Ph.D. thesis, UCL-Université Catholique de Louvain (2022)
14. Gillard, X., Coppé, V., Schaus, P., Cire, A.A.: Improving the filtering of branch-and-bound mdd solver. In: Stuckey, P.J. (ed.) International Conference on Integration of Constraint Programming, Artificial Intelligence, and Operations Research, pp. 231–247. Springer, Heidelberg (2021). DOI: https://doi.org/10.1007/978-3-030-78230-6_15
15. Gillard, X., Schaus, P., Coppé, V.: Ddo, a generic and efficient framework for mdd-based optimization. In: Proceedings of the Twenty-Ninth International Conference on International Joint Conferences on Artificial Intelligence, pp. 5243–5245 (2021)
16. Haahr, J.T., Pisinger, D., Sabbaghian, M.: A dynamic programming approach for optimizing train speed profiles with speed restrictions and passage points. Transport. Res. Part B: Methodol. **99**, 167–182 (2017)
17. Held, M., Karp, R.M.: A dynamic programming approach to sequencing problems. J. Soc. Ind. Appl. Math. **10**(1), 196–210 (1962)
18. Horn, M., Raidl, G.R.: A*-based compilation of relaxed decision diagrams for the longest common subsequence problem. In: Stuckey, P.J. (ed.) International Conference on Integration of Constraint Programming, Artificial Intelligence, and Operations Research, pp. 72–88. Springer, Heidelberg (2021). DOI: https://doi.org/10.1007/978-3-030-78230-6_5

19. Ibaraki, T.: The power of dominance relations in branch-and-bound algorithms. J. ACM (JACM) **24**(2), 264–279 (1977)

20. Kohler, W.H., Steiglitz, K.: Characterization and theoretical comparison of branch-and-bound algorithms for permutation problems. J. ACM (JACM) **21**(1), 140–156 (1974)

21. Kuroiwa, R., Beck, J.C.: Domain-independent dynamic programming: generic state space search for combinatorial optimization. In: Proceedings of the International Conference on Automated Planning and Scheduling, vol. 33, pp. 236–244 (2023)

22. Langevin, A., Desrochers, M., Desrosiers, J., Gélinas, S., Soumis, F.: A two-commodity flow formulation for the traveling salesman and the makespan problems with time windows. Networks **23**(7), 631–640 (1993)

23. Lee, J.H., Zhong, A.Z.: Exploiting functional constraints in automatic dominance breaking for constraint optimization. J. Artif. Intell. Res. **78**, 1–35 (2023)

24. Lieder, A., Briskorn, D., Stolletz, R.: A dynamic programming approach for the aircraft landing problem with aircraft classes. Eur. J. Oper. Res. **243**(1), 61–69 (2015)

25. Mears, C., De La Banda, M.G.: Towards automatic dominance breaking for constraint optimization problems. In: Twenty-Fourth International Joint Conference on Artificial Intelligence (2015)

26. Morrison, D.R., Jacobson, S.H., Sauppe, J.J., Sewell, E.C.: Branch-and-bound algorithms: a survey of recent advances in searching, branching, and pruning. Disc. Optim. **19**, 79–102 (2016)

27. Nikolic, B., Kartelj, A., Djukanovic, M., Grbic, M., Blum, C., Raidl, G.: Solving the longest common subsequence problem concerning non-uniform distributions of letters in input strings. Mathematics **9**(13), 1515 (2021)

28. Pesant, G., Gendreau, M., Potvin, J.Y., Rousseau, J.M.: An exact constraint logic programming algorithm for the traveling salesman problem with time windows. Transp. Sci. **32**(1), 12–29 (1998)

29. Pisinger, D.: Where are the hard knapsack problems? Comput. Oper. Res. **32**(9), 2271–2284 (2005)

30. Potvin, J.Y., Bengio, S.: The vehicle routing problem with time windows part ii: genetic search. INFORMS J. Comput. **8**(2), 165–172 (1996)

31. Righini, G., Salani, M.: Decremental state space relaxation strategies and initialization heuristics for solving the orienteering problem with time windows with dynamic programming. Comput. Oper. Res. **36**(4), 1191–1203 (2009)

32. Shyu, S.J., Tsai, C.Y.: Finding the longest common subsequence for multiple biological sequences by ant colony optimization. Comput. Oper. Res. **36**(1), 73–91 (2009)

33. Smith-Miles, K., Christiansen, J., Muñoz, M.A.: Revisiting where are the hard knapsack problems? via instance space analysis. Comput. Oper. Res. **128**, 105184 (2021)

ViolationLS: Constraint-Based Local Search in CP-SAT

Toby O. Davies$^{(\boxtimes)}$, Frédéric Didier, and Laurent Perron

Google Research, Paris, France
tobyodavies@google.com, fdid@google.com, lperron@google.com

Abstract. We introduce ViolationLS, a Constraint-Based Local Search (CBLS) solver for arbitrary MiniZinc models based on the Feasibility Jump Mixed-Integer Programming Heuristic. ViolationLS uses a novel approach for finding multi-variable moves without requiring any "implicit" or "one-way" constraints traditionally used by CBLS solvers. It significantly outperforms other state-of-the-art CBLS solvers on the MiniZinc challenge benchmarks, both with and without multi-variable moves. We demonstrate that integrating ViolationLS into the parallel portfolio framework of the state-of-the-art CP-SAT Lazy Clause Generation solver improves performance on more than half of the instances in the MiniZinc challenge benchmarks suite. To our knowledge this is the first instance of such an integration of CBLS and Lazy Clause Generation inside the same solver.

1 Background

1.1 Constraint-Based Local Search

Constraint-Based Local Search (CBLS) [6] is an approach to local search that takes inspiration from Constraint Programming. In particular, both exploit the fact that many combinatorial sub-structures occur in many different problems to build models using reusable "constraints" based on these structures.

A CBLS model consists of an objective function, a set of variables, and a set of constraints. Constraints in CBLS each define a "violation" function given an assignment to the variables that takes the value 0 if the constraint is satisfied, and a positive value otherwise. A CBLS solver tries to find an assignment such that all constraints have a violation of 0, while secondarily minimising the objective.

CBLS solvers typically either require explicit neighbourhoods to be provided by the modeller, or provide "implicit constraints" [2] that act as neighbourhoods and define moves and initialisation logic that guarantee that the implicit constraint is initially satisfied and remains satisfied after each move. To create an overall neighbourhood from a set of implicit constraints, each variable may participate in at most one implicit constraint, and any variable in an implicit constraint must only be initialised and modified by moves generated by that constraint.

© The Author(s), under exclusive license to Springer Nature Switzerland AG 2024
B. Dilkina (Ed.): CPAIOR 2024, LNCS 14742, pp. 243–258, 2024.
https://doi.org/10.1007/978-3-031-60597-0_16

T. O. Davies et al.

Additionally CBLS solvers usually use "one way constraints" [2], these must form a Directed Acyclic Graph of functional definitions, used to reduce the number of decision variables. These allow other soft constraints and the objective to depend on the auxiliary variables while reducing the search space.

In order to support arbitrary models, CBLS solvers must define a "soft" version of implicit constraints, as the structure of the problem may contain variables that participate in multiple implicit constraints. Solvers must also provide a soft version of all one-way constraints as arbitrary constraint graphs may contain cycles that must be broken using soft constraints.

1.2 CP-SAT

The CP-SAT solver (also called CP-SAT-LP) [10] is a state-of-the-art Lazy Clause Generation (LCG)-based solver [11]. CP-SAT has won multiple medals in the MiniZinc Challenge [12] every year since it replaced the "CP" solver as the OR-tools entry in 2013.

In its multi-threaded configuration, CP-SAT implements an information-sharing portfolio of diverse subsolvers. There is a distinction between 3 different kinds of subsolvers: "full" subsolvers; "first solution" subsolvers; and "incomplete" subsolvers.

"Full" subsolvers exist for the full duration of the solve. Most "full" workers implement depth-first searches using various branching and restart heuristics, using a Lazy Clause Generation propagation and explanation engine to prune variable domains after each decision and avoid repeated search when back-jumping.

"First solution" workers terminate as soon as the first solution is found, and are replaced with "incomplete" workers. Most "first solution" workers are variants of one of the "full" workers, with additional randomness. Most "incomplete" workers perform Large Neighbourhood Search (LNS), using various neighbourhoods. These neighbourhoods fix part of the solution to be the same as a base solution sampled from a shared solution pool, and are then solved by creating a new CP-SAT instance, presolving the smaller problem, and solving the problem using a single-thread LCG worker.

CP-SAT expects there to be more incomplete solvers than threads to run them, so whenever a thread becomes available, an incomplete subsolver is chosen to generate and run a short-lived "task". The incomplete subsolvers are chosen in a round-robin pattern.

Workers share information between one another by periodically synchronising with various thread-safe shared state classes. Two such classes are relevant to this paper. First, the Shared Solution Manager, which maintains a small pool of the best 3 solutions seen, and the upper and lower bounds on the objective. Second, the Shared Bounds Manager, which allows full workers to share root-level reductions of variable domains with other workers.

Another important component of CP-SAT is its presolver which runs before any subsolvers start, modifying the model in various ways. A full description of these transformations is beyond the scope of this paper, but they are summarised here. The presolver replaces some complex constraints with simpler ones (often a set of linear constraints); adds symmetry breaking constraints; and substitutes equivalent variables. The CBLS solvers we have implemented and described in this paper all use CP-SAT's presolver.

The presolver may act similarly to limited "one-way" constraints in this final regard as many CP-SAT constraints accept affine transformations of the form $ax + b$ in place of a variable, where a and b are constants and x is a variable. This allows some variables to be removed, but is still far more restrictive than the general functional dependencies enabled by one-way constraints in other CBLS solvers, which could for example remove the variable y in $y = x^2$ from the set of decision variables.

2 Violation Functions

Each constraint in a CBLS solver defines a violation that is 0 when the constraint is satisfied and strictly positive otherwise. Our solver defines a decomposition or violation function for every type of constraint supported natively by CP-SAT.

Constraints in our solver have a simple API, they need to be able to: compute their initial violation given a new assignment; compute the violation delta when a single variable changes value; and update their internal state when a single variable changes value. We use the notation $\delta_G(v, j)$ to denote the function returning a sparse vector containing the change in violation of each constraint in the constraint graph G when variable v changes its value to j.

These three operations on linear constraints are identical to those in the paper introducing Feasibility Jump [8] (modulo half-reification, explained below [3]).

In the implementation tested in this paper all non-linear constraints compute their violation delta by computing their violation, temporarily changing the value of the variable in question, recomputing their violation from scratch, then resetting the variable and returning the difference between the two violations. We expect significant improvements in performance on larger scheduling and routing instances by improving this aspect of the implementation as the computation of violation for these types of constraints in particular is expensive.

To maximise the benefit of making linear constraint violation efficient and incremental to compute, many constraints in our implementation are partially or fully decomposed into a set of linear constraints instead of or in addition to defining a general constraint. Our implementation generally fully linearises a constraint whenever a complete encoding of the constraint requires no new variables and at-most a linear number of additional constraints. Additionally it will partially linearise the violation function whenever doing so requires no new variables, at most a linear number of new constraints, and the linear constraints form a sparser constraint graph than the global would.

CP-SAT has native support for half-reified versions of many constraints. Half-reified versions of the constraints defined below are the same, except have violation defined to be 0 if the constraint is disabled in the current assignment.

Linear constraints define violation as in the original Feasibility Jump paper: the absolute value of the difference between the left and right hand side of the constraint if violated, or 0 if the constraint is satisfied.

Boolean constraints are transformed into equivalent linear expressions, except XOR which is defined to have violation 0 if satisfied, and 1 otherwise.

Min (and max) constraints where $t = min(xs)$ $(t = max(xs))$ define violation by decomposing into a $t \leq x$ $(t \geq x)$ constraints for each variable $x \in xs$. Thus the linear submodel enforces that t is a lower (upper) bound on the other arguments, and we also add a non-linear constraint $t \geq min(xs)$ $(t \leq max(cs))$. This definition creates a sparser graph in the linear submodel, so fewer variables will be considered when repairing its violation if few variables are close to t.

Circuit constraints in CP-SAT are slightly unusual compared to other CP solvers in that they are defined in terms of a binary variable for each arc, rather than a "next" array of integers. The violation of this constraint is decomposed into the flow-conservation linear constraints (in-arcs and out-arcs for each node must sum to 1), plus a subcircuit-elimination constraint. The subcircuit-elimination constraint considers the graph containing only arcs whose controlling literals are 1. Its violation is defined to be 0 if all Strongly-Connected Components (SCCs) are singletons (i.e. contain only a single node). Otherwise, the violation is the number of non-singleton strongly connected components (SCCs), minus 1 (as a valid circuit contains at most 1 non-singleton SCC), plus the number of non-singleton SCCs that do not have an active arc from any node in that SCC to any other node in a different non-singleton SCC, minus 1. An important property of this definition is that removing arcs from a valid tour leaves the subcircuit-elimination constraint satisfied (though flow constraints will become violated), only introducing a second cycle causes it to become violated.

Route constraints are nearly identical to circuit constraints, but the subcircuit-elimination constraint also adds 1 to its violation if the size of the SCC containing the depot is one, unless there are no non-unit-size SCCs at all. Also the in and out arcs at the depot must merely equal one another, rather than equalling exactly 1.

Cumulative and Disjunctive constraints violations are defined to be the total area of above the capacity of the cumulative (above 1 for Disjunctive constraints).

2D Packing constraints are implemented using a pairwise decomposition, and the violation of each pairwise constraint is the area of overlap of the two rectangles.

3 Generalised Feasibility Jump

Our two closely-related CBLS solvers are both based on extending the Feasibility Jump heuristic [8] designed for Mixed-Integer Programming to work with general constraints. Feasibility Jump is not described as nor compared to CBLS, we believe only because CBLS is not well-known in the mathematical programming community. For all intents and purposes Feasibility Jump is a type of CBLS solver restricted to Linear Constraints.

Feasibility Jump is novel compared to existing CBLS solvers in that it does not require constraints to define separate implicit, one-way, and soft versions. Nor for the solver to have to choose which of these implementations each particular instance of a constraint should use. In Feasibility Jump, all constraints are soft. Using only soft constraints has been shown to have a significant negative effect on the performance of another CBLS solver [2], but in Sect. 6 we show that our Generalised Feasibility Jump outperforms other CBLS solvers by a significant margin.

Generalised Feasibility Jump and ViolationLS both maintain very similar state, to simplify presentation we define a state tuple that we use extensively below and in our algorithm listings.

$$S = \langle G, X, W, V, Q, J \rangle$$

Where G is a constraint graph containing variables and constraint, G_c denotes the set of variables in constraint c, and G_v is the set of constraints containing variable v. Each variable v in G has a finite domain of possible values it may take, denoted $D(v)$. X is a valuation of the variables in G, where $X[v] \in D(v)$. W is a vector with a strictly positive weight for each constraint c. V is the subset of constraints in G that are violated in the assignment X. Q the set of variables that may be valid moves, and J is a "jump table" containing either a "jump value" and "score" (explained below) or nil for each variable in G.

Feasibility Jump repeatedly selects a variable v and sets its value to the value $j \in D(v)$ (excluding the current value of v) that minimises the weighted sum of violations over all constraints. This value j is called the variable's jump value, and is cached in J to maximise incrementality. The reduction in weighted violation from assigning a variable to its jump value is the variable's score, $-W \cdot \delta_G(v, j)$. Weights (W) are initialised to 1, and the weights of violated constraints are increased by 1 whenever no variable has positive score. We also implement a variant where weights are decayed by $\rho \in (0, 1)$ before weights are updated.

Note that our implementation of Generalised Feasibility Jump differs from that described in the original paper [8] in 2 key ways when applied to a model containing only linear constraints. First, we re-compute the jump values and scores of a variable v whenever any variable that shares a constraint with v changes, in the original, only v's jump value would be recomputed, and others would be recomputed at fixed point. Second, we apply the best jump of 5 with positive score, whereas the original algorithm selects 25 random variables that participate in a violated constraint, and if none have negative score then weights of all violated constraints are increased.

To compensate for some of the extra work due to the first difference above, we use the fact the jump value cannot change for binary variables to only consider the constraints that have been updated when computing the score, using the same logic used for all variables in the original algorithm. For other variables, we compute jumps and scores lazily, whenever a variable is updated in Algorithm 1 we invalidate the jump for adjacent variables, and recompute as needed in Algorithm 2.

Feasibility Jump exploits the structure of linear constraints to compute a subset of values for a variable that must include one with minimal score. This structure does not apply to general constraints, so in our generalisation we instead assume a slightly weaker property: that each violation function is convex when we change a single variable within its domain. This allows us to use a convex minimisation algorithm like ternary search to compute a variable's jump value and score, since a positive-weighted sum of convex functions is convex.

Note that it does not actually impact the correctness of any of our algorithms if violation functions are not in fact convex. The only difference is that the search is more restricted, as we may incorrectly conclude that a variable has no good jump values.

Our implementation maintains a set of variables to scan Q, initialised with all variables participating in any violated constraint. The main loop repeatedly samples from that set, setting variables with positive score to their jump value, then adding any variable in any constraint touched by the changed variable into the set to scan. When no more improving jumps can be found, it increases the weights of all violated constraints by 1, re-initialises Q and repeats. This is a type of Guided Local Search (GLS) [13], shown in Algorithm 3.

Our generalisation first finds a solution to the linear subset of constraints, by performing the GLS using only the linear submodel, then applies the same algorithm to the full model, starting from the solution found in the first phase. This allows us to minimize the impact of relatively inefficient global constraint evaluation. This simple algorithm alone finds a surprising number of solutions to problems in the MiniZinc Challenge instances, as shown in Sect. 6.

4   Novelty Jump

Local-search solvers' moves usually affect more than one variable at a time in order to maintain desirable invariants (notably, maintaining the satisfaction of all implicit constraints in CBLS). Many algorithms to generate such moves search for a sequences of smaller component changes, where later components repair invariants broken earlier in the sequence. The Lin-Kerighan TSP heuristic [5] can be viewed as using a neighbourhood that chains dependent sequences of changes. Similarly, the "Ejection Chain" family of heuristics explicitly attempt to find sequences of changes [4].

To take advantage of the incremental updates used by Feasibility Jump, we search for such sequences using a depth-first search, using modified constraint weights to find values likely to allow chains to continue. Our approach to modifying the weights used in the search is inspired by the concept of Novelty used in AI planning [7], in particular Novelty Jump is very similar to a sequence of calls to a depth-first variant of $IW(1)$.

Algorithm 1: UpdateVar(S, v)

Input:
  S: A state tuple $\langle G, X, W, V, Q, J \rangle$
  v: The variable to update
Output: S is modified

1 $j, s \leftarrow J[v]$
2 $x_0 \leftarrow X[v]$
3 $X[v] \leftarrow j$
4 **foreach** $c \in G_v$ **do**
5 | Update the internal state of c with the new value of v
6 | **if** $v(X, c) > 0$ **then**
7 | | $V \leftarrow V \cup \{c\}$
8 | **else**
9 | | $V \leftarrow V \setminus \{c\}$

10 **foreach** $c \in G_v$ **do**
11 | **foreach** $v' \in G_c \setminus \{v\}$ **do**
12 | | **if** v' *is binary and* c *is linear* **then**
13 | | | Update the score of v' in J as in [8]
14 | | **else**
15 | | | $J[v'] \leftarrow$ nil
16 | | **if** $V \cap G_{v'} \neq \emptyset$ **then**
17 | | | $Q \leftarrow Q \cup \{v'\}$

$IW(1)$ performs a structured exploration of the state-space, requiring each transition to set at least one state variable to a "novel" value not seen since the current best state. Analogously in Novelty Jump, we use modified weights to make it significantly cheaper to increase the violation of any constraint that has not been violated since the best state seen so far.

These weights prioritise jump values that fix constraints that are initially broken, in the hope that constraints that have not yet been broken can be fixed later, by further changes. If backtracking is necessary, subsequent children will prefer to avoid breaking the same constraints as were broken in prior exploration, making the search more structured than a purely random exploration.

Novelty Jump initialises the weights of currently violated constraints to the same value as used in the Guided Local Search metaheuristic, and sets non-violated constraint weights to ϵ times the GLS weight ($\epsilon = 2^{-10}$ in our implementation). Whenever we take a move, the weight of any newly violated constraints are set to the GLS weight. Jump values used in Novelty Jump are computed using these "novelty weights", to avoid confusion with scores using the original weights we will use "score" to refer to the score with respect to the original weights unless otherwise stated, and "novelty score" of a jump value to be the score computed with the two weights.

Algorithm 2: ApplyJump(S)

Input: S: A state tuple $\langle G, X, W, V, Q, J \rangle$
Output: S is modified, returns False if no move could be found.

1   $v, j, s \leftarrow$ nil, nil, 0
2   $n \leftarrow 0$
    // Sample 5 vars from Q and keep the one with the best score
3   **while** Q *is not empty* **do**
4 | $v' \leftarrow$ a random variable from Q
5 | **if** $J[v'] = nil$ **then**
6 | | $j' \leftarrow \underset{j' \in D(v')}{\text{ConvexArgMin}}(W \cdot \delta_G(v', j'))$
7 | | $J[v'] \leftarrow j', -W \cdot \delta_G(v', j')$
8 | $j', s' \leftarrow J[v']$
9 | **if** $s' <= 0$ **then**
10 | | $Q \leftarrow Q \setminus \{v\}$
11 | | **continue**
12 | **if** $s' ¿ s$ **then**
13 | | $v, j, s \leftarrow v', J[v'], s'$
14 | $n \leftarrow n + 1$
15 | **if** $n \geq 5$ **then**
16 | | **break**

17   **if** $v = nil$ **then return** False
18   UpdateVar(S,v)
19   **return** True

During depth-first search if we can't find any way to extend the compound move given the filtering mechanisms described below, we backtrack. When backtracking we revert the last assignment, but not the changes to any novelty weights, which we update only when we fail to find any move at the root.

Note that whenever a constraint is violated, its novelty weight is equal to the original weight. Thus reductions in violation of currently violated constraints are responsible for all positive terms in the computation of a variable's score. Consequently the novelty score of a jump cannot be lower than its score, as only the negative terms can have smaller weights. So long as the depth-first search always explores at least one positive score move at the root (if one exists), any state that is a local minimum for Novelty Jump must also be a local minimum for Feasibility Jump.

Our preliminary investigation showed that exploring all moves with negative novelty score explores too much, taking too long to detect a local minimum and increase the GLS weights. To combat this, Novelty Jump limits exploration in three ways. First, we do not change the same variable twice in the same compound move.

Second we limit the number of backtracks along any path, once this is exceeded we backtrack until we reach a node with remaining backtrack bud-

---

Algorithm 3: GLS(G, X, W, M, ρ)

Input:
    G: A bipartite graph of variables and constraints,
    X: An assignment from variables to values
    W: A weight per constraint
    M: A "move" function that modifies X or returns False
    ρ: A discount factor for weights
Output:
    X, and W are modified

1 $V \leftarrow \{c \,|\, c(X) > 0, \quad \forall c \in \text{Constraints}(G)\}$
2 $Q \leftarrow \bigcup_{c \in V} G_c$
3 $J[v] \leftarrow \text{nil} \qquad \forall v \in Variables(G)$
4 **while** *effort limit is not reached* **do**
5   **if** $M(\langle G, X, W, V, Q, J\rangle) = False$ **then**
6   **if** $V = \emptyset$ **then**
7   **return** FEASIBLE
8      $W \leftarrow \rho W$
9   **foreach** $c \in V$ **do**
10        $W[c] \leftarrow W[c] + 1$
11   **foreach** $v \in G_c$ **do**
12          $S[v] \leftarrow \text{nil}$

13 **return** UNSOLVED

---

get $b > 0$. If we backtrack to the root, the limit is increased, to a maximum of 2. If the search fails with its maximum limit, we increment weights like in feasibility jump, re-initialise the novelty weights and continue.

Third, we track the cumulative score of individual moves along the path (s_m), and for each node on the stack we track the best score of any immediate child move that has previously been explored (s_c, initially 0). To be considered in the search, a move must either have sufficient novelty that the novelty score plus s_m is positive, or the score is greater than the s_c value of the current search node. This is defined by the filter predicate F, used in Algorithm 5:

$$F(G, W, W', s_m, s_c, v, j) \leftrightarrow (s_m - W' \cdot \delta_G(v, j) > 0) \vee (-W \cdot \delta_G(v, j) > s_c)$$

This definition allows 2 categories of move to be explored. The first criterion allows chains of values to be swapped between variables. The second allows multiple small changes to combine to repair a large change that was allowed by the first criterion.

For example if the first criterion allowed search to explore $x = 1$ with a score of –1, the constraint $x \rightarrow y + z = 0$ would require at least 2 moves to repair if $y = z = 1$, but these two moves would both have positive score so either jump would be allowed at the first node, and the other would be allowed as a child of that node (assuming its score was still positive). However if the search later backtracked over these moves, the sequence would *not* be explored again, as the score would not be higher than the best score explored previously (s_c).

The second criterion alone would obviously not be sufficient to perform exploration as this would only explore positive score moves which would be immediately committed, making this algorithm nearly equivalent to Feasibility Jump. However it does guarantee that Novelty Jump will always explore at least one positive score move (if one exists) at the root node of the tree, ensuring its local minima are at least as good as Feasibility Jump.

Another property of this filtering mechanism to note is that no two moves causing the same violation delta can be explored as children of the same node. After a move $X[v] = j$ has been explored W' is updated so that $W \cdot \delta_G(v, j) = W' \cdot \delta_G(v, j)$, so the first criterion cannot apply to the new move if $\delta_G(v', j') = \delta_G(v, j)$, and the new move must also have the same score as the previously explored move, so cannot have a better score than all previously explored moves, thus the second criterion cannot apply.

If, at any point in the search, the sum of scores of single-variable moves on the stack is positive, we commit the compound move, and reset the per-node backtrack limit to 0. This resetting means that the algorithm limits exploration further when moves are easy to find.

We provide pseudo-code for this algorithm in Algorithm 4, and the recursive helper function shown in Algorithm 5. Our actual implementation is iterative, with an explicit stack of moves instead of using the call stack, which is somewhat more efficient, but more cumbersome to present.

Algorithm 4: ApplyNoveltyJump(S)

Input: S: A state tuple $\langle G, X, W, V, Q, J \rangle$
Output: S and W' are modified, returns False if no move could be found.

1   $b \leftarrow 0$
2   **while** $b \leq 2$ **do**
3      $W' \leftarrow \epsilon W$
4   **foreach** $c \in V$ **do**
5         $W'[c] \leftarrow W[c]$
6   **while** $NoveltyJumpSearch(S, W', 0, b, \emptyset)$ **do**
7   **if** $V = \emptyset$ **then return** True
8         $b \leftarrow 0$
9      $b \leftarrow b + 1$
10 **return** False

Algorithm 5: NoveltyJumpSearch(S, W', s_m, b, T)

Input:

    S: A state tuple $\langle G, X, W, V, Q, J \rangle$

    W': A "novelty" weight per constraint

    s_m: The score of the partial move so far

    b: The remaining backtrack limit for the search

    T: The set of variables on the stack, not eligible for moves.

Output: S, and W' are modified, returns False if no move can be applied.

1   **if** $b < 0$ **then return** False

2   $s_c \leftarrow 0$

3   **while** $Q \supset T$ **do**

4   **while** $b \geq 0$ **do**

5       $v \leftarrow$ Best of 3 sampled vars in $Q \setminus T$ satisfying $F(G, W, W', s_m, s_c, v, j)$

6   **if** $v = nil$ **then return** False

7       $Q \leftarrow Q \setminus \{v\}$

8       $j \leftarrow \underset{j' \in D(v')}{\text{ConvexArgMin}}(W' \cdot \delta_G(v', j'))$

9       $s \leftarrow -W \cdot \delta_G(v, j)$

10      $s_c \leftarrow \min(s, s_c)$

11      $x_0 \leftarrow X[v]$

12   UpdateVar(S,v)

13   **foreach** $c \in V \cap G_v$ **do**

14   **if** $W'[c] \neq W[c]$ **then**

15         $W'[c] \leftarrow W[c]$

16         $Q \leftarrow Q \cup G_c$

17   **if** $s_m + s > 0$ **then return** True

18   **if** $NoveltyJumpSearch(S, W', s_m + s, b, T \cup \{v\})$ **then**

19   **return** True

20      $J[v] \leftarrow x_0, -s$

21      $UpdateVar(S, v)$

22      $b \leftarrow b - 1$

23 **return** False

---

5   ViolationLS

Our overall algorithm, ViolationLS (Algorithm 6) picks either Feasibility Jump or Novelty Jump (in our experiments, with 50% probability each), and runs it for a "batch". The stopping criteria for a batch is complicated and is not particularly scientifically interesting, as choosing a good value mostly depends on the other solvers in the "incomplete" and "first solution" worker pools rather than any property of the algorithm itself.

At the start of each batch, if there is a new best solution in the solution pool or we have seen no solutions for 100 batches, we choose the algorithm again. Otherwise we continue with the same algorithm as last batch, reusing the same weights from the end of the last batch.

In our experiments below, where we report the performance of "Feasibility Jump", we use a 100% probability of choosing Feasibility Jump in this algorithm.

Algorithm 6: ViolationLS(G)

Input: G: A bipartite graph of variables and constraints
1 $W[c] \leftarrow 1 \quad \forall c \in \text{Constraints}(G)$
2 Initialize X to the value in each variable's domain closest to 0
3 $A \leftarrow \text{RandomChoice}(\{FJ, NJ\})$
4 **while** *The problem is not solved* **do**
5   **if** *If a new best solution S is available in the shared pool* **then**
6          $X \leftarrow S$
7   Tighten the RHS of the objective constraint.
8          $W[c] \leftarrow 1 \quad \forall c \in \text{Constraints}(G)$
9          $\rho \leftarrow \text{RandomChoice}(\{0.95, 1.0\})$
10          $A \leftarrow \text{RandomChoice}(\{FJ, NJ\})$
11   **if** *No new solutions found or imported for 100 iterations* **then**
12   Perturb X, randomising each variable's value with probability 0.1
13          $W[c] \leftarrow 1 \quad \forall c \in \text{Constraints}(G)$
14          $\rho \leftarrow \text{RandomChoice}(\{0.95, 1.0\})$
15          $A \leftarrow \text{RandomChoice}(\{FJ, NJ\})$
16   **if** $A = FJ$ **then**
17          $G' \leftarrow$ the linear submodel of G
18   **if** *GLS(G', X, W, ApplyJump, ρ) = FEASIBLE* **then**
19   **if** *GLS(G, X, W, ApplyJump, ρ) = FEASIBLE* **then**
20   Add X to the Shared Solution Pool
21   **else**
22   **if** *GLS(G, X, W, ApplyNoveltyJump, ρ) = FEASIBLE* **then**
23   Add X to the Shared Solution Pool
24   Yield to allow other incomplete solvers to run

---

6 Experimental Results

We compare our Feasibility Jump and ViolationLS purely local search solvers against 2 local search solvers that have previously won medals in the local search category of the MiniZinc challenge: Yuck [9], and fzn-oscar-cbls [2]. In addition we show the impact of adding our CBLS solvers to CP-SAT's "first solution" and "incomplete" subsolver pool by comparing several configurations of CP-SAT with and without these algorithms. Each solver was run 5 times with 5 different random seeds and a 5 min time limit, using 8 worker threads, as CP-SAT is designed for multi-threaded solving, and 8 cores is the minimum number of threads that CP-SAT recommends. Note that fzn-oscar-cbls is not multi-threaded, so can only take advantage of one core, but is included for completeness as it is better-described in the literature than Yuck. The experiments were performed on 64-cores machines (n2d-standard-64 Google Compute Engine instances), each running 8 solves concurrently.

In Table 1, we show the mean number of points scored by each solver, averaged across 5 different random seeds for each solver and instance. A solver scores a point for every other solver and instance where it solved that instance better than the other solver. For the "Complete" column, the better solver either closed the problem faster; found a better solution; or found the same quality solution faster (if neither solver closed the problem). For the "Incomplete" column, the better solver either found a better solution, or found the same quality solution faster. The percentage columns show what percentage of the maximum possible score each solver achieved.

Table 1. MiniZinc Challenge Scores, averaged over 5 random seeds

Solver	Complete		Incomplete	
	Points	%	Points	%
CP-SAT+ViolationLS	**5841.04**	**80.06%**	**5630.04**	**77.17%**
CP-SAT+Feasibility Jump	5812.60	79.67%	5583.20	76.52%
CP-SAT (No LS)	5643.76	77.35%	5406.68	74.10%
ViolationLS	2632.60	36.08%	3003.92	41.17%
Feasibility Jump	1788.08	24.51%	2050.24	28.10%
Yuck	1180.92	16.19%	1200.08	16.45%
fzn-oscar-cbls	487.32	6.68%	511.76	7.01%

Figure 1 shows that while ViolationLS helps substantially more often than it hinders time to prove optimality, the magnitude of improvement is often small, and it can occasionally cause substantial slowdown. We suspect two possible causes. First, some global constraints are slow to evaluate, this may cause CP-SAT to allocate disproportionate CPU-time to ViolationLS that would be better spent on LNS. Improving the incrementality of global constraints may mitigate this, and is likely to lead to better performance generally, there is significant room for improvement in our implementation.

Second, ViolationLS finds many (quite similar) solutions in quick succession. We suspect the solution pool becomes full of similar solutions before LNS workers have the opportunity to explore neighbourhoods around earlier solutions. However quickly finding nearby solutions is the goal of ViolationLS, so mitigating this would require changes to encourage diversity in the solution pool.

Table 2 shows the average number of times that each of the pure local-search solvers tested found a better solution faster than each other local-search solver. Similarly, Table 3 shows how often adding the two local search algorithms introduced in this paper outperformed the other CP-SAT configurations.

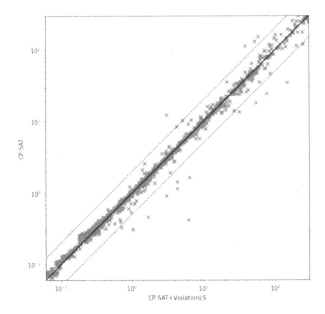

Fig. 1. Shifted geometric mean time to optimality for CP-SAT with and without Violation LS on each the MiniZinc challenge benchmark instances, averaged over 5 seeds, shifted by 10 s. The 609 instances (50.1%) above the diagonal are faster with ViolationLS, and 397 (32.6%) below are slower, 209 (17.2%) were consistently unsolved by either solver within 5 min. Dotted lines show 2x increase or decrease in solve time.

Table 2. Pairwise incomplete MiniZinc Challenge scores. Each cell shows the average number of instances where the solver in that row was better than the other solver in that column, using the "incomplete" definition of better. Averaged over 5 random seeds. The highest score in each column is highlighted.

Solver	ViolationLS	Feasibility Jump	Yuck	fzn-oscar-cbls
ViolationLS	–	**650.6**	**652.28**	**734.64**
Feasibility Jump	182.00	–	523.84	617.24
Yuck	**209.52**	268.96	–	435.72
fzn-oscar-cbls	101.96	135.36	147.6	–

Table 3. Pairwise complete MiniZinc Challenge scores for CP-SAT variants. Each cell shows the average number of instances where CP-SAT plus the heuristic in that row was better than CP-SAT plus the heuristic in that column, using the "complete" definition of better. Averaged over 5 random seeds. The highest score in each column is highlighted.

CP-SAT+	ViolationLS	Feasibility Jump	Baseline
ViolationLS	–	**609.20**	**659.04**
Feasibility Jump	**588.36**	–	651.80
Baseline	538.32	545.48	–

7   Conclusions and Further Work

Both our generalisation of Feasibility Jump and ViolationLS improve substantially on the current state-of-the-art local search solver for general models expressed in MiniZinc. ViolationLS finds better solutions faster compared to "Yuck" more than 3x as frequently as Yuck outperformed ViolationLS.

Including these algorithms in the incomplete subsolver framework inside the state-of-the-art CP-SAT improves performance in 20% more instances than it degrades it. However, it does slow solve time significantly on some instances. Mitigating this will make using ViolationLS more effective as a primal heuristic, but further investigation into the causes and possible solutions is required. We suspect improving the incrementality of global constraint violation computation to improve this significantly.

Global constraints can also make the constraint graph more densely connected, causing ViolationLS to needlessly recompute jumps for variables that cannot have positive score because they do not actually participate in the violation detected by the global constraint. For example if 2 intervals overlap in an otherwise satisfied Disjunctive constraint, there is no need to consider changing the value of any variables in any other intervals (unless these variables participate in other violations). It may be possible to extend the constraint API to allow ViolationLS to cheaply skip computing jumps for such variables, or to skip enqueing them entirely.

Another possible avenue for improvement would be to revisit the definition of the violations of global constraints. The theoretical basis of jump values relies on the convexity of violation functions, however several of our constraint definitions are non-convex (notably scheduling constraints). While this has no impact on correctness, there may be opportunities for improved or more consistent performance if the violation definitions of more constraints were convex.

The parameters in ViolationLS were chosen based only on very limited preliminary experiments. There will almost certainly be a benefit to automatically tuning this during search based on which configurations find solutions faster on a specific instance.

Finally, the Novelty Jump heuristic was inspired by treating the search for an improving neighbour as a domain-independent planning problem. There may be other promising avenues of research based on adapting other planning algorithms to create other effective, problem-independent CBLS neighbourhoods.

8   Related Work

Previous work [1] has explored adding local-search to CP-based MiniZinc solvers. This approach requires annotations to be written by the modeller to define the neighbourhood, so cannot be used with an arbitrary MiniZinc model.

References

1. Björdal, G., Flener, P., Pearson, J., Stuckey, P.J., Tack, G.: Declarative local-search neighbourhoods in minizinc. In: 2018 IEEE 30th International Conference on Tools with Artificial Intelligence (ICTAI), pp. 98–105. IEEE (2018)
2. Björdal, G., Monette, J.N., Flener, P., Pearson, J.: A constraint-based local search backend for MiniZinc. Constraints **20**, 325–345 (2015)
3. Feydy, T., Somogyi, Z., Stuckey, P.J.: Half reification and flattening. In: Lee, J. (ed.) CP 2011, vol. 6876, pp. 286–301. Springer, Heidelberg (2011). https://doi.org/10.1007/978-3-642-23786-7_23.pdf
4. Glover, F., Rego, C.: Ejection chain and filter-and-fan methods in combinatorial optimization. 4OR **4**, 263–296 (2006)
5. Helsgaun, K.: An effective implementation of the Lin-Kernighan traveling salesman heuristic. Eur. J. Oper. Res. **126**(1), 106–130 (2000)
6. Hentenryck, P.V., Michel, L.: Constraint-Based Local Search. The MIT press, Cambridge (2009)
7. Lipovetzky, N., Geffner, H.: Width and serialization of classical planning problems. In: European Conference on Artificial Intelligence (ECAI), pp. 540–545. IOS Press (2012)
8. Luteberget, B., Sartor, G.: Feasibility Jump: an LP-free Lagrangian MIP heuristic. Math. Program. Comput. **15**(2), 365–388 (2023)
9. Marte, M.: Yuck. https://github.com/informarte/yuck. Accessed 01 Nov 2023
10. Perron, L., Didier, F., Gay, S.: The CP-SAT-LP solver. In: Yap, R.H.C. (ed.) 29th International Conference on Principles and Practice of Constraint Programming (CP), Dagstuhl, Germany (2023)
11. Stuckey, P.J.: Lazy clause generation: combining the power of sat and cp (and mip?) solving. In: Lodi, A., Milano, M., Toth, P. (eds.) Integration of AI and OR Techniques in Constraint Programming for Combinatorial Optimization Problems, pp. 5–9. Springer, Heidelberg (2010). https://doi.org/10.1007/978-3-642-13520-0_3
12. Stuckey, P.J., Feydy, T., Schutt, A., Tack, G., Fischer, J.: The minizinc challenge 2008–2013. AI Mag. **35**(2), 55–60 (2014)
13. Voudouris, C., Tsang, E.P., Alsheddy, A.: Guided local search. In: Marti, R., Panos, P., Resende, M. (eds.) Handbook of Metaheuristics, pp. 321–361. Springer, Heidelberg (2010). https://doi.org/10.1007/978-3-319-07153-4_2-1

ULD Build-Up Scheduling
with Logic-Based Benders Decomposition

Ricardo Euler[1]([✉])[iD], Ralf Borndörfer[1][iD], Christian Puchert[2][iD],
and Tuomo Takkula[2][iD]

[1] Zuse Institute Berlin, Takustraße 7, 14195 Berlin, Germany
{euler,borndoerfer}@zib.de
[2] Ab Ovo Deutschland GmbH, Prinzenallee 9, 40549 Düsseldorf, Germany
{christian.puchert,tuomo.takkula}@ab-ovo.com

Abstract. We study a complex planning and scheduling problem arising from the build-up process of air cargo pallets and containers, collectively referred to as unit load devices (ULD), in which ULDs must be
assigned to workstations for loading. Since air freight usually becomes
available gradually along the planning horizon, ULD build-ups must be
scheduled neither too early to avoid underutilizing ULD capacity, nor
too late to avoid resource conflicts with other flights. Whenever possible, ULDs should be built up in batches, thereby giving ground handlers
more freedom to rearrange cargo and utilize the ULD's capacity efficiently. The resulting scheduling problem has an intricate cost function
and produces large time-expanded models, especially for longer planning
horizons. We propose a logic-based Benders decomposition approach that
assigns batches to time intervals and workstations in the master problem, while the actual schedule is decided in a subproblem. By choosing appropriate intervals, the subproblem becomes a feasibility problem
that decomposes over the workstations. Additionally, the similarity of
many batches is exploited by a strengthening procedure for no-good cuts.
We benchmark our approach against a time-expanded MIP formulation
from the literature on a publicly available data set. It solves 15% more
instances to optimality and decreases run times by more than 50% in the
geometric mean. This improvement is especially pronounced for longer
planning horizons of up to one week, where the Benders approach solves
over 50% instances more than the baseline.

Keywords: Air Cargo · Unit Load Device · Build-Up Scheduling ·
Batch Scheduling · Logic-Based Benders Decomposition

1 Introduction

Getting air cargo off the ground is a complex operation that requires solving a
sequence of challenging optimization problems. Brandt & Nickel [7] identify four

Supported by the Research Campus MODAL funded by the German Federal Ministry
of Education and Research (BMBF) [grant number 05M20ZBM].

main problems: aircraft configuration, build-up scheduling, air cargo palletiza-
tion, and finally the weight and balancing problem. In this paper, we consider a
variant of the build-up scheduling problem with an additional batching objective.

Many flights contain stop-overs at which ULDs may be discharged. A single
flight hence services multiple connections, which we call its *flight segments*. ULD
build-up is the process of loading a pallet or container, collectively referred to as
unit load devices (ULD), with cargo for a single segment of an outbound flight.

These build-up processes must be performed on dedicated workstations,
which are blocked until the build-up is completed. As many outbound flights
compete for a limited number of workstations, a schedule of build-up processes
maximizing the amount of cargo delivered in time is sought. Cargo for a single flight segment is not available at some specific release time, but gradually
becomes available in the days or hours leading up to the flight's departure as
corresponding break-down processes for inbound ULDs release it to the airport's
storage unit. Hence, scheduling a build-up process too early risks building up
ULDs with underutilized capacity, potentially leading to capacity problems for
shipments that are broken down later. Thus, it is usually beneficial to build up
ULDs as close to the flight's departure time as possible [6].

Additionally, we aim to build up ULDs in batches, that is, sets of ULDs
that are built up at the same time and in spatial proximity. The rationale for
this is twofold. First, it offers cargo handlers more degrees of freedom in loading
shipments into ULDs, thereby utilizing freight capacity more efficiently. Second,
it reduces the cost incurred by the repeated transport of cargo from the storage unit to the build-up area. Batch building is an operational objective, while
failing to actually transport cargo usually incurs contractual penalties. We combine ULD in large batches only if it does not result in offloaded cargo or ULDs.
Batches, therefore, cannot be considered an input, but must be found as part
of the build-up scheduling problem. Build-up scheduling is hence a challenging
problem encompassing a machine assignment, a scheduling, and a set partitioning component. Additionally, we consider planning horizons of up to a whole
week of operations, which results in huge time-expanded models.

We address these difficulties with a logic-based Benders (LBB) decomposition
approach that assigns batches to time intervals and workstations in the master
problem, while the schedule is determined in a subproblem. Here, the main innovation is an appropriate choice of the time intervals that allows the subproblem
to become a feasibility problem that decomposes over the workstations.

In the remainder of this section, we give a short literature review and formally define the build-up scheduling problem with batching. In Sect. 2, we briefly
present a time-expanded formulation from the literature. The LBB approach is
introduced in Sect. 3. We evaluate both approaches computationally in Sect. 4
and conclude in Sect. 5.

Related Literature. Scheduling problems arising from ULD build-up processes
have been considered in the literature predominantly with a focus on crew
scheduling [16, 20, 22]. These models do not consider the scheduling of ULDs,
the assignment of freight to ULDs, or of ULDs to workstations.

Cargo availability and workstation assignments were considered for the first time by Brandt [6] who employed a rolling horizon approach to calculate heuristic build-up schedules that minimized offloaded cargo and the number of crews employed per shift. They also introduced the first publicly available data set for build-up scheduling problems [5]. In contrast to [6], we aim to build up ULD in batches. In our application, a fixed number of crews is provided by a staffing agency. Therefore, the available workforce is known in advance and influences our model only in so far as it determines the number of concurrently available workstations. Heuristic methods for scheduling batch build-ups and personnel have also been studied by Emde et al. [8]. They, however, do not take the loading of ULDs with cargo into account. Instead, they aim to minimize the maximum amount of workforce employed concurrently. Their batches are given as an input, whereas in our model, they are part of the output.

The build-up scheduling problem studied here has previously been considered in [10], as part of a network design model that also incorporates break-down scheduling. When considering only build-ups, the network flow reduces to simple inventory constraints. We restate this simplified network design model as a scheduling problem in Sect. 2.

Logic-Based Benders decomposition (LBB) [13] generalizes Benders decomposition [4] to cases in which the subproblem is no longer a linear program. Most importantly, it allows the subproblem to be tackled by constraint programming (CP) techniques, making it, besides constraint integer programming [1], one of the main techniques for the integration of mixed integer programming (MIP) and CP approaches. In contrast to classical Benders decomposition, efficient cuts for the master cannot be derived from the LP dual, but must be handcrafted from specific problem knowledge [15]. For a recent survey on applications of classical Benders decomposition, see [19]; for an overview of LBB, see [14]. Early on, LBB has been applied predominantly to scheduling problems [12], with the number of applications ever-growing since [15]. Our approach is inspired by a common technique for planning and scheduling problems by Hooker [12] in which jobs are assigned to machines (planning) in a master problem, usually a MIP. The scheduling of jobs is then shifted into one CP subproblem per machine that provides feasibility and optimality cuts to the master problem.

Çoban and Hooker [23,24] present an LBB variant for single machine scheduling problems, in which jobs are assigned to time intervals in the master problem. While LBB has occasionally been applied to scheduling problems with batching [9], there is, to the best of our knowledge, no previous work applying it to build-up scheduling problems.

Problem Definition. In the following, we formally define the build-up scheduling problem with batching. In it, ULDs need to be built up for a set of flights K in time for each flight's $k \in K$ departure time δ_k. Using the notational convention $[n] := \{1, \ldots, n\}$, the overall time horizon is hence $T := [\max_{k \in K} \delta_k]$.

For simplicity, we do not distinguish between flights and their segments, as a flight with multiple segments can easily be modeled as multiple flights. During build-up, a ULD is loaded with available shipments for flight k. These shipments are modeled as an inflow of cargo volume with d_{kt} denoting the cargo for flight k arriving at time point $t \in [\delta_k]$. A ULD can only be loaded with cargo that is available at the start of the build-up process, and each volume unit of cargo that is not loaded before the departure of flight k is not shipped at all. In this case, an offload penalty θ_k has to be paid.

ULDs come in different types $u \in U$ characterized by their capacity c_u and build-up time β_u. The number of ULDs of each type $u \in U$ planned per flight $k \in K$, denoted by p_{ku}, is an input determined in the aforementioned aircraft configuration phase. Occasionally, it might not be possible to build up all ULDs for a given flight k. If this happens, a penalty ϕ_{ku} applies per offloaded ULD of type u. Each ULD build-up must be performed on a workstation that is blocked during the build-up time. Build-up times are independent of the chosen workstation. As all workstations are identical, we can partition them into disjoint *workstation groups* W of workstations that are near each other in the cargo terminal. The capacity c_w of such a group $w \in W$ is the number of workstations it contains. In this framework, a batch is then defined as a set of ULDs of the same type u, assigned to the same flight k starting the build-up process at the same time point $t \in T$ in the same workstation group w.

As ULDs of the same type are indistinguishable, it can be represented as a five-tuple (k, u, w, t, n) with n being the number of ULDs in the batch. The aim of build-up scheduling with batching is to minimize the sum of penalty terms for offloaded cargo volume θ_k and offloaded planned ULDs ϕ_{ku} and a flat penalty $\rho > 0$ for constructing a batch, which incentivizes the build-up of large batches. Batch building is a secondary objective, i.e., saving batches should never lead to offloaded cargo or ULDs. The specific value of ρ is hence irrelevant as long as it is chosen significantly smaller than the cargo and ULD offload penalties. In contrast to many other batch scheduling problems [18], we do not conserve resources by batch processing. It has an effect solely on the objective value.

2   Time-Indexed MIP Formulation

We present a time-indexed MIP formulation for build-up scheduling with batching. It is identical to the model in [10] if break-down decisions are fixed.

Let $\mathcal{J}_B$ be an index set whose members $j \in \mathcal{J}_B$ are four-tuples (k, u, w, t) representing the start of the build-up of a batch of ULD type u for flight k in workstation group w at time point t. The maximum number of ULDs that such a batch can contain is $M_j := \min(p_{ku}, c_w)$. For all $j \in \mathcal{J}_B$ and $n \in [M_j]$, we introduce a binary variable x_{jn} representing the decision of building up batch j comprised of n ULDs. Hence, $x_{jn} = 1$ implies $x_{jn'} = 0$ for all $n' \in [M_j] \backslash \{n'\}$. The objective coefficient of x_{jn} is the sum of the batch building and ULD offload penalties, i.e., $\rho - n\phi_{ku}$. We also add variables $l_{k,t}$ for $k \in K$ and $t \in [\delta_k]$, representing the cargo volume for flight k in storage at time point t. The variable $l_{k\delta_k}$ contains the offloaded cargo, penalized with the objective coefficient θ_k.

For ease of notation, we treat $\mathcal{J}_B$ as a selector function, e.g., we write $\mathcal{J}_B(w') := \{(k, u, w, t) \in \mathcal{J}_B : w' = w\}$ to denote all possible batches in workstation group w and $\mathcal{J}_B(k', t') := \{(k, u, w, t) \in \mathcal{J}_B : k' = k \wedge t' = t\}$ for all possible batches for flight k' starting build-up at time point t'. With this, the time-indexed model *(TI)* can be stated as follows.

$$(TI) \quad \min \sum_{k \in K} \sum_{u \in U} \sum_{j \in \mathcal{J}_B(k,u)} \sum_{n \in [M_j]} (\rho - n\phi_{ku})x_{jn} + \sum_{k \in K} \theta_k l_{k\delta_k} \tag{1a}$$

$$\text{s.t.} \quad \sum_{u \in U} \sum_{\substack{t' \in T: \\ t \in [t', t' + \beta_u)}} \sum_{j \in \mathcal{J}_B(u,w,t')} \sum_{n \in [M_j]} n x_{jn} \leq c_w \qquad \forall w \in W \, \forall t \in T \tag{1b}$$

$$\sum_{j \in \mathcal{J}_B(k,u)} \sum_{n \in [M_j]} n x_{jn} \leq p_{ku} \qquad \forall k \in K \, \forall u \in U \tag{1c}$$

$$l_{kt-1} - \sum_{u \in U} \sum_{j \in \mathcal{J}_B(k,u,t)} \sum_{n \in [M_j]} c_u n x_{jn} - l_{kt} \leq -d_{kt} \qquad \forall t \in [\delta_k] \, \forall k \in K \tag{1d}$$

$$\sum_{n \in [M_j]} x_{jn} \leq 1 \qquad \forall j \in \mathcal{J}_B \tag{1e}$$

$$x_{jn} \in \{0,1\} \qquad \forall n \in [M_j] \, \forall j \in \mathcal{J}_B \tag{1f}$$

$$l_{kt} \in \mathbb{Q} \qquad \forall t \in [\delta_k] \, \forall k \in K. \tag{1g}$$

Constraints (1b) ensure that workstation capacity is never exceeded on any workstation group, while constraints (1c) guarantee that at most as many ULDs are scheduled as planned. For each flight $k \in K$, constraints (1d) ensure that the net freight change in storage between time points $t - 1$ and t is no larger than the total capacity of all ULD built up at time point t. This ensures that $l_{k\delta_k}$ contains at least the amount of offloaded cargo volume for all $k \in K$. In any optimal solution, it is precisely the offloaded cargo volume. Finally, constraints (1e) force that for each flight k, ULD type u, workstation group w and time point t, at most one batch is constructed. This constraint is not necessary since the condition is already ensured by letting $\rho > 0$ in the objective. It serves, however, to strengthen the model's LP relaxation.

3   Logic-Based Benders Decomposition

The time-indexed formulation both exhibits a weak LP relaxation and leads to large models. We therefore propose an alternative approach based on logic-based Benders decomposition. The decomposition is specifically designed such that its subproblem decomposes into one scheduling problem per workstation group. These problems can then be solved independently. In planning and scheduling problems, this is usually achieved by moving the planning, which here corresponds to building batches and assigning them to workstation groups, into the master problem [cf. 12].

In the case of build-up scheduling, this approach does, however, not suffice. A ULD can only accommodate cargo that is in storage when the build-up process begins. After the build-up is completed, it will be closed and never reopened.

The contribution of a batch build-up to the objective value hence depends on the amount of cargo in storage at the build-up start time, which, in turn, depends on the cargo that has already arrived, and on the cargo capacity of all ULDs for the same flight that were built up earlier, including build-ups in other workstation groups. Hence, only assigning batches to workstation groups in the master problem does not result in subproblem that decomposes along W: Batches remain linked across workstation groups via inventory constraints.

We make, however, the following observation: Shifting the start of a batch build-up from time point t to $t + \epsilon$ can only influence the objective value if new cargo becomes available in $(t, t + \epsilon]$. By decomposing the planning horizon into intervals based on the cargo arrival times, we therefore find that the objective value of any schedule is not dependent on the precise scheduling of batches but only on their assignments to time intervals. We use this observation to derive our master problem: It decides on the number and size of batches and assigns them to workstation groups and time intervals. The actual scheduling is then delegated to a pure feasibility subproblem. As cargo assignments can already be fixed in the master problem, the subproblem decomposes into $|W|$ cumulative scheduling problems with deadlines and release times.

3.1 Master Problem

For each flight $k \in K$, let $t_{k_1} < \cdots < t_{k_{r-1}}$ be the set of all time points $t \in [\delta_k]$ for which $d_{k,t} > 0$ and also let $t_{k_0} := 0$ and $t_{k_r} := \delta_k$. From these, we construct a set of intervals $\mathcal{I}_k := \{[t_{k_0}, t_{k_1}), [t_{k_1}, t_{k_2}), \dots, [t_{k_2}, t_{k_3}), [t_{k_{r-1}}, t_{k_r}), [t_{k_r}, t_{k_r} + 1)\}$, to which batches for k will be assigned in the master problem. This means that a batch's build-up start time lies in the interval but not necessarily the end time. Note that any ULD assigned to interval $[t_{k_0}, t_{k_1})$ must be empty, as no cargo for k becomes available before t_{k_1}. Such a build-up might still appear in an optimal solution if workstation capacity is tight and $\phi_{ku} > 0$. Especially when considering long planning horizons, this interval can be large while remaining unlikely to contain build-ups. Aggregating it, therefore, removes many binary variables that are very likely at zero in an optimal solution. Additionally, no batch build-up can ever happen in $[t_{k_r}, t_{k_r} + 1)$; cargo assigned to this interval counts as offloaded.

For any flight k, ULD type u and workstation group w, we can compute an upper bound on the number of batches of size $n \in \{1, \dots, c_w\}$ that can be built up in interval $I = [t_1, t_2) \in \mathcal{I}_k$ as

$$G_{kuwIn} := \sum_{i=1}^{\min(\lfloor \frac{c_w}{n} \rfloor, \beta_u)} \left\lfloor \frac{\min(t_2 + \beta_u, \delta_k + 1) - t_1 - i}{\beta_u} \right\rfloor. \tag{2}$$

Here, G_{kuwIn} is the closed form result of a simple Greedy procedure that schedules as many batches in I and before δ_k as possible. This scheduling takes into account, that, in an optimal solution, no two batches of the same flight and type can be scheduled in the same workstation group at the same time, as the

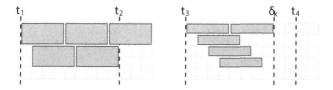

Fig. 1. Two examples of bounds on the maximum number of batches starting in an interval, obtained by Greedy scheduling. We can schedule a maximum of five batches of size two to start build-up in interval $I_1 = [t_1, t_2)$ (left) and a maximum of five batches of size one in $I_2 = [t_2, t_3)$ (right). The fifth workstation on the right remains unoccupied, as all feasible start times are already blocked.

objective value could be improved by merging them. Such schedules can be seen in Fig. 1. Considering the number of planned ULDs, we get the stronger bound $M_{kuwIn} := \min\left(G_{kuwIn}, \left\lfloor \frac{p_{ku}}{n} \right\rfloor\right)$.

For long intervals and flights with many planned ULDs, M_{kuwIn} can be quite large. We hence further reduce the number of binary variables by introducing $\mathcal{O}(\log_2 M_{kuwIn})$ variables x_{kuwInm} representing the build-up of 2^{m-1} batches for $m = 1, \ldots, \lfloor \log_2 M_{kuwIn} \rfloor + 1$. The objective coefficient of x_{kuwInm} is hence $o_{kuwInm} = \rho - s_j \phi_{ku}$ with $s_j := 2^{m-1} n$.

The index set $\mathcal{J}_L$ of all binary variables is then

$$\mathcal{J}_L := \bigcup_{k \in K} \bigcup_{u \in U} \bigcup_{w \in W} \bigcup_{I \in \mathcal{I}_k} \bigcup_{n=1}^{c_w} \bigcup_{m=1}^{\lfloor \log_2 M_{kuwIn} \rfloor + 1} (k, u, w, I, n, m). \tag{3}$$

The master problem *(LBB)* can now be stated as

$$(LBB) \quad \min \sum_{j \in \mathcal{J}_L} o_j x_j + \sum_{k \in K} \theta_k l_{k[\delta_k, \delta_k + 1)} \tag{4a}$$

$$\text{s.t.} \qquad \sum_{j \in \mathcal{J}_L(k,u)} s_j x_j \leq p_{k,u} \qquad \forall k \in K \; \forall u \in U \tag{4b}$$

$$l_{k\text{pred}(I)} - \sum_{u \in U} \sum_{j \in \mathcal{J}_L(k,u,t)} c_u s_j x_j - l_{kI} \leq -d_{kI} \qquad \forall I \in \mathcal{I}_k \; \forall k \in K \tag{4c}$$

$$\sum_{j \in C} x_j \leq |C| - 1 \quad \forall C \in \mathcal{C} \tag{4d}$$

$$\sum_{j \in \mathcal{J}_L(w)} e_{jE} x_j \leq c_w |E| \qquad \forall E \in \mathcal{E} \; \forall w \in W \tag{4e}$$

$$x_j \in \{0, 1\} \qquad \forall j \in \mathcal{J}_L \tag{4f}$$

$$l_{kI} \in \mathbb{Q} \qquad \forall I \in \mathcal{I}_k \; \forall k \in K, \tag{4g}$$

with $\text{pred}(I)$ denoting the interval before $I \in \mathcal{I}_k$ in $\mathcal{I}_k$. Constraints (4b) ensure that no more ULDs are scheduled than have been planned; constraints (4c) guarantee flow conservation of the inventory. For each flight k, $l_{k[\delta_k, \delta_k + 1)}$ hence contains the offloaded cargo volume, which is penalized in the constraints (4a) with θ_k. Constraints (4d) apply all no-good cuts $\mathcal{C}$ obtained from the subproblem.

Finally, constraints (4e) enforce an energetic reasoning-based relaxation of the scheduling subproblem. Here, $\mathcal{E}$ is a set of intervals $E \subseteq [T]$ and e_{jE} is a lower bound on the minimum energy used by $j = (k, u, w, I, n, m)$ in E. We let $\mathcal{E} := \{[t_1, t_2 - 1 + \beta_u] : (k, u, w, I = [t_1, t_2), n, m) \in \mathcal{J}_L\}$. The rational is that any batch for index $j = (k, u, w, I = [t_1, t_2), n, m)$ must start and finish build-up in $[t_1, t_2 - 1 + \beta_u]$. The coefficient is hence maximized as $e_{jE} = s_j\beta_u$. For all other indices, a bound can be computed by Algorithm 1. It calculates a lower bound on the minimum energy of an index j in E by left- and right-shifting the corresponding batches to their earliest and latest start times. As no two batches defined by j can be scheduled at the same time point in any optimal solution, Algorithm 1 successively increases the left- and right-shift values.

Input: Index $j = (k, u, w, I, n, m)$ with $I = [t_1, t_2)$, Interval $E = [\tau_1, \tau_2]$
Output: Energy e_{jE}
$e_{jE} \leftarrow 0$;
rshift $\leftarrow \tau_2 - \min(t_2 - 1, \delta_k - \beta_u)$;
lshift $\leftarrow t_1 + \beta_u - \tau_1$;
for $i = 1$ **to** 2^{m-1} **do**
    $e_{jE} \leftarrow e_{jE} + \max(0, \min(\text{rshift}, \text{lshift}, \beta_u, \tau_2 - \tau_1))$;
    if lshift $<$ rshift **then** lshift $\leftarrow$ lshift $+ 1$ **else** rshift $\leftarrow$ rshift $+ 1$;
end

Algorithm 1: Lower Bound on Energy.

3.2 Subproblem

For any optimal solution (x^*, l^*) to *(LBB)*, x^* induces one cumulative scheduling subproblem for each workstation group $w \in W$. Each variable $x_j^* = 1$ with $j = (k, u, w, I, n, m)$ and $I = [t_{k_i}, t_{k_{i+1}})$ induces a set $\mathcal{J}_S(k, u, w, I, n)$ of 2^{m-1} identical jobs. Each such job $j \in \mathcal{J}_S(k, u, w, I, n)$ has a release time $\mathbf{r}_j := t_{k_i}$, deadline $\mathbf{d}_j := t_{k_{i+1}} + \beta_u - 1$, processing time $\mathbf{p}_j := \beta_u$ and resource consumption $\mathbf{c}_j := n$. We denote by $\mathcal{J}_S$ the set of all jobs and follow the same convention as for previous index sets, e.g., we write $\mathcal{J}_S(k, u)$ to denote all jobs for some $k \in K$ and $u \in U$. Using the general constraints *cumulative* and *alldifferent*, the subproblem *(CS)* on workstation group $w \in W$ can then be stated as

$$\begin{align}
\textit{(CS)} \quad & \mathbf{r}_j \leq y_j \leq \mathbf{d}_j - \mathbf{p}_j & \forall j \in \mathcal{J}_S & \quad (5a) \\
& \text{cumulative}(y, \mathbf{p}, \mathbf{c}, c_w) & & \quad (5b) \\
& \text{alldifferent}(y_j, j \in \mathcal{J}_S(k, u, w)) & \forall k \in K, u \in U. & \quad (5c)
\end{align}$$

The variable y_j indicates the starting time of job j. Constraints (5a) ensure that all scheduled ULD build-up processes start in the correct time interval and do not violate the corresponding flight's departure time. Constraints (5b) guarantee that no more ULDs are built up than there is workstation capacity available. Constraints (5c) guarantee that no two batch build-ups for the same flight and

ULD type are started at the same time in the same workstation group. Again, they are redundant because two such batches could be combined to improve the objective, thereby contradicting the optimality of (x^*, l^*). Their additional domain reduction rules might, however, aid in detecting infeasibility faster.

The constraint program *(CS)* is a feasibility problem. If it is feasible for all $w \in W$, the optimal master solution (x^*, l^*) is feasible. If it is infeasible for any $w \in W$, the variables indexed by $C := \{j \in \mathcal{J}_L(w) | x_j^* = 1\}$ can not all be set to one. The infeasibility can be removed by introducing the standard no-good cut

$$\sum_{j \in C} x_j \leq |C| - 1. \tag{6}$$

Note that the number of variables and constraints in both *(LBB)* and *(CS)* does not depend on the chosen time discretization but rather on the number of cargo arrival events. This is an important advantage over the time-indexed formulation, allowing us to theoretically solve problems with arbitrary precision. In practice, however, this will depend on the effectiveness of the CP solver's domain propagation rules, as finer discretization leads to larger variable domains.

3.3   Strengthening No-Good Cuts

Pure no-good cuts are commonly known to be weak, as they only remove few solutions [14]. Instead, one strives to extract the reason for the infeasibility. Since CP solvers provide no dual information, a usual strategy is to search for a reason heuristically. To do so, we follow a procedure by Hooker [12] and strengthen a no-good $\sum_{j \in C} x_j \leq |C| - 1$ by tentatively removing elements from C and resolving *(CS)*. If the result remains infeasible, the element is permanently removed from C. We apply this procedure to every element of C, thereby solving $|C|$ additional subproblems. We then strengthen this cut again by adding back some variables. Consider an index $i = (k, u, w, I, n, m) \in C$ that is part of the reduced no-good cut. It contributes to the infeasibility by introducing 2^{m-1} jobs that must start build-up on w during I. Hence, replacing i in C with any index that assigns more jobs that are of larger or equal size and duration to workstation group w and any interval $I' \subseteq I$ maintains the infeasibility. We thus let

$$C_i := \{(k', u', w, I', n', m') \in \mathcal{J}_L(w) \backslash \{C\}|$$
$$n' \geq n \wedge m' \geq m \wedge I' \subseteq I \wedge \beta_{u'} \geq \beta_u\}. \tag{7}$$

Theorem 1. *Let C be a reduced no-good cut and $i = (k, u, w, I, n, m) \in C$. With $K := \min(|C_i| + 1, \lfloor G_{kuwIn}/2^{m-1} \rfloor)$, a valid feasibility cut for* (LBB) *is given by*

$$x_i + \sum_{j \in C_i} x_j + K \sum_{j \in C \backslash \{i\}} x_j \leq K(|C| - 1). \tag{8}$$

For a proof, see Appendix A. This technique can be applied only to a single $j \in C$. To add the maximum number of variables, we hence choose $i := \text{argmax}_{j \in C} |C_j|$.

Equation (8) defines the knapsack polytope

$$K(C) := \text{conv}\left\{z \in \{0,1\}^{|C| \times |C_i|} \,\middle|\, z_i + \sum_{j \in C_i} z_j + K \sum_{j \in C \setminus \{i\}} z_j \leq K(|C| - 1)\right\}.$$

(9)

For $h \in C_i \cup \{i\}$, the set $C \setminus \{i\} \cup h$ defines a *minimal cover* of $K(C)$. Moreover, if $K \geq 2$, $\sum_{j \in C \setminus \{i\}} z_j + z_h \leq |C| - 1$ holds with equality for $|C| + |C_i|$ affinely independent points $z \in K(C) \cap \{0,1\}^{|C| \times |C_i|}$. These inequalities are hence facet-defining for $K(C)$. This indicates that, while Eq. (8) excludes more infeasible solutions from *(LBB)*, a reduced no-good of the form Eq. (6) provides a tighter LP relaxation. In Sect. 4, we investigate which cut is more useful in practice.

4   Computational Experiments

We compared the performance of the logic-based Benders decomposition against the time-indexed baseline model *(TI)* on an anonymized data set of real flights and shipments. As the master problem *(LBB)* remains hard to solve, we implemented not a classical logic-based Benders decomposition, but a Branch&Check algorithm [2,21]. In Branch&Check, *(LBB)* is solved once using Branch&Bound. When a solution (x^*, l^*) with x^* integer is found, its feasibility is checked by *(CS)*. If (x^*, l^*) is infeasible, cuts are introduced to *(LBB)* and the current tree node is resolved; if (x^*, l^*) is feasible, we terminate with an optimal solution.

Instances. Our experiments are based on the ACLPP data set used in the dissertation of Brandt [6]. It contains a real flight schedule of 82 flights with 158 flight segments between 11/23/15 and 11/30/15. ULDs belong to one of four types (three types of pallets and one container type). Cargo for each segment is given as a list of shipments that were sourced from real booking data, anonymized and randomly assigned to flight segments. For a description of the procedure, see [5].

The ACLPP data set contains three different scenarios: a scenario *base* where no flight is overbooked (regarding the aircraft's weight capacity), a scenario *fast* that halves the time between the arrival of a shipment and its flight segment's deadline compared to *base*, and a scenario *high* that artificially overbooks all flights. In the *base* scenario, all flight segments are assigned an average of 13.24 ULDs. This results in 299 ULDs per day on average. For scenarios *fast* and *high*, these numbers are 13.05 (295) and 27.14 (613), respectively. In most, but not all, instances, all cargo can be shipped. This is roughly similar to practice where long-term planning is done to avoid frequent offloads. From each scenario, we obtain 28 different planning horizons ranging from single days to the whole week by considering all possible sequences of consecutive days. Combining each planning horizon with one of two workstation setups results in 56 instances per scenario and 168 instances in total. The parameters of each instance are derived in the following way: Flight departures δ_k, planned ULDs p_{ku}, ULD break-down times β_u and ULD capacities c_u are taken directly from the ACLPP data. Following a recommendation in [6], we apply a factor of 0.66 to the ULD capacities, accounting for the fact that full capacity utilization is unachievable in practice. We calculate the cargo demand d_k of a flight segment as the volume of all its shipments. Cargo demand per flight and time point d_{kt} is derived from the shipments' arrival times. The loss parameter θ_k per flight segment k is calculated by taking the average of the shipments' offload penalties normalized by their volume. Following [6], the offload penalty per ULD of type $u \in U$ for segment $k \in K$ is set to $\phi_{ku} := \theta_k c_u$. We chose a small batch penalty of $\rho := 1$ and hence strongly prioritize the shipping of cargo, which is a contractual obligation, over the operational benefits of batching. For each instance, we use a time discretization of five minutes in accordance with the precision requirements of our industry partner's application. As [5] does not contain workstation data, we developed two artificial but realistic setups: the first contains two workstation groups; the second contains four workstation groups. Each workstation group contains six workstations, resulting in a total of 12 and 24 workstations, respectively.

MIP Warm Starts. For *(TI)*, we find that a large portion of the solver's initial progress is made via primal heuristics. These are far less effective for *(LBB)* as they don't take the subproblem into account, and hence often produce infeasible solutions that are then discarded. We address this difficulty by first running an initial round of *(TI)* with a time discretization of 30 min. This model is smaller and easier to solve but provides only an upper bound, as the coarse discretization might cut off the optimal solution. The discretization of 30 min was chosen as the largest common divisor of the build-up durations of all ULD types. We applied the warm start procedure to both *(TI)* and *(LBB)*. It was run until either a gap of under 2.5% or the time limit of 20 min was reached.

Computational Environment. All MIP models were solved using the Python interface of Gurobi 9.5.2 [11]; the subproblem *(CS)* was solved using the CP-SAT Solver in Google's ortools 9.6 [17]. We addressed its C++ interface via Cython [3] as it proved to be considerably faster than the Python interface. All experiments were run on PowerEdge C6520 machines using Intel(R) Xeon(R) Gold 6338 CPUs with 2.00 GHz. Each query was allocated 128 GB of RAM and a run time limit of four hours. We used Gurobi with default settings, apart from the following two changes: All root relaxations were solved using the dual Simplex algorithm. Each call to Gurobi was allowed to allocate up to eight threads. Both settings were chosen to conserve memory for *(TI)*. Other settings led to up to 33% of the instances violating the generous memory limit. Due to its considerably smaller size, *(LBB)* did not exhibit any memory issues. For *(TI)*, we additionally set Gurobi's parameter *Presolve* to *conservative* as the automatic setting led to excessive time spent on presolving, thereby severely degrading the overall performance. A similar effect could not be observed for *(LBB)*.

Table 1. Computational results. Reported are the number of instances solved (*Opt*), the gap (*Gap*, %), the time spent on the algorithm (*Time*, *s*), the number of integer variables and constraints (*Vars*, *Cons*), and, for *(LBB)*, the time spent in the sub-problem (*SubT*, *s*) and the number of cuts (*Cuts*). Parameters are warm starts (*W*), reduced no-good cuts (*N*), and strengthened cuts (*S*).

Scen	Works	Method	W	N	S	Opt	Gap	Time	Vars	Cons	SubT	Cuts
Base	12	LBB	○	●	○	16	3.56	1766	20296	6032	568	474
		LBB	●	●	○	18	0.19	888	20296	6032	34	67
		LBB	●	○	●	20	0.14	869	20296	6032	35	70
		LBB	●	●	●	**21**	**0.11**	**784**	20296	6032	29	119
		TI	○	–	–	17	0.71	1859	398448	114986	–	–
		TI	●	–	–	17	0.60	2503	398448	114986	–	–
	24	LBB	○	●	○	**28**	**0.00**	**251**	40592	10098	8	48
		LBB	●	●	○	**28**	**0.00**	455	40592	10098	0	2
		LBB	●	○	●	**28**	**0.00**	453	40592	10098	0	2
		LBB	●	●	●	**28**	**0.00**	464	40592	10098	0	4
		TI	○	–	–	23	0.31	1523	796894	166587	–	–
		TI	●	-	–	25	0.19	1902	796894	166587	-	-
Fast	12	LBB	○	●	○	13	9.26	2580	17580	5471	815	421
		LBB	●	●	○	19	0.17	1017	17580	5471	30	81
		LBB	●	○	●	**22**	0.29	**797**	17580	5471	22	65
		LBB	●	●	●	21	**0.13**	916	17580	5471	30	148
		TI	○	-	-	14	1.18	2620	324124	95157	–	–
		TI	●	–	–	15	0.82	2889	324124	95157	–	–
	24	LBB	○	●	○	27	0.18	**308**	35160	9054	12	78
		LBB	●	●	○	27	**0.00**	468	35160	9054	1	3
		LBB	●	○	●	27	**0.00**	459	35160	9054	1	2
		LBB	●	●	●	27	**0.00**	444	35160	9054	0	4
		TI	○	–	–	23	0.52	1487	648248	138786	–	–
		TI	●	–	–	24	0.33	2261	648248	138786	–	–
High	12	LBB	○	●	○	1	81.65	12727	51609	8300	1513	424
		LBB	●	●	○	3	24.40	10504	51609	8300	433	454
		LBB	●	○	●	3	24.67	10023	51609	8300	360	377
		LBB	●	●	●	4	24.27	**9078**	51609	8300	363	686
		TI	○	-	-	**5**	4.16	10408	692966	226730	–	–
		TI	●	–	–	3	**1.20**	11292	692966	226730	–	–
	24	LBB	○	●	○	16	5.19	4514	103217	13655	114	339
		LBB	●	●	○	16	**0.19**	5315	103217	13655	5	22
		LBB	●	○	●	16	**0.19**	4789	103217	13655	4	20
		LBB	●	●	●	18	**0.19**	4463	103217	13655	4	40
		TI	○	-	-	18	1.59	**4261**	1385931	293991	–	–
		TI	●	–	–	19	1.20	4461	1385931	293991	–	–
Total		LBB	○	●	○	101	4.52	1652	37345	8362	145	218
		LBB	●	●	○	111	0.87	1486	37345	8362	13	29
		LBB	●	○	●	116	0.89	1381	37345	8362	11	26
		LBB	●	●	●	**119**	0.83	**1348**	37345	8362	12	51
		TI	○	-	-	100	1.15	2807	632577	154019	–	–
		TI	●	–	–	103	**0.68**	3408	632577	154019	–	–

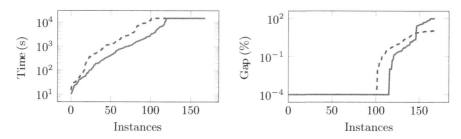

Fig. 2. Performance profile for the total run time and gap over all 168 instances. The gap for all instances solved to optimality is plotted as 10^{-4}. The dashed line represents *(TI)*; the continuous line represents *(LBB)*. Each time, we compare *(LBB)* to the better performing baseline, that is, we show *(TI)* without warm starts for the run time and *(TI)* with warm starts for the gap.

Results. In Table 1, we summarize the computational results for the *base*, *fast* and *high* scenarios for both workstation configurations. Experiments were conducted with the time-indexed formulation *(TI)* and logic-based Benders decomposition *(LBB)*. For *(LBB)*, we studied the effect of reduced no-good cuts (column N), the strengthened cuts (S) and a combination where both cuts were added (N and S). We also investigate the effect of the warm start procedure (W). All values, apart from the number of instances solved, are calculated as the shifted geometric mean. For one instance, *(LBB)* produced huge gaps in the *high* scenario, as the subproblem stalled in presolving and no meaningful lower bound was produced. These gaps were truncated to a value of 100% to not skew the results.

The strengthening procedure for no-good cuts improves slightly on the reduced no-good cuts in terms of instances solved and run times, as fewer subproblems are solved. As expected, however, they provide weaker LP relaxations, leading to a slightly larger gap. Adding both strengthened and reduced no-goods, however, works best with 119 instances out of 168 solved to optimality, compared to only 111 for the reduced no-goods. Additionally, both run times and gap are the best among all experiments. We also see that warm starts significantly improve the performance of *(LBB)*. In the following, if we compare *(LBB)* to the baseline, we always refer to the variant using both types of cuts and warm starts.

Overall, *(LBB)* solves 16 instances more to optimality than the best time-indexed variant, *(TI)* with warm starts, while reducing run times by over 52%. The master problem *(LBB)* contains, on average, only 5.9% as many variables and 5.5% as many constraints, including lazy cuts, as *(TI)*. The average gap, however, remains slightly higher. This somewhat unintuitive result is illuminated by considering the scenarios individually. We find that *(LBB)* outperforms the time-indexed formulations consistently and substantially in the *base* and *fast* scenarios on all metrics, including the gap. The difference is especially pronounced in scenarios with 24 workstations, where all configurations of *(LBB)* were able to

Table 2. Aggregated computational results for *short* (one to three days) *long* (four to seven days) planning horizons.

Time Horizon	Method	W	N	S	Opt	Gap	Time	Vars	Cons	SubT	Cuts
Short	LBB	○	●	○	74	2.65	837	25466	5722	97	139
	LBB	●	●	○	82	0.70	716	25466	5722	8	19
	LBB	●	○	●	86	0.68	629	25466	5722	7	17
	LBB	●	●	●	**87**	0.68	**622**	25466	5722	7	32
	TI	○	–	–	84	0.34	1261	324138	79961	–	–
	TI	●	–	–	82	**0.14**	1718	324138	79961	–	–
Long	LBB	○	●	○	27	10.62	5612	74394	16552	301	492
	LBB	●	●	○	29	1.21	5527	74394	16552	29	58
	LBB	●	○	●	30	1.33	5676	74394	16552	27	56
	LBB	●	●	●	**32**	**1.14**	**5417**	74394	16552	30	111
	TI	○	–	–	16	4.01	11854	2107671	501215	–	–
	TI	●	–	–	21	2.33	11685	2107671	501215	–	–

solve all but one instances to optimality, even when no warm start was applied. On the *high* scenario with 24 workstations, *(LBB)* and *(TI)* achieve comparable run times and solve roughly the same number of instances, while *(LBB)* is still able to achieve a significantly better gap. This situation reverses when considering only 12 workstations. Here, *(TI)* achieves the best gap with 1.20%, while the best gap we could achieve with an LBB approach jumps to 24.27%. Upon closer examination, we find that the higher amount of cargo, combined with fewer available workstations, leads to a considerable amount of offloaded cargo. In contrast, there are no offloads in the *base* and *fast* scenarios and the objective value is solely defined by the number of batches. In cases where offloads occur, the incumbents produced by *(LBB)* are usually not far from those of *(TI)*. The issue lies instead in a weak dual bound. This indicates that either our subproblem relaxation or the cuts are not strong enough to deal with extreme competition for workstation resources. Figure 2 provides performance profiles that reveal that the bad gap is indeed caused by a small set of instances.

Due to the aggregation of the planning horizon, we suspect that *(LBB)* performs better on longer horizons. To investigate this effect, we separated our instances into two sets: *short* contains 108 instances of a planning horizon of up to three days, *long* contains 60 instances of between four and seven days. All scenarios and workstation setups were included. The results are reported in Table 2. We find *(LBB)*'s model size to be far less sensitive to changes in the planning horizon. Between *short* and *long*, the number of variables and constraints increases less than threefold, whereas *(TI)* sees more than six times as many variables and constraints. While *(LBB)* solves more instances than *(TI)* for short and long planning horizons, the difference is far larger for the long horizon where *(LBB)* solves 53% more instances compared to only 6% for the short horizon. Also, *(LBB)* achieves a significantly tighter gap for *long* than *(TI)*.

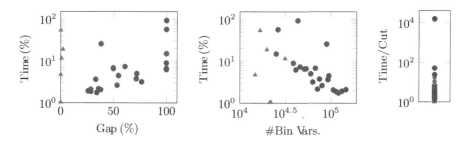

Fig. 3. Left/Middle: Fraction of total run time spent on the subproblem for the *high* scenario on twelve workstations by experiment *(LBB)* with both types of cuts and warm starts, mapped against the achieved gap (left) and problem size (middle). Right: Average subproblem time spend per cut generation normalized to multiples of the lowest value. Each data point is a single instance. Instances with offloaded cargo are marked with circles; instances where a solution shipping all cargo and all ULDs was found are marked with triangles.

In general, *(LBB)* is able to generate hundreds of cuts in a few minutes, spending only a fraction of its run time on the subproblem. However, we see a high variability for scenario *high* on twelve workstations, with one subproblem taking up nearly 91% of the run time for one instance. There is no strong relation of this behavior to the gap, problem size or the amount of offloaded cargo (Fig. 3). Instead, the average time spent per cut varies by up to four orders of magnitude, hinting at performance variability in the CP solver as the underlying issue.

5   Conclusion

We proposed a logic-based Benders decomposition formulation for build-up scheduling with batching. Based on a careful partition of the planning horizon, the subproblem decomposes into one cumulative scheduling problem per workstation group. This approach compares favorably to time-indexed formulations both theoretically, as the model size no longer depends on the time discretization and in practice, as LBB solves 15% more instances in less than 52% of the time-indexed model's run time. This advantage is more pronounced for longer planning horizons, where LBB solves roughly 50% more instances to optimality.

While significant, this improvement appears comparatively modest compared to what LBB achieved for other scheduling problems [14]. A reason may be the imbalance between the master and subproblem as conventional wisdom is that LBB performs best when roughly equal effort is spent on each. It may, therefore, seem worthwhile to seek a more balanced LBB decomposition. However, we believe that this would likely prevent the subproblems decomposition or turn it into an optimization problem less suited for CP solvers. Considering that a few high-load instances already spent large amounts of time on the subproblem, such a change might turn out unproductive, at least for this most difficult scenario.

Acknowledgments. We thank Felix Prause and Enrico Bortoletto for constructive criticism of the manuscript.

Disclosure of Interests. The authors have no competing interests to declare that are relevant to the content of this article.

A   Proof of Theorem 1

Proof. Consider a reduced no-good cut C obtained from an infeasible solution (x', l') and some index $i \in C$ with $i = (k, u, w, I, n, m)$. First note, that Eq. (8) does indeed cut off (x', l'):

$$x_i' + \sum_{j \in C_i} x_j' + K \sum_{j \in C \setminus \{i\}} x_j' = 1 + \sum_{j \in C_i} x_j' + K(|C| - 1) > K(|C| - 1). \quad (10)$$

In fact, it cuts off any solution (x, l) that has $x_j = 1$ for all $j \in C \setminus \{i\}$ and either $x_i = 1$ or $x_j = 1$ for at least one $j \in C_i$. None of these solutions are feasible as i consumes at most as many resources on w during I as any $j \in C_i$. The quantity G_{kuwIn} is an upper bound on the number of batches of size $n' \geq n$ in an interval $I' \subseteq I$ with build up time $\beta_u' \leq \beta_u$ that can be build-up on w. Each $j \in C_i$ represents at least 2^{m-1} such batches. Hence, for any feasible master solution (x, l), we have

$$2^{m-1} \left(x_i + \sum_{j \in C_i} x_j \right) \leq G_{kuwIn}. \quad (11)$$

Then, we have that either $x_i + \sum_{j \in C_i} x_j = 0$ in which case Eq. (8) is trivially fulfilled, or if $x_i + \sum_{j \in C_i} x_j > 0$, we must have $\sum_{j \in C \setminus \{i\}} x_j \leq |C| - 2$ and thus

$$x_i + \sum_{j \in C_i} x_j + K \sum_{j \in C \setminus \{i\}} x_j \leq$$

$$\underbrace{\min \left(C_i + 1, \left\lfloor \frac{G_{kuwIn}}{2^{m-1}} \right\rfloor \right)}_{K} + K \sum_{j \in C \setminus \{i\}} x_j \leq K(|C| - 1). \quad (12)$$

$\square$

References

1. Achterberg, T.: Constraint Integer Programming. Ph.D. thesis, TU Berlin (2007). https://doi.org/10.14279/depositonce-1634
2. Beck, J.C.: Checking-up on branch-and-check. In: Cohen, D. (ed.) CP 2010. LNCS, vol. 6308, pp. 84–98. Springer, Berlin, Heidelberg (2010). https://doi.org/10.1007/978-3-642-15396-9_10. ISBN 978-3-642-15396-9

3. Behnel, S., Bradshaw, R., Citro, C., Dalcin, L., Seljebotn, D.S., Smith, K.: Cython: the best of both worlds. Comput. Sci. Eng. **13**(2), 31–39 (2011). https://doi.org/10.1109/MCSE.2010.118

4. Benders, J.F.: Partitioning procedures for solving mixed-variables programming problems. Numer. Math. **4**(1), 238–252 (1962). https://doi.org/10.1007/BF01386316

5. Brandt, F.: Aclpp instances. https://github.com/fbrandt/ACLPP. commit: 3516c2b

6. Brandt, F.: The Air Cargo Load Planning Problem. Ph.D. thesis, Karlsruher Institut für Technologie (KIT) (2017). https://doi.org/10.5445/IR/1000075507

7. Brandt, F., Nickel, S.: The air cargo load planning problem - a consolidated problem definition and literature review on related problems. Eur. J. Oper. Res. **275**(2), 399–410 (2019). https://doi.org/10.1016/j.ejor.2018.07.013. https://www.sciencedirect.com/science/article/pii/S0377221718306180. ISSN 0377-2217

8. Emde, S., Abedinnia, H., Lange, A., Glock, C.H.: Scheduling personnel for the build-up of unit load devices at an air cargo terminal with limited space. OR Spect. **42**(2), 397–426 (2020). https://doi.org/10.1007/s00291-020-00580-2

9. Emde, S., Polten, L., Gendreau, M.: Logic-based benders decomposition for scheduling a batching machine. Comput. Oper. Res. **113**, 104777 (2020). https://doi.org/10.1016/j.cor.2019.104777

10. Euler, R., Borndörfer, R., Strunk, T., Takkula, T.: Uld build-up scheduling with dynamic batching in an air freight hub. In: Trautmann, N., Gnägi, M. (eds.) Operations Research Proceedings 2021, pp. 254–260. Springer, Cham (2022). https://doi.org/10.1007/978-3-031-08623-6_38

11. Gurobi Optimization, LLC: Gurobi Optimizer Reference Manual (2023). https://www.gurobi.com

12. Hooker, J.N.: Planning and scheduling by logic-based benders decomposition. Oper. Res. **55**(3), 588–602 (2007). https://doi.org/10.1287/opre.1060.0371

13. Hooker, J.N., Ottosson, G.: Logic-based benders decomposition. Math. Program. **96**(1), 33–60 (2003). https://doi.org/10.1007/s10107-003-0375-9

14. Hooker, J.: Logic-based benders decomposition: theory and applications. In: Synthesis Lectures on Operations Research and Applications. Springer, Heidelberg (2024). https://doi.org/10.1007/978-3-031-45039-6

15. Hooker, J.N.: Logic-based benders decomposition for large-scale optimization. In: Velásquez-Bermúdez, J.M., Khakifirooz, M., Fathi, M. (eds.) Large Scale Optimization in Supply Chains and Smart Manufacturing: Theory and Applications, pp. 1–26. Springer, Cham (2019). https://doi.org/10.1007/978-3-030-22788-3_1

16. Nobert, Y., Roy, J.: Freight handling personnel scheduling at air cargo terminals. Transport. Sci. **32**(3), 295–301 (1998). https://doi.org/10.1287/trsc.32.3.295

17. Perron, L., Didier, F.: OR-Tools CP-SAT v9.6. https://developers.goo-gle.com/optimization/cp/cp_solver/

18. Pinedo, M.L.: Scheduling, vol. 29. Springer, Cham (2012). https://doi.org/10.1007/978-3-319-26580-3

19. Rahmaniani, R., Crainic, T.G., Gendreau, M., Rei, W.: The benders decomposition algorithm: a literature review. Eur. J. Oper. Res. **259**(3), 801–817 (2017). https://doi.org/10.1016/j.ejor.2016.12.005

20. Rong, A., Grunow, M.: Shift designs for freight handling personnel at air cargo terminals. Transport. Res. Part E: Logist. Transport. Rev. **45**(5), 725–739 (2009). https://doi.org/10.1016/j.tre.2009.01.005

21. Thorsteinsson, E.S.: Branch-and-check: a hybrid framework integrating mixed integer programming and constraint logic programming. In: Walsh, T. (ed.) Principles and Practice of Constraint Programming—CP 2001, pp. 16–30. Springer, Heidelberg (2001). https://doi.org/10.1007/3-540-45578-7_2

22. Yan, S., Chen, C.H., Chen, M.: Stochastic models for air cargo terminal manpower supply planning in long-term operations. Appl. Stochastic Models Bus. Ind. **24**(3), 261–275 (2008). https://doi.org/10.1002/asmb.710

23. Çoban, E., Hooker, J.: Single-facility scheduling by logic-based Benders decomposition. Ann. Oper. Res. **210**(1), 245–272 (2013). https://doi.org/10.1007/s10479-011-1031-z

24. Çoban, E., Hooker, J.N.: Single-facility scheduling over long time horizons by logic-based benders decomposition. In: Lodi, A., Milano, M., Toth, P. (eds.) Integration of AI and OR Techniques in Constraint Programming for Combinatorial Optimization Problems, pp. 87–91. Springer, Heidelberg (2010). https://doi.org/10.1007/978-3-642-13520-0_11

A Benders Decomposition Approach for a Capacitated Multi-vehicle Covering Tour Problem with Intermediate Facilities

Vera Fischer[1]([⊠])[iD], Antoine Legrain[2,3][iD], and David Schindl[1,4][iD]

[1] University of Fribourg, Fribourg, Switzerland
vera.fischer@unifr.ch
[2] Polytechnique Montréal, 6079 Station Centre-Ville, Montréal, QC, Canada
[3] CIRRELT & GERAD, 2920 Chemin de la Tour, Montréal, QC, Canada
[4] Haute Ecole de Gestion de Genève, HES-SO University of Applied Sciences and Arts Western Switzerland, Geneva, Switzerland

Abstract. We consider a waste collection problem with intermediate disposal facilities to accommodate the usage of smaller, but more sustainable vehicles, with less capacity than the traditional waste collecting trucks. The optimization problem consists in determining the locations to place the collection points and the routes of a capacitated collection vehicle that visits these locations. We first present a mixed-integer linear programming formulation that exploits the sparsity of the road network. To efficiently solve practical instances, we propose a Benders decomposition approach in which a set covering problem to select the collection points is solved in the master problem and a capacitated vehicle routing problem with intermediate facilities to determine the routes and price the set covering solution is solved in the subproblem. We show a way to derive valid Benders cuts when solving the Benders subproblem with column generation and propose a heuristic Benders approach that provides better solutions for larger instances and approximated lower bounds to assess the quality of the obtained solutions.

Keywords: covering tour problem · intermediate facilities · Benders decomposition · column generation · waste collection

1 Introduction

In this article, we investigate a waste collection problem involving vehicles that are smaller, but more sustainable than the traditional collecting trucks. Such vehicles can be efficient in residential areas with family houses or in city centers with narrow streets. However, due to their limited capacity (about 0.5 tons while usual trucks have a capacity of 10 tons) and driving speed, the installation of intermediate disposal facilities is required to avoid long trips to the final disposal facility. We formulate the problem as a capacitated multi-vehicle covering tour problem with intermediate facilities (Cm-CTP-IF), which is a variation of the

B. Dilkina (Ed.): CPAIOR 2024, LNCS 14742, pp. 277–292, 2024.
https://doi.org/10.1007/978-3-031-60597-0_18

capacitated multi-vehicle covering tour problem on a road network (Cm-CTP-R; [7]). This problem is motivated by a collaboration with Schwendimann AG, a company operating in waste management, but our approach also applies to, for instance, the location of mobile health-care teams in rural areas [8], the location of boxes for overnight mail service [9], and the planning of routes for urban patrolling [11]. For an overview of relevant research in the context of multi-vehicle covering tour problems (m-CTP; [8]) and vehicle routing problems (VRP; [4]) with road-network information, we refer the reader to [7] and [2].

Given the road network with the locations of a vehicle depot, candidate collection points and intermediate facilities, the decisions consist in selecting the collection points where residents must leave their bags and determining the routes of a fleet of uniform vehicles. As a consequence, the latter is equivalent to determining the routes of a single vehicle that performs exactly one rotation starting and ending at the depot, and carries out a sequence of collections with disposals at the available intermediate facilities [5]. For each residential building, a given rank defines and sorts the candidate collection points eligible for that building in compliance with some criterion (e.g., walking distance) and we assume that residents will always consider the highest-ranked eligible selected collection point for leaving their bags. The goal of the Cm-CTP-IF is to determine routes of minimum total travel time that visit a subset of candidate collections points, such that all residents are covered (their waste is picked-up at a collection point from their rank) and the capacity of the vehicle is respected.

The contributions of this paper are the following:

1. We define and formulate a variation of the Cm-CTP-R [7] that includes intermediate facilities and that has not been defined yet in the literature.
2. To handle larger instances derived from real-life data [7], we develop a Benders decomposition approach [3] in which a set covering problem (SCP) to select the collection points is solved in the master problem and a capacitated vehicle routing problem with intermediate facilities (CVRP-IF) to determine the routes and price the set covering solution is solved in the subproblem.
3. We discuss the consequences of solving the Benders subproblem with column generation and show a way to derive valid Benders cuts. Finally, we propose a heuristic Benders approach for which approximated lower bounds are computed to assess the quality of the obtained solutions.

The remainder of the paper is organized as follows. Section 2 formally defines the problem. Section 3 presents the road-network based mixed-integer linear programming (MILP) formulation. Section 4 presents the Benders decomposition approach. Section 5 reports the computational experiments, and Sect. 6 gives some concluding remarks.

2   Problem Definition

In this section, we formally define the Cm-CTP-IF using the notation and the modeling assumptions introduced in [7]. We are given a directed strongly connected graph $G = (V \cup W, A)$ with two node sets V and W, and an arc set A

representing the road network. Following the notation proposed in [8], W is a set of nodes with positive demand and V includes nodes that represent candidate collection points, intermediate facilities and road intersections. For each demand node $i \in W$, its demand d_i must be satisfied at exactly one node from its rank $V_i^{\text{rank}} \subseteq V$, which contains candidate collection points eligible for i. We assume that V_i^{rank} is totally ordered based on the assumed criterion to sort candidate locations (in our case, in increasing order of walking distance). We assume that d_i must be satisfied at the first node in V_i^{rank} at which a vehicle stops. This assumption is based on real-world observations that a resident will simply go to their closest collection point rather than abide by specific assignment decisions. Then, the set of candidate locations that can be visited by the vehicle is $V^{\text{stop}} = \cup_{i \in W} V_i^{\text{rank}}$. Finally, we define $\text{rank}(i, j)$ as the index of node j in V_i^{rank}.

The set $V^{\text{fac}} \subset V$ contains the available intermediate facilities where the vehicle can dump the load. Note that some of them might not be used in a solution. Let $c_{ll'}$ be the non-negative length of arc $(l, l') \in A$ representing the travel time of (l, l') that can be asymmetric, i.e., $c_{ll'} \neq c_{l'l}$. To capture the time needed to dump the load, a fixed cost is added to each arc entering an intermediate facility $h \in V^{\text{fac}}$. We consider a single vehicle that is located at a distinguished depot node $\sigma \in V$ and is characterized by its capacity Q. It performs exactly one rotation that starts and ends at the depot and is composed of a set of single-facility (i.e., starting and ending at the same intermediate facility) and inter-facility (i.e., connecting two different intermediate facilities) routes with a mandatory visit to a final intermediate facility $h \in V^{\text{fac}}$ before going back to the depot. The concepts of rotation was introduced in [5] as the set of all routes assigned to a vehicle. A route is a part of the rotation that starts and ends at an intermediate facility or the depot. In contrast to [7], we do not allow for split collection because it is assumed that intermediate facilities are closer to the residential buildings and therefore, overall costs are not significantly improved by this approach. To this end, we assume $d_i \leq Q, \forall i \in W$.

A solution of the Cm-CTP-IF is specified by a sequence of the selected collection points $V^{\text{sel}} \subseteq V^{\text{stop}}$ and the given intermediate facilities (i.e., which are not necessarily distinct), and it covers a demand node $i \in W$ if $V_i^{\text{rank}} \cap V^{\text{sel}} \neq \phi$. A solution is feasible if V^{sel} covers all demand nodes (i.e., V^{sel} is a set cover of W) and the vehicle performs exactly one connected rotation collecting the demands d_i at the highest-ranked nodes in $V_i^{\text{rank}} \forall i \in W$, while respecting its capacity Q in all routes. Figure 1 displays an example of a potential solution. The objective of the Cm-CTP-IF is to find a feasible solution with minimum total travel time which is computed as the sum of the lengths of the shortest paths between consecutive nodes in the sequence plus a penalty t^{sto} for each stop performed by the vehicle.

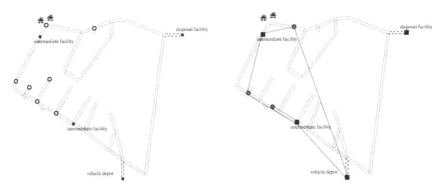

(a) Street network of a small part of a Swiss municipality.

(b) A solution to the C m-CTP-IF.

Fig. 1. Street network of a small part of a Swiss municipality and a possible solution to the Cm-CTP-IF.

3 Road-Network-Based Formulation

We extend the road-network-based formulation presented in [7] in which decision variables are introduced for each road segment [10]. It was shown in [7] that such a formulation is better suited as the set of candidate collection points is very large. Even though including intersections leads to more nodes, the road-network graph has much less arcs than a complete graph resulting in a smaller model. Let $\mathcal{M} = \{1, \ldots, m\}$ be the set of the m routes, where m defines an upper bound on the number of routes that have to be performed. Let $x_{ll'k}$ be an integer variable indicating the number of traversals on arc $(l, l') \in A$ in route $k \in \mathcal{M}$ and y_j be a binary variable taking value 1 if the vehicle stops at node $j \in V^{\text{stop}}$. Two non-negative continuous variables q_j and q_{jk} define the total amount of demand satisfied at node $j \in V^{\text{stop}}$ and the amount collected in each route k at node j, respectively, and a continuous variable z_{ij} indicates if the demand of $i \in W$ is satisfied at $j \in V_i^{\text{rank}}$. Finally, we introduce a non-negative continuous variable $f_{ll'k}$ to capture the flow passing through arc $(l, l') \in A$ in route $k \in \mathcal{M}$, and a non-negative continuous variable $g_{hh'}$ to indicate the flow between each pair $\{h, h'\} : h, h' \in V^{\text{fac}} \cup \{\sigma\}$.

$$\min \quad \sum_{k \in \mathcal{M}} \sum_{(l,l') \in A} c_{ll'} x_{ll'k} + \sum_{j \in V^{\text{stop}}} t^{\text{sto}} y_j \tag{1a}$$

The objective function (1a) expresses the total cost, which is computed as the sum of the total travel times plus t^{sto} times the total number of stops.

$$\text{s.t.} \quad \sum_{j \in V_i^{\text{rank}}} z_{ij} = 1 \qquad \forall i \in W \tag{1b}$$

$$\sum_{\substack{j' \in V_i^{\text{rank}}: \\ \text{rank}(i,j') > \text{rank}(i,j)}} z_{ij'} \leq 1 - y_j \qquad \forall i \in W, j \in V_i^{\text{rank}} \tag{1c}$$

$$\sum_{i\in W: j\in V_i^{\text{rank}}} d_i z_{ij} = q_j \qquad\qquad \forall j \in V^{\text{stop}} \qquad\qquad (1d)$$

$$q_j \leq D_j y_j \qquad\qquad \forall j \in V^{\text{stop}} \qquad\qquad (1e)$$

$$y_j \leq q_j \qquad\qquad \forall j \in V^{\text{stop}} \qquad\qquad (1f)$$

$$z_{ij} \leq y_j \qquad\qquad \forall i \in W, j \in V_i^{\text{rank}} \qquad\qquad (1g)$$

Constraints (1b) ensure that the demand of node $i \in W$ is satisfied at exactly one candidate collection point from its rank. Constraints (1c) state that this demand is satisfied at the first node in V_i^{rank} at which the vehicle stops. Constraints (1d) guarantee that the total demand that is assigned to node $j \in V^{\text{stop}}$ is equal to the total amount that is collected at that node. Constraints (1e) link the variables q with the variables y and impose an upper bound $D_j = \min\{(\sum_{i\in W: j\in V_i^{\text{rank}}} d_i), Q\}$ $\forall j \in V^{\text{stop}}$ on the total amount that can be satisfied at node j. In particular, they prevent the demand of a node to exceed the capacity Q of the vehicle. Constraints (1f) link the same variables in the opposite direction such that the vehicle can stop at a node j only if some demand is assigned to it. This prevents from suboptimal solutions in which selected nodes are not needed to satisfy demand. Constraints (1g) link the variables z with the variables y by imposing that demand can only be assigned to a node $j \in V_i^{\text{rank}}$ $\forall i \in W$ which is visited by the vehicle. Notice that constraints (1c) and (1g) together enforce variables z_{ij} to be binary.

$$\sum_{k\in\mathcal{M}}\sum_{l\in V:(l,j)\in A} x_{ljk} \geq y_j \qquad\qquad \forall j \in V^{\text{stop}} \qquad\qquad (1h)$$

$$\sum_{k\in\mathcal{M}}\sum_{l\in V:(l,h)\in A} x_{lhk} = \sum_{k\in\mathcal{M}}\sum_{l\in V:(h,l)\in A} x_{hlk} \quad \forall h \in V^{\text{fac}} \qquad\qquad (1i)$$

$$\sum_{k\in\mathcal{M}}\sum_{l\in V:(\sigma,l)\in A} x_{\sigma lk} = 1 \qquad\qquad\qquad (1j)$$

$$\sum_{l'\in V:(l',l)\in A} x_{l'lk} = \sum_{l'\in V:(l,l')\in A} x_{ll'k} \qquad \forall l \in V\setminus V^{\text{fac}}, k \in \mathcal{M} \quad (1k)$$

$$\sum_{h\in V^{\text{fac}}}\sum_{l\in V:(h,l)\in A} x_{hlk} \leq 1 \qquad\qquad \forall k \in \mathcal{M} \qquad\qquad (1l)$$

Constraints (1h)–(1l) force the variables x to take values that make up a valid rotation. More precisely, constraints (1h) specify that the vehicle can stop at a node $j \in V^{\text{stop}}$ only if it traverses an incoming arc of node j at least once. Constraints (1i) state the degree constraints for each facility $h \in V^{\text{fac}}$. Constraint (1j) ensures that the depot is visited exactly once in a rotation. Constraints (1k) define that the vehicle enters and leaves any node $h \in V \setminus V^{\text{fac}}$ the same number of times in route $k \in \mathcal{M}$. Constraints (1l) ensure that in each route $k \in \mathcal{M}$ at most one facility is visited.

$$W^{\text{tot}} x_{ll'k} \geq f_{ll'k} \qquad\qquad \forall (l, l') \in A, k \in \mathcal{M} \qquad\qquad (1m)$$

$$f_{ll'k} = 0 \qquad \forall (l, l') \in A :$$
$$l \in V^{\text{fac}} \cup \{\sigma\}, k \in \mathcal{M} \qquad (1n)$$

$$\sum_{h' \in V^{\text{fac}} \cup \{\sigma\}} g_{hh'} - \sum_{h' \in V^{\text{fac}} \cup \{\sigma\}} g_{h'h}$$
$$= \sum_{k \in \mathcal{M}} \sum_{l \in V:(l,h) \in A} f_{lhk} \qquad \forall h \in V^{\text{fac}} \qquad (1o)$$

$$\sum_{h \in V^{\text{fac}}} g_{\sigma h} = 0 \qquad (1p)$$

Constraints (1m) link the variables f with the variables x. If $x_{ll'k} = 0$, i.e., arc (l, l') is not traversed in route k, then the flow $f_{ll'k}$ does not pass through it. Constraints (1n) impose a 0 outflow for $l \in V^{\text{fac}} \cup \{\sigma\}$. Constraints (1o) link the variables f with the variables g by passing the flow on the route level f to the rotation level g. Constraint (1p) defines a 0 outflow at the depot.

$$\sum_{h \in V^{\text{fac}}} g_{h\sigma} = W^{\text{tot}} \qquad (1q)$$

$$\sum_{l' \in V:(l,l') \in A} f_{ll'k} - \sum_{l' \in V:(l',l) \in A} f_{l'lk}$$
$$= \begin{cases} q_{lk} & \forall l \in V^{\text{stop}} \\ 0 & \forall l \in V \setminus (V^{\text{stop}} \cup V^{\text{fac}} \cup \{\sigma\}) \end{cases} \qquad \forall k \in \mathcal{M} \qquad (1r)$$

$$\sum_{k \in \mathcal{M}} q_{jk} = q_j \qquad \forall j \in V^{\text{stop}} \qquad (1s)$$

$$\sum_{j \in V^{\text{stop}}} q_{jk} \leq Q \qquad \forall k \in \mathcal{M} \qquad (1t)$$

$$y_j \in \{0, 1\}, q_j \geq 0 \qquad \forall j \in V^{\text{stop}} \qquad (1u)$$

$$z_{ij} \geq 0 \qquad \forall i \in W, j \in V_i^{\text{rank}} \qquad (1v)$$

$$q_{jk} \geq 0 \qquad \forall j \in V^{\text{stop}}, k \in \mathcal{M} \qquad (1w)$$

$$x_{ll'k} \in \mathbb{Z}_{\geq 0}, f_{ll'k} \geq 0 \qquad \forall (l, l') \in A, k \in \mathcal{M} \qquad (1x)$$

$$g_{hh'} \geq 0 \qquad \forall h, h' \in V^{\text{fac}} \cup \{\sigma\} \qquad (1y)$$

Constraint (1q) enforces that the total amount going into the depot equals the total amount satisfied in the rotation. To ensure a last visit to an intermediate facility before going back to the depot, arcs entering the depot are defined in G only for pairs $\{l, \sigma\} : l \in V^{\text{fac}}$. Constraints (1r) define that the net outflow of any node $l \in V^{\text{stop}}$ must be the amount q_{lk} satisfied at node l in route $k \in \mathcal{M}$. For any other node, these constraints impose a 0 net outflow. Constraints (1s) link the variables q with each other such that the total demand satisfied at node $j \in V^{\text{stop}}$ corresponds to the total demand satisfied in all routes. Constraints (1t) ensure the respect of the vehicle's capacity Q. Finally, constraints (1u) - (1y) define the domain of the decision variables.

4   A Benders Decomposition Approach

To solve large instances derived from real-life data, we propose a *Benders decomposition approach* in which a SCP is solved in the Benders master problem (BMP; Sect. 4.1) and a CVRP-IF is solved with column generation in the Benders subproblem (BSP; Sect. 4.2). Given an optimal solution to the linear relaxation BSP, a Benders optimality cut (Sect. 4.3) is derived from the dual solution. This approach was proposed by [3], with the main objective of tackling problems with complicating variables, which, when temporarily fixed, yield a subproblem significantly easier to handle. The solution of the BSP provides a new constraint to incorporate into the BMP which is then solved again. This process is repeated iteratively until the optimal solution is found or another stopping criterion is met. We refer the reader to [12] for a state-of-the-art survey on the Benders decomposition approach.

In our case, the BMP (2) consists in selecting a subset of candidate collection points with minimum total cost. Given a solution to the BMP, which is a set cover V^{sel}, the goal of the linear relaxation BSP is to generate routes for V^{sel} and to provide lower bounds on the routing cost. From the dual BSP (4), Benders cuts (2b) are derived which are added to the BMP. Note that all feasible BMP solutions lead to feasible BSP solutions which is why only optimality cuts are considered. To keep track of the best solution, an integer routing solution is created by solving the BSP with all generated columns set as binary variables.

4.1   Benders Master Problem

The BMP formulation is presented in (2). The objective function (2a) represents the total cost of visiting the collection points plus an underestimator θ corresponding to the routing cost of the vehicle. Let B be the set of BSP solutions generated in the solving process. Each such solution corresponds to an extreme point of the dual of BSP. Then constraints (2b) represent the respective Benders cuts added to the BMP, with ξ^b, γ^b, and κ^b being the associated dual variables of $b \in B$. In general, (2) contains more constraints than (1), but many of the cuts are inactive in an optimal solution and will not be generated.

$$\min \quad \sum_{j \in V^{\text{stop}}} t^{\text{sto}} y_j + \theta \qquad (2a)$$

$$\text{s.t.} \quad \theta \geq \sum_{j \in V^{\text{stop}}} \xi_j^b q_j + \gamma^b + W^{\text{tot}} \kappa^b \qquad \forall b \in B \qquad (2b)$$

$$(1b) - (1g) \qquad (2c)$$

$$y_j \in \{0,1\} \qquad \forall j \in V^{\text{stop}} \qquad (2d)$$

$$q_j \geq 0 \qquad \forall j \in V^{\text{stop}} \qquad (2e)$$

$$z_{ij} \geq 0 \qquad \forall i \in W, j \in V_i^{\text{rank}} \qquad (2f)$$

4.2   Benders Subproblem

The BSP formulation corresponds to a Dantzig-Wolfe decomposition of the formulation (1h)–(1r) and is solved with column generation, which is a technique to solve problems with a large number of variables by iteratively adding some of the variables to the model [6]. Let $\bar{q}_j \ \forall j \in V^{\mathrm{stop}} : \bar{q}_j = 0 \ \forall j \notin V^{\mathrm{sel}}$ represent a solution to the BMP and define the amount of demand assigned to each node j. Let R be the set of all feasible routes starting and ending at the intermediate facilities or the depot $V^{\mathrm{fac}} \cup \{\sigma\}$, with $R_h^+ \subseteq R$ and $R_h^- \subseteq R$ representing the subsets of routes that start and end at h, respectively. Note that R is different from $\mathcal{M}$, thus we use index $r \in R$ instead of index $k \in \mathcal{M}$. Let $W_r = \sum_{j \in V^{\mathrm{sel}}} q_{jr}$ be the amount of demand picked up in route $r \in R$, with q_{jr} indicating the amount of demand collected at node j in route r, and let $c_r \ \forall r \in R$ be the travel cost of route r. We introduce a continuous variable x_r indicating if route $r \in R$ is selected. To ensure feasibility of the problem, artificial variables ν_j for nodes $j \in V^{\mathrm{stop}}$ that are not satisfied and λ for proportion of demand that is not collected are introduced and highly penalized in the objective function (3a) by c^{pick} and c^{coll}, respectively. Similarly as in formulation (1), we define a non-negative continuous variable $g_{hh'}$ to indicate the flow between each pair $\{h, h'\} : h, h' \in V^{\mathrm{fac}} \cup \{\sigma\}$. The goal is to select a subset of routes $R^{V^{\mathrm{sel}}} \subseteq R$ that collect the demand from the set cover nodes V^{sel} at minimum total cost.

$$\min \quad \sum_{r \in R} c_r x_r + \sum_{j \in V^{\mathrm{stop}}} c^{\mathrm{pick}} \nu_j + c^{\mathrm{coll}} \lambda \tag{3a}$$

$$\text{s.t.} \quad \sum_{r \in R} q_{jr} x_r + \nu_j = \bar{q}_j \qquad\qquad \forall j \in V^{\mathrm{stop}} \quad (\xi_j) \tag{3b}$$

$$\sum_{r \in R_h^-} x_r - \sum_{r \in R_h^+} x_r = 0 \qquad \forall h \in V^{\mathrm{fac}} \cup \{\sigma\} \quad (\beta_h) \tag{3c}$$

$$\sum_{r \in R_\sigma^+} x_r = 1 \qquad\qquad (\gamma) \tag{3d}$$

$$W^{\mathrm{tot}} \sum_{r \in R_h^+ \cap R_{h'}^-} x_r - g_{hh'} \geq 0 \qquad \forall (h, h') \in A^{\mathrm{H}} \quad (\delta_{hh'}) \tag{3e}$$

$$\sum_{(h,h') \in A^{\mathrm{H}}} g_{hh'} - \sum_{(h',h) \in A^{\mathrm{H}}} g_{h'h}$$

$$\leq \sum_{r \in R_h^-} W_r x_r \qquad\qquad \forall h \in V^{\mathrm{fac}} \quad (\zeta_h) \tag{3f}$$

$$\sum_{(\sigma,h) \in A^{\mathrm{H}}} g_{\sigma h} = 0 \qquad\qquad (\eta) \tag{3g}$$

$$\sum_{(h,\sigma) \in A^{\mathrm{H}}} g_{h\sigma} + W^{\mathrm{tot}} \lambda = W^{\mathrm{tot}} \qquad\qquad (\kappa) \tag{3h}$$

$$x_r \geq 0 \qquad\qquad \forall r \in R \tag{3i}$$

$$\nu_j \geq 0 \qquad\qquad \forall j \in V^{\mathrm{stop}} \tag{3j}$$

$$\lambda \geq 0 \tag{3k}$$

$$g_{hh'} \geq 0 \qquad\qquad \forall (h, h') \in A^{\mathrm{H}} \tag{3l}$$

Constraints (3b) enforce the total demand to be collected at each demand node $j \in V^{\mathrm{stop}}$, unless ν_j fills the gap. Constraints (3c) ensure that the number of routes entering each intermediate facility or the depot is the same as the number of routes leaving it, while constraint (3d) sets this number to one for the depot. Constraints (3e) force the existence of at least one route between two facilities if there is a strictly positive flow between them, where A^{H} consists of all the feasible arcs of the graph with only the depot and the intermediate facilities. Constraints (3f) are the flow conservation constraints at each facility. Constraint (3g) ensures that the flow leaving the depot is 0 and constraint (3g) computes the demand λ that is not collected. Finally, constraints (3i) to (3l) set the domains of the variables.

The routing subproblems, which for each pair $(h, h'), h \in V^{\mathrm{fac}} \cup \{\sigma\}, h' \in V^{\mathrm{fac}}$ aim at finding negative reduced cost routes with respect to the routing master problem (3), are also called the pricing problems (PP). We follow the standard procedure in which iteratively the restricted master problem (RMP) and the PP are solved until no route with negative reduced cost can be found. Note that the names RMP and BSP are used interchangeably depending on the context.

To solve the PP in an efficient way, we rely on dynamic programming to solve a resource constrained shortest path problem and the concept of ng-path [1]. The goal is to find a shortest path from $h \in V^{\mathrm{fac}} \cup \{\sigma\}$ to $h' \in V^{\mathrm{fac}}$ on the road-network graph $G = (V, A)$, such that the path respects the vehicle capacity and its cost corresponds to the reduced cost. Ng-paths are a compromise between elementary and non-elementary paths and are built according to customer sets (i.e., ng-sets) which are associated with each customer and often contain neighbours within a short travelling distance. Those ng-sets allow to enforce locally the elementary constraint only on those neighbours.

Using the dual variables of the RMP (i.e., depicted in parentheses next to their constraints), the reduced cost is computed as

$$\sum_{(l,l') \in A} c_{ll'} a_{ll'r} - \sum_{j \in V^{\mathrm{stop}}} (\xi_j + \zeta_{h'}) q_{jr} + \beta_h - \beta_{h'} - \mathbb{1}_{\{\sigma\}}(h)\gamma - W^{\mathrm{tot}}\delta_{hh'}$$

where $\mathbb{1}_{\{\sigma\}}(h)$ indicates if the starting node h corresponds to σ and $a_{ll'r}$ define whether arc $(ll') \in A$ is used in route r. Note that, $c_r = \sum_{(l,l') \in A} c_{ll'} a_{ll'r}$ and $\sum_{k \in \mathcal{M}} x_{ll'k} = \sum_{r \in R} a_{ll'r} x_r$.

The RMP (3) is well-known to provide strong dual bounds as it is based on a Dantzig-Wolfe decomposition [6], thus allowing the derivation of tighter Benders cuts which, however, introduce some other difficulties that are discussed in the following. From the definition of the Cm-CTP-IF, we assume that a node $j \in V^{\mathrm{stop}}$ is only visited once in a solution to collect its assigned demand, such that $\bar{q}_j = \sum_{r \in R} q_{jr} x_r \ \forall j \in V^{\mathrm{stop}}$ for a given set cover solution V^{sel} of the BMP. Following that assumption, the BSP only considers routes that collect the total assigned demand $\bar{q}_j$ from the set cover nodes $j \in V^{\mathrm{sel}}$, and $\sum_{r \in R} q_{jr} =$

$0 \ \forall j \in V^{\text{stop}} \backslash V^{\text{sel}}$. This restriction results from the column generation procedure where only such columns (i.e., with $q_j = \bar{q}_j \ \forall j \in V^{\text{stop}}$) are generated in the PP and added to the BSP, leading to a restricted BSP (RBSP) that only considers a subset of all feasible routes. As a consequence, we do not take all constraints (4b) of the dual BSP (4) into account, which leads to a dual solution that could be infeasible and consequently to an invalid Benders cut. The dual BSP (4) is presented below and can be deduced directly from the primal BSP (3), where each dual variable is indicated next to its corresponding constraint.

$$\max \quad \sum_{j \in V^{\text{stop}}} \bar{q}_j \xi_j + \gamma + W^{\text{tot}} \kappa \tag{4a}$$

$$\text{s.t.} \quad \sum_{j \in V^{\text{stop}}} q_{jr} \xi_j + \beta_{h'} - \beta_h$$
$$+ \mathbb{1}_{\{\sigma\}}(h)\gamma + W^{\text{tot}} \delta_{hh'}$$
$$+ \mathbb{1}_{V^{\text{fac}}}(h')\zeta_{h'} \sum_{j \in V^{\text{stop}}} q_{jr} \leq c_r \quad \forall (h, h') \in A^{\text{H}}, r \in R_{h+} \cap R_{h-} \tag{4b}$$

$$\xi_j \leq c^{\text{pick}} \qquad\qquad \forall j \in V^{\text{stop}} \tag{4c}$$

$$W^{\text{tot}} \kappa \leq c^{\text{coll}} \tag{4d}$$

$$-\delta_{hh'} + \mathbb{1}_{V^{\text{fac}}}(h)\zeta_h - \mathbb{1}_{V^{\text{fac}}}(h')\zeta_{h'}$$
$$+ \mathbb{1}_{\{\sigma\}}(h)\eta + \mathbb{1}_{\{\sigma\}}(h')\kappa \leq 0 \quad \forall (h, h') \in A^{\text{H}} \tag{4e}$$

$$\zeta_h \geq 0 \qquad\qquad \forall h \in V^{\text{fac}} \tag{4f}$$

$$\delta_{hh'} \geq 0 \qquad\qquad \forall (h, h') \in A^{\text{H}} \tag{4g}$$

To overcome this issue, we would need to solve the real BSP (not the restricted one) including all feasible routes that exist between each pair $\{h, h'\}, h \in V^{\text{fac}} \cup \{\sigma\}, h' \in V^{\text{fac}}$, such that demands $q_j \neq \bar{q}_j \ \forall j \in V^{\text{stop}}$ can be collected. As defined in (1), q_j could take any value between 0 and $D_j \ \forall j \in V^{\text{stop}}$ which, however, would lead to a dramatic increase in computation time as all the nodes in V^{stop} instead of V^{sel} would be considered and even the smallest instances in our dataset could not be handled anymore. Instead, starting from formulation (4), we derive a set of weaker, but valid inequalities by allowing for only some variability in the PP. More precisely, the PP considers all feasible routes with $0 \leq q_j \leq D_j \ \forall j \in V^{\text{sel}}$ (instead of all $j \in V^{\text{stop}}$). Even though this simplifies the problem a lot, it is still more time consuming than the PP without any variability and thus, only few instances can be solved (see Sect. 5.2). Therefore, we also consider the Benders decomposition approach without any variability in the PP as a heuristic solution method for the Cm-CTP-IF and refer to it as *BendersH* in the remainder of the paper.

4.3   Valid Benders Cuts

To derive valid cuts from the ones obtained in formulation (4), we are going to consider the routing problem on the rotation level instead of the route level.

Let P be the set of all feasible rotations. For some rotation p, let R_p represent the set of routes $r \in R$ that are in rotation p, with q_{jp} being the amount of demand picked at node j in rotation p, and let $c_p = \sum_{r \in R_p} c_r$ be its cost. We introduce a continuous variable x_p indicating if rotation $p \in P$ is selected and a continuous variable ν_j representing the amount of demand that is not satisfied at node $j \in V^{\text{stop}}$, and present the BSP on the rotation level (BSP-P) in (5).

$$\min \quad \sum_{p \in P} c_p x_p + \sum_{j \in V^{\text{stop}}} c^{\text{pick}} \nu_j \tag{5a}$$

$$\text{s.t.} \quad \sum_{p \in P} q_{jp} x_p + \nu_j = \bar{q}_j \qquad \forall j \in V^{\text{stop}} \quad (\xi'_j) \tag{5b}$$

$$\sum_{p \in P} x_p = 1 \qquad (\gamma') \tag{5c}$$

$$x_p \geq 0 \qquad \forall p \in P \tag{5d}$$

$$\nu_j \geq 0 \qquad \forall j \in V^{\text{stop}} \tag{5e}$$

The objective function (5a) computes the total routing costs plus the penalty costs for uncollected demand. Constraints (5b) are the analogous of constraints (3b) and allow to ensure feasibility. Constraint (5c) ensures that exactly one rotation is selected, and constraints (5d) and (5e) define the variables domains.

The dual BSP-P (6) is based on the respective dual variables depicted in (5) next to their constraints.

$$\max \quad \sum_{j \in V^{\text{stop}}} \bar{q}_j \xi'_j + \gamma' \tag{6a}$$

$$\text{s.t.} \quad \sum_{j \in V^{\text{stop}}} q_{jp} \xi'_j + \gamma' \leq c_p \qquad \forall p \in P \tag{6b}$$

$$\xi'_j \leq c^{\text{pick}} \qquad \forall j \in V^{\text{stop}} \tag{6c}$$

Instead of solving the BSP-P directly to generate the Benders cuts, we derive them from the dual BSP solution (4). As explained above, if we fix a set of nodes V^{sel} and restrict to rotations p only stopping at nodes in V^{sel} to pick demands $q_{jp} \geq 0$, we are only dealing with a subset of possible rotations. Let us denote $P^{V^{\text{sel}}}$ this restricted subset of rotations. For a given route r, we denote h_r^+ and h_r^- the intermediate facility (or depot) where r starts and ends respectively. Consider a rotation $p \in P^{V^{\text{sel}}}$. We know that $c_p = \sum_{r \in R_p} c_r$ and $q_{jp} = \sum_{r \in R_p} q_{jr} \ \forall j \in V^{\text{stop}}$, and that all the routes in $P^{V^{\text{sel}}}$ are considered in the RBSP, therefore:

$$\sum_{r \in R_p} \Big(\sum_{j \in V^{\text{stop}}} q_{jr} \xi_j + \beta_{h_r^-} - \beta_{h_r^+} + \mathbb{1}_{\{o\}}(h_r^+)\gamma + W^{\text{tot}} \delta_{h_r^+ h_r^-}$$

$$+ \mathbb{1}_{V^{\text{fac}}}(h_r^-)\zeta_{h_r^-} \sum_{j \in V^{\text{stop}}} q_{jr} \Big) \leq \sum_{r \in R_p} c_r = c_p \quad \forall p \in P^{V^{\text{sel}}}. \tag{7}$$

As a rotation enters and leaves an intermediate facility (or the depot) the same number of times (3c), $\sum_{r \in R_p}(\beta_{h_r^-} - \beta_{h_r^+}) = 0$. Since the vehicle depot σ is only visited once in a rotation (3d), $\sum_{r \in R_p} \mathbb{1}_{\{\sigma\}}(h_r^+)\gamma = \gamma$. Therefore:

$$\sum_{j \in V^{\text{stop}}} q_{jp}\xi_j + \gamma + \sum_{r \in R_p}(W^{\text{tot}}\delta_{h_r^+ h_r^-} + \mathbb{1}_{V^{\text{fac}}}(h_r^-)\zeta_{h_r^-} \sum_{j \in V^{\text{stop}}} q_{jr}) \leq c_p \quad \forall p \in P^{V^{\text{sel}}}. \quad (8)$$

Moreover, by summing up constraints (4e) over all routes of p, we obtain

$$\sum_{r \in R_p}(-\delta_{h_r^+ h_r^-} + \mathbb{1}_{V^{\text{fac}}}(h_r^+)\zeta_{h_r^+} - \mathbb{1}_{V^{\text{fac}}}(h_r^-)\zeta_{h_r^-} s + \mathbb{1}_{\{\sigma\}}(h_r^+)\eta + \mathbb{1}_{\{\sigma\}}(h_r^-)\kappa) \leq 0$$

$$\Leftrightarrow \quad \sum_{r \in R_p}(-\delta_{h_r^+ h_r^-}) + \eta + \kappa \leq 0$$

$$\Leftrightarrow \quad \eta + \kappa \leq \sum_{r \in R_p} \delta_{h_r^+ h_r^-}$$

We can then relax our inequalities (8), and by defining

$$\xi_j' = \xi_j + \min_{h \in V^{\text{fac}}} \zeta_h \qquad\qquad \forall j \in V^{\text{stop}} \quad (9a)$$

$$\gamma' = \gamma + W^{\text{tot}}(\eta + \kappa) \qquad\qquad (9b)$$

we obtain

$$\sum_{j \in V^{\text{stop}}} q_{jp}\xi_j' + \gamma' \leq c_p \quad \forall p \in P^{V^{\text{sel}}}. \quad (10)$$

To derive cuts that are valid not only for rotations in $P^{V^{\text{sel}}}$ but for any rotation in P, we need to further relax this constraint. To do so, we set $\xi_j' = 0 \ \forall j \in V^{\text{stop}} \setminus V^{\text{sel}}$ (i.e., $\xi_j = -\min_{h \in V^{\text{fac}}} \zeta_h \ \forall j \in V^{\text{stop}} \setminus V^{\text{sel}}$). Indeed, let $p \in P$ be any rotation with set cover V^p containing the nodes at which demand is collected in rotation p. We define $\tilde{p}$ as the rotation that only stops at nodes that are in both V^{sel} and V^p, and collects there the demands induced by the rotation p and its nodes V^p (i.e., $q_{j\tilde{p}} = q_{jp} \ \forall j \in V^{\text{sel}} \cap V^p$, 0 otherwise). Since $\tilde{p}$ is a subtour of p and does not collect more demand than p, we have $c_{\tilde{p}} \leq c_p$. Furthermore, by construction $\tilde{p}$ belongs to the set of restricted rotations $P^{V^{\text{sel}}}$. Therefore, it satisfies inequality (10), and with the above assumptions we have

$$\sum_{j \in V^{\text{stop}}} q_{jp}\xi_j' + \gamma' = \sum_{j \in V^{\text{stop}}} q_{j\tilde{p}}\xi_j' + \gamma' \leq c_{\tilde{p}} \leq c_p. \quad (11)$$

As a consequence, constraints (6b) are respected by $p \in P$, and the valid Benders cut b that is added to (2) is defined as

$$\theta \geq \sum_{j \in V^{\text{stop}}}(\xi_j^b + \min_{h \in V^{\text{fac}}} \zeta_h^b)q_j + \gamma^b + W^{\text{tot}}(\eta^b + \kappa^b) \quad \forall b \in B \quad (12)$$

5   Computational Experiments

In this section, we present the results of the computational experiments. The solution approaches have been implemented in Java. To solve the MILP, the BMP and the RMP we used the Gurobi 9.5.2 MIP solver via its Java API. The instances were tested on a computer with a Intel(R) Xeon(R) Gold 6148 CPU @ 2.40 GHz processor, 32 GB of RAM, operating under Linux, and a time limit (TL) of 12 h was set.

5.1   Problem Instances

We use the problem instances introduced by [7]. They are defined with respect to the number of demand nodes $|W| \in \{15, 50, 100, 200\}$ and are labelled accordingly. The three main parameters that characterize an instance are the maximum walking distance $w \in \{100, 200, 300\}$, the number of routes $m \in \{5, 10, 20\}$ for the datasets $W50, W100$ and $W200$, and $m \in \{2, 3, 5, 10\}$ for the dataset $W15$, and the number of intermediate facilities $f \in \{1, 2, 3\}$, whose locations are given. This results into 27 instances for $W50, W100$ and $W200$, and 36 instances for $W15$, which yields 117 instances in total. Finally, we assume a stop penalty value of $t^{sto} = 5$ s. We refer the reader to [7] for more details about these instances. Since we do not allow to split the demand in the collection process, we preprocess each demand node $i \in W$ with $d_i > Q$, by distributing the excess to its closest nodes $l \in V \setminus W$ and add each such l to the set of demand nodes W.

5.2   Validation of the Benders Decomposition Approach

This section aims to validate the Benders decomposition approach with respect to the MILP formulation. Figure 2 presents the gap to the best solution (i.e., $(UB^{best} - LB)/UB^{best}$ in %) for the MILP formulation, the *Benders* and the *BendersH* approaches and datasets $W15$ and $W50$. We observe in Fig. 2a that for the smaller dataset the MILP formulations reports a large range of gaps with an average value of 25.15%, while the two Benders approaches have a smaller range of gaps with average values of 25.74% and 18.71%, respectively. For the larger dataset in Fig. 2b, the gaps vary more for the *Benders* and are higher than for the MILP formulation, while the *BendersH* has smaller gaps, with average gap values of 43.89%, 64.18% and 39.22%, respectively.

Looking at the *Benders*, we observe that no lower bound could be computed for some of the instances (i.e., 4/27 for $W50$). This is due to the variability required in the PP to solve the real BSP (i.e., not a restricted one) which leads to increased computation time, taking on average 78.85% of the total time to solve the LP. Furthermore, the relaxations performed to derive valid cuts from the dual BSP, result in optimality cuts that are weaker and on average have a small gap with respect to the primal solutions. The cuts underestimate the LP routing cost of the BMP solution, so that optimality can no longer be proven as the *Benders* may never converge to an optimal solution. More precisely, for a given BMP solution, the cut generally also underestimates the LP routing cost associated with that solution. Nevertheless, these gaps are in general relatively small with an average value of 1.75%, supporting the quality of the derived cuts.

The *BendersH* approach, which does not allow for variability in the PP, and can therefore only be considered as an approximation, is faster and can handle on average more set covers within the TL (i.e., 9.75% more set covers). As a consequence, more time is spent in the BMP which needs on average 95% of the computation time (i.e., 40% more than the *Benders*). The positive gap values indicate that this heuristic did never report approximated lower bounds that are higher than the best upper bound which would have shown the invalidity of the approximated lower bounds. Moreover, for instances where the MILP formulation has proven optimality (i.e., 8/36 for $W15$), the heuristic returned valid lower bounds with an average gap value of 17.61%.

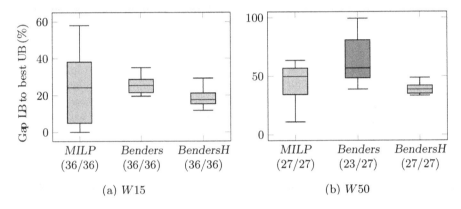

(a) $W15$ (b) $W50$

Fig. 2. Boxplots of the gaps to the best solution (i.e., $(UB^{best} - LB)/UB^{best}$ in %) for the MILP formulation, the *Benders* and *BendersH* approaches and datasets $W15$ and $W50$.

To assess the performance of the *BendersH* with respect to the MILP formulation, Table 1 presents the number of instances for which each method provided the best solution within the TL, the average valid optimality gaps computed with respect to the lower bounds of the MILP formulation, and the average approximated optimality gaps computed with respect to the approximated lower bounds of the *BendersH*. The average number of Benders cuts, number of column generation iterations per cut, and objective value (i.e., w/o dumping cost) are also depicted. We observe that the MILP formulation cannot solve most of the larger instances and that the valid and approximated optimality gaps are similar on average. The *BendersH* can provide better solutions for larger instances, mainly for those with a higher number of tours m, leading to shorter routes and computing times in the PP of the column generation approach. As a consequence, more set covers can be treated within the TL resulting in more cuts added and thus higher approximated bounds. From a managerial perspective, we point out that more routes (i.e., smaller vehicles) could lead to reduction of operational costs, as the routing cost increases are mild for most instances (especially for those with lower population density: W15 and W200). These results confirm the difficulty of the Cm-CTP-IF and the potential of the *BendersH* to solve and evaluate instances of larger size. However, we recall that the *BendersH* approach can only be considered as an approximation and could lead to false lower bounds.

Table 1. Performance of the MILP formulation and the *BendersH* approach for instances of datasets $W15, W50, W100$ and $W200$.

| Instance | | # solutions | | # is best | | Averages | | | | # cuts | # CG iter. | obj. val. |
| | | | | | | valid opt. gap % | | approx. opt. gap % | | | | |
		MILP	BendersH	MILP	BendersH	MILP	BendersH	MILP	BendersH	BendersH	BendersH	BendersH
W15	m = 2	9	9	9	0	**0.58%**	4.69%	**17.32%**	20.75%	356	8	2776
	m = 3	9	9	9	2	**11.98%**	13.28%	**15.70%**	16.91%	628	6	2523
	m = 5	9	9	7	2	**35.29%**	35.90%	**25.15%**	25.87%	736	5	2765
	m = 10	9	9	8	1	**52.90%**	53.10%	**16.87%**	17.22%	1484	16	2926
W50	m = 5	9	9	8	1	**25.74%**	30.25%	**43.35%**	46.67%	112	22	2882
	m = 10	9	9	3	6	52.08%	**51.74%**	41.09%	**40.44%**	132	16	3139
	m = 20	9	9	0	9	58.21%	**54.78%**	39.60%	**34.47%**	163	14	4357
W100	m = 5	9	9	7	2	**57.20%**	58.46%	**79.15%**	79.78%	103	20	3615
	m = 10	7	9	0	9	78.20%	**73.56%**	86.34%	**84.29%**	101	15	3637
	m = 20	1	9	0	9	84.94%	**84.76%**	83.17%	86.10%	99	12	4401
W200	m = 5	6	9	2	7	47.48%	**42.90%**	63.01%	**58.98%**	91	30	6938
	m = 10	1	9	0	9	63.16%	**60.19%**	72.05%	**67.94%**	88	23	6671
	m = 20	–	9	0	9	–	**74.69%**	–	**75.26%**	86	17	7902

6 Conclusion

In this article, we formulated the Cm-CTP-IF which is inspired by a waste collection problem with intermediate disposal facilities to accommodate the usage of smaller vehicles. It is a variation of the Cm-CTP-R ([7]) which is closely related to the m-CTP. We extended the proposed road-network-based MILP formulation and developed a Benders decomposition approach in which a SCP is solved in the master problem and a CVRP-IF is solved with column generation in the subproblem to handle large instances derived from real-life data. The results show that the Cm-CTP-IF is a difficult problem and the MILP formulation fails to solve most instances of the largest dataset and could only prove optimality for some instances of the smallest dataset. The *BendersH* approach provides better solutions for larger instances and allows to evaluate the solutions by providing approximated lower bounds that are close to the valid bounds. To derive valid Benders cuts, one should remember that the real Benders subproblem needs to be solved, and not a restricted one. We believe that this is an important aspect to consider when combining these two concepts, which are both widely used in location and routing problems. Further research will be needed to improve the BMP so that less time is spent on that problem and more set covers can be evaluated within the TL. For instance, valid inequalities that revolve around the residents' ranks could be derived to strengthen the linear relaxation BMP.

Acknowledgement. The authors gratefully acknowledge the support of Innosuisse under grant 36157.1 IP-EE. They also thank the Hasler Foundation for their support and they highly value the involvement of Schwendimann AG in conducting practical experiments and providing the associated data.

References

1. Baldacci, R., Mingozzi, A., Roberti, R.: New route relaxation and pricing strategies for the vehicle routing problem. Oper. Res. **59**(5), 1269–1283 (2011)
2. Ben Ticha, H., Absi, N., Feillet, D., Quilliot, A.: Vehicle routing problems with road-network information: state of the art. Networks **72**(3), 393–406 (2018)
3. Benders, J.: Partitioning procedures for solving mixed-variables programming problems '. Numer. Math. **4**(1), 238–252 (1962)
4. Braekers, K., Ramaekers, K., Van Nieuwenhuyse, I.: The vehicle routing problem: state of the art classification and review. Comput. Ind. Eng. **99**, 300–313 (2016)
5. Crevier, B., Cordeau, J.F., Laporte, G.: The multi-depot vehicle routing problem with inter-depot routes. Eur. J. Oper. Res. **176**(2), 756–773 (2007)
6. Dantzig, G.B., Wolfe, P.: Decomposition principle for linear programs. Oper. Res. **8**(1), 101–111 (1960)
7. Fischer, V., Paneque, M.P., Legrain, A., Bürgy, R.: A capacitated multi-vehicle covering tour problem on a road network and its application to waste collection. Eur. J. Oper. Res. (2023)
8. Hachicha, M., Hodgson, M.J., Laporte, G., Semet, F.: Heuristics for the multi-vehicle covering tour problem. Comput. Oper. Res. **27**(1), 29–42 (2000)
9. Labbé, M., Laporte, G.: Maximizing user convenience and postal service efficiency in post box location. JORBEL-Belgian J. Oper. Res. Stat. Comput. Sci. **26**(2), 21–36 (1986)
10. Letchford, A.N., Nasiri, S.D., Theis, D.O.: Compact formulations of the Steiner traveling salesman problem and related problems. Eur. J. Oper. Res. **228**(1), 83–92 (2013)
11. Oliveira, W.A.D., Moretti, A.C., Reis, E.F.: Multi-vehicle covering tour problem: building routes for urban patrolling. Pesquisa Operacional **35**, 617–644 (2015)
12. Rahmaniani, R., Crainic, T.G., Gendreau, M., Rei, W.: The benders decomposition algorithm: a literature review. Eur. J. Oper. Res. **259**(3), 801–817 (2017)

Don't Explain Noise: Robust Counterfactuals for Randomized Ensembles

Alexandre Forel[1]([✉])[iD], Axel Parmentier[2][iD], and Thibaut Vidal[1][iD]

[1] Department of Mathematics and Industrial Engineering, Polytechnique Montreal, Montreal, Canada
{alexandre.forel,thibaut.vidal}@polymtl.ca
[2] CERMICS, Ecole des Ponts, Marne-la-Vallée, France
axel.parmentier@enpc.fr

Abstract. Counterfactual explanations describe how to modify a feature vector in order to flip the outcome of a trained classifier. Obtaining robust counterfactual explanations is essential to provide valid algorithmic recourse and meaningful explanations. We study the robustness of explanations of randomized ensembles, which are always subject to algorithmic uncertainty even when the training data is fixed. We formalize the generation of robust counterfactual explanations as a probabilistic problem and show the link between the robustness of ensemble models and the robustness of base learners. We develop a practical method with good empirical performance and support it with theoretical guarantees for ensembles of convex base learners. Our results show that existing methods give surprisingly low robustness: the validity of naive counterfactuals is below 50% on most data sets and can fall to 20% on problems with many features. In contrast, our method achieves high robustness with only a small increase in the distance from counterfactual explanations to their initial observations.

1 Introduction

Counterfactual explanations provide a course of action to change the outcome of a classifier and reach a target class. Since the seminal work of [35], counterfactual explanations have received a large amount of attention [19,34]. Their applications can be divided into two main categories: (1) to provide algorithmic recourse to users or customers so that they can react to a given outcome (e.g., a customer applies for a loan and is rejected), and (2) to explain the recommendation of complex classifiers to stakeholders (e.g., a medical diagnosis based on a large set of features).

In high-stakes environments, such as the loan application example, algorithmic recourse must remain valid over time. That is, the loan should be approved when the customer returns after having acted upon the explanation received. Yet, the classification model might be retrained in the meantime, which alters its prediction function. The robustness of explanations when a model is retrained has been studied when additional data is observed, possibly affected by a shift in the distribution of the data-generating process [8,13,28,32]. However, a fundamental source of uncertainty has been overlooked so far: when the classifier is a randomized ensemble, *the random training procedure always leads to algorithmic uncertainty* when retraining the model, even if the training data is fixed.

From an explainability perspective, a lack of robustness due to the random training procedure when the training data is fixed is already very problematic. It raises the question of whether explanations that are not robust to model retraining allow any meaningful interpretation of a classifier's decisions. This issue is akin to the concept of predictive multiplicity, in which models from different classes have similar average performance but wildly different predictions for certain samples [16]. By design, nearest counterfactual explanations identify the minimum change that flips the classifier's label. Thus, they might be attracted to regions of the feature space that are particularly vulnerable to predictive multiplicity. In that case, the explanations obtained are only noise: artifacts of the random training procedure that exploits the closest region with high predictive multiplicity. Such non-robust explanations are equivalent to adversarial examples, which have no explainability value but are optimized to fool a given model [17,27].

In this paper, we study the robustness of counterfactual explanations of randomized ensembles to algorithmic uncertainty. Ensemble learning is a powerful technique that aggregates several models to reduce the risk of over-fitting and achieve high generalization power [15,29]. Common approaches to building an ensemble of learners revolve around the use of randomization (e.g., the bootstrap aggregating procedure of random forests) and are thus susceptible to algorithmic uncertainty. A key result of our paper is to show that naive explanations that ignore the algorithmic uncertainty of random ensembles are not robust to model retraining even when the training data is fixed. Hence, naive explanations provide neither robust algorithmic recourse nor explainability.

To bridge this gap, we develop methods to obtain counterfactual explanations that are robust to algorithmic uncertainty. We make the following contributions:

1. We show that naive methods to generate counterfactual explanations fail to provide robust explanations on common data sets—the validity is often below 50% and falls below 20% on the most complex data set.
2. We derive an efficient method to generate robust explanations by identifying a robust threshold on the ensemble's score. Our approach is flexible in the sense that it can be combined with any counterfactual explanation method and applies to all ensembles made of independently trained base learners (e.g., random forests, deep ensembles with random weights initialization). We demonstrate the value of our approach by obtaining explanations of random forest ensembles that are robust to algorithmic uncertainty on real-world data.
3. We support our practical results with theoretical guarantees that hold for ensembles of convex learners, such as random forests made of trees with a single decision split or ensembles of input-convex neural networks.

4. Finally, we study the connection between the predictive importance of features and the robustness of counterfactual explanations. We show that generating robust counterfactuals is more challenging for data sets with many features with high predictive importance.

2   Problem Statement and Background

We consider the standard binary classification setting. A training set $z_n = \{(x_i, y_i)\}_{i=1}^n$ of size n is available where $x_i \in \mathcal{X}$ is a feature vector and $y_i \in \{0, 1\}$ is a binary label. We assume throughout the paper that the training samples are i.i.d. observations of an unknown distribution P_{XY}.

2.1   Classification Ensembles

We focus on classification ensembles made of base learners trained independently and identically. This encompasses for instance the random forests of [5] or ensembles of neural networks. The randomness of random forests stems from two factors: (1) each base learner uses a re-sampling of the original training set (most commonly a bootstrap sample), and (2) a random subset of features is selected at each node of the tree to identify the best split. Interestingly, bootstrapped ensembles of neural networks do not benefit from the performance improvements of their tree-based counterparts. Instead, deep ensembles are often made of base learners trained independently on the same training set with random initial weights [20]. Randomization improves the generalization performance of ensembles as it reduces the correlation between base learners and thus significantly reduces variance (see Chap. 15, Sect. 15.4 of [15]).

Formally, we denote by $t(\cdot; \xi) : \mathcal{X} \to \{0, 1\}, x \mapsto t(x; \xi)$ the prediction function of a base learner with random training procedure parameterized by ξ. A classification ensemble $T(\cdot; \boldsymbol{\xi}) : \mathcal{X} \to \{0, 1\}, x \mapsto T(x; \boldsymbol{\xi})$ consists of N base learners $\{t(\cdot; \xi_i)\}_{i=1}^N$, where $\boldsymbol{\xi} = \{\xi_i\}_{i=1}^N$ parameterizes the training procedure of the ensemble. We denote by $h_N(\cdot; \boldsymbol{\xi}) : \mathcal{X} \to [0, 1]$ the score function of an ensemble, defined as the average of the class predictions:

$$h_N(x; \boldsymbol{\xi}) = \frac{1}{N} \sum_{i=1}^N t(x; \xi_i), \qquad (1)$$

For any observation $x \in \mathcal{X}$, a trained ensemble returns a class prediction as a majority vote among the base learners, so that $T(x; \boldsymbol{\xi}) = 1$ if $h_N(x; \boldsymbol{\xi}) \geq 1/2$ and $T(x; \boldsymbol{\xi}) = 0$ otherwise.

2.2   Counterfactual Explanations of Ensembles

Let T^0 be a classification ensemble trained on z_n with score function h_N^0, and let x_{n+1} be a new observation with predicted class $T^0(x_{n+1}; \boldsymbol{\xi}^0)$. We want to find the nearest counterfactual explanation of this class prediction. We assume, w.l.o.g, that the predicted class of x_{n+1} is $T^0(x_{n+1}; \boldsymbol{\xi}^0) = 0$ so that the target class of its counterfactual explanation is 1. Multi-class problems can be converted to this setting by assigning label 1 to the target class and label 0 to all other classes.

We focus on nearest counterfactual explanations that minimize a distance function $f(\cdot, x_{n+1})$, such as the l_1-norm that encourages sparse explanations for which only a few features are modified. Counterfactual explanations should satisfy two essential conditions: they should be actionable and plausible. Actionability ensures that users can act upon the algorithmic recourse by making sure that immutable features are not modified or respect a specific structure [33]. For instance, an individual cannot decrease their age. Plausibility ensures that the counterfactual explanation is not an outlier of the distribution of the target class. This can be done by ensuring that the likelihood of an explanation is larger than a desired threshold, where the likelihood can be estimated for instance using density estimation [3], local outlier factor [18], or local-neighborhood search [21]. An open question is whether plausibility can also help in obtaining explanations that are robust to the random training procedure. We show in our experiments that this is not the case.

The majority of existing counterfactual methods modify the given sample until the target class is attained [see e.g., 23, 31]. Thus, they work in a heuristic fashion and do not guarantee that the explanation found is the closest one to the original sample. Conversely, approaches based on integer programming can determine counterfactuals that are optimal for the distance metric under consideration and readily integrate constraints that reflect the actionability of the feature changes [18]. Such methods have proved especially relevant for generating counterfactual explanations over tree ensembles [11]. In particular, [26] provide an efficient formulation for cost-optimal counterfactuals in tree ensembles. Plausibility constraints are integrated using isolation forests, a tree-based method to estimate the likelihood of an explanation.

We can now state the problem of finding the nearest counterfactual explanation of x_{n+1} in a general way as:

$$\min_{x} \quad f(x, x_{n+1}) \tag{2a}$$

$$\text{s.t.} \quad h_N^0(x; \boldsymbol{\xi}^0) \geq 1/2 \tag{2b}$$

$$x \in \mathcal{X}^a \cap \mathcal{X}^p. \tag{2c}$$

Constraint (2b) ensures that the counterfactual explanation reaches the target class according to the ensemble T^0. Constraint (2c) specifies that the counterfactual explanation belongs to both the actionable domain $\mathcal{X}^a$ and the plausible domain $\mathcal{X}^p$. We call counterfactual explanations obtained by solving Problem (2) *naive* since they ignore the algorithmic uncertainty caused by the random training procedure of the ensemble.

2.3   Algorithmic Uncertainty, Validity and Robustness

Let T^0 and T^1 be two classification ensembles of size N trained on the same training set z_n. Due to the random training procedure, the classifiers have different prediction functions. Let $\mathcal{C}(\,\cdot\,; T^0) : \mathcal{X} \to \mathcal{X}$ be an algorithm that maps any observation x to a counterfactual explanation $\hat{x}$ for the classifier T^0. We now formalize the concepts of validity and robustness of counterfactual explanations.

Definition 1 (Validity). *The counterfactual explanation $\hat{x} = \mathcal{C}(x; T^0)$ is valid for classifier T^1 if $T^1(\hat{x}, \boldsymbol{\xi}^1) = 1$.*

Definition 2 (Algorithmic robustness). *A counterfactual algorithm $\mathcal{C}(\cdot; T^0)$ is robust with tolerance α if, for any new observation x_{n+1}, the probability that its counterfactual explanation is valid for a new classifier T trained on z_n is greater than $(1 - \alpha)$, that is:*

$$\mathbb{P}_{\boldsymbol{\xi}}\left(T\left(\mathcal{C}\left(x_{n+1}; T^0\right); \boldsymbol{\xi}\right) = 1\right) \geq 1 - \alpha, \tag{3}$$

where the uncertainty is taken with regard to the random training procedure of $T(\cdot; \boldsymbol{\xi})$ on the fixed training set z_n.

Definition 2 introduces robustness in a probabilistic sense over all possible ensembles of size N trained on the fixed set z_n. Intuitively, the tolerance parameter α controls the trade-off between the expected robustness of the counterfactual algorithm and the average distance between counterfactual explanations and the original observations. One of the goals of this work is to investigate the trade-off between these two objectives and to provide robust counterfactuals that remain close to their original observations.

3 Robust Counterfactual Explanations

Due to the random training procedure of randomized ensembles, the class prediction of an ensemble trained on z_n is a random variable. To generate a counterfactual explanation robust to algorithmic uncertainty, Problem (2) has to be augmented with the probabilistic constraint introduced in Definition 2, which can be equivalently expressed as:

$$\mathbb{P}_{\boldsymbol{\xi}}\left(h_N(x; \boldsymbol{\xi}) \geq 1/2\right) \geq 1 - \alpha. \tag{4}$$

Main Result. The core result of our paper is to show that, given a trained ensemble with score function h_N^0, we can obtain robust counterfactual explanations by replacing the complex probabilistic condition in Eq. (4) with the much simpler deterministic condition:

$$h_N^0(x; \boldsymbol{\xi}^0) \geq \tau(N, \alpha), \tag{5}$$

where $\tau(N, \alpha) \in [1/2, 1]$ is a well-defined threshold. The computational complexity of finding robust explanations is thus the same as the one of the naive Problem (2).

Deriving this result consists of two steps: (i) reformulating the robustness constraint on ensembles into an equivalent constraint on base learners, and (ii) showing that the probability of any base learner to output the target class is well approximated by the score function of the initial ensemble. The remainder of this section details these two steps.

3.1 Reformulating the Robustness Constraint

We start by formalizing the statistical properties of the class predicted by a randomized ensemble when the training data z_n is fixed. For any observation $x \in \mathcal{X}$, the event that a base learner trained on z_n outputs the target class at x can be seen as a random event with probability $p(x)$. Since we consider a binary classification setting, the class predicted by a base learner trained on z_n thus follows a Bernoulli distribution: $t(x; \xi) \sim \text{Bernoulli}(p(x))$.

A key observation is that, given a randomized ensemble trained on z_n, its base learners are independent and identically distributed observations of any base learner trained on z_n. Consequently, the score of a randomized ensemble with N base learners follows a binomial distribution:

$$N \cdot h_N(x, \xi) \sim \text{Bin}(N, p(x)). \tag{6}$$

Denote by $B(k; N, p)$ the cumulative distribution function (c.d.f.) of the binomial distribution $\text{Bin}(N, p)$ evaluated at k. The following property can be identified.

Lemma 1. *Given $N \in \mathbb{N}$, the map $g_N : [0, 1] \rightarrow [0, 1], p \mapsto B(N/2; N, p)$ is decreasing and invertible.*

The proof is based on the link between the c.d.f. of the binomial distribution and the c.d.f. of the beta distribution. It is given in Appendix A.1[1]. Using Lemma 1, we can reformulate the probabilistic condition in Eq. (4) by an equivalent condition on the prediction of a single (random) base learner.

Proposition 1. *Let h_N be the score function of an ensemble of N base learners trained on z_n. The robustness condition $\mathbb{P}_\xi (h_N(x, \xi) \geq 1/2) \geq (1 - \alpha)$ is satisfied if $\mathbb{P}_\xi (t(x, \xi) = 1) \geq g_N^{-1}(\alpha)$. The two conditions are equivalent when N is odd.*

The proof is given in Appendix A.2. Proposition 1 provides a criterion to generate robust counterfactual explanations based only on the probability that a base learner predicts the target class. We denote by $p_{N,\alpha}^* = g_N^{-1}(\alpha)$ this key robustness threshold and provide two properties. They describe how the robustness threshold $p_{N,\alpha}^*$ varies with the tolerance α and the ensemble size N.

Proposition 2. *Let $N \in \mathbb{N}$, the threshold $p_{N,\alpha}^*$ is monotonic increasing with the robustness target $(1 - \alpha)$.*

The proof follows directly from Lemma 1. This proposition confirms the intuitive property that a counterfactual explanation algorithm that is robust with a robustness level $(1 - \alpha)$ is also robust at a robustness level lower than $(1 - \alpha)$. We now study how increasing the number of base learners impacts robustness. We will need the following lemma.

---

[1] Appendix is available in the full version of the paper: https://arxiv.org/abs/2205.14116.

Lemma 2. *Let $m \in \mathbb{N}$, the following relationships hold:*

(a) $\forall \alpha \leq 1/2,\ p^*_{2m+3,\alpha} \leq p^*_{2m+1,\alpha},$
(b) $\forall \alpha \leq 1/2,\ p^*_{2(m+1),\alpha} \leq p^*_{2m,\alpha},$
(c) $\forall \alpha,\ p^*_{2m+1,\alpha} \leq p^*_{2m,\alpha}.$

The proof is based on studying the variations of $B(k; N, p)$ with fixed p and is given in Appendix A.3.

Proposition 3. *A counterfactual algorithm robust to algorithmic uncertainty for an ensemble of size N with N even is also robust for any ensemble of larger size.*

The proof follows directly from Lemma 2. Proposition 3 generalizes the robustness of counterfactual explanations for an ensemble of size N. We illustrate the results of Proposition 2 and 3 in Fig. 1, which shows how the robustness threshold $p^*_{N,\alpha}$ varies as a function of the robustness target $(1 - \alpha)$ and the ensemble size N.

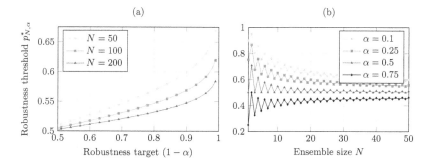

Fig. 1. Sensitivity of the robustness threshold $p^*_{N,\alpha}$.

All the above results hold as long as the ensembles are made of base learners trained independently and identically on the fixed data set z_n. For instance, they apply to random forests, deep ensembles, or any ensemble built using the bootstrap aggregating procedure. Another observation is that $\alpha = 1/2$ recovers the naive condition presented in Problem (2) when N is odd. Thus, explanations satisfying exactly the naive constraint (2b) have an average validity of only 50% when N is odd (regardless of the ensemble size!), or when N is even and the ensemble size is large as illustrated in Fig. 1. Therefore, even if the ensemble size is large, naive algorithms should fail to provide explanations robust to the noise of the training procedure.

3.2   Sample-Average Approximations

The robustness condition in Proposition 1 remains a probabilistic constraint and, as such, cannot be directly integrated into any solution algorithm. Further, the probability $\mathbb{P}_\xi\left(t(x,\xi)=1\right)$ that a base learner trained on z_n predicts the target class at x is unknown. We now present two methods to approximate the probability $\mathbb{P}_\xi\left(t(x,\xi)=1\right)$ that use only a given trained ensemble T^0.

Background. Optimization problems with a probabilistic constraint such as Eq. (5) belong to the class of chance-constrained problems. These problems have been studied extensively in the stochastic optimization literature since the seminal work of [10]. In particular, sampling approaches, such as the sample-average approximation (SAA), are well-known techniques to solve chance-constrained problems with strong theoretical foundations and good empirical results [25].

To approximate the probability $\mathbb{P}_\xi\left(t(x,\xi)=1\right)$, a possible sampling-based approach is to repeatedly train base learners on z_n and evaluate them for a given observation x. Yet, in counterfactual explanation applications, we already have "sampled" observations of the base learner: given the trained ensemble T^0, the base learners $\{t^0(x,\xi_i^0)\}_{i=1}^N$ are i.i.d. observations of $t(x,\xi)$. Thus, we do not need to train any additional learners. This is very desirable in practice since it means that robust explanations can be obtained without having access to the training data when a trained ensemble is available.

Direct SAA. Given the trained learners $\{t^0(x,\xi_i^0)\}_{i=1}^N$, the Direct SAA of the probabilistic robustness constraint is:

$$h_N^0(x,\boldsymbol{\xi}^0) \geq p_{N,\alpha}^*. \tag{7}$$

Thus, we recover the simple condition presented in Eq. (5) by taking $\tau(N,\alpha) = p_{N,\alpha}^*$. Note also that, when $p_{N,\alpha}^* = 1/2$, we recover the naive condition of Problem (2).

Sample-average approximations of chance constraints have been shown to give good performance in numerous applications. However, the Direct SAA does not guarantee that the probabilistic constraint is satisfied in general with a finite ensemble size N. Hence, we provide a more robust approach, that still uses only the initial ensemble and does not need to train any additional learners.

Robust SAA. This approach is motivated by a statistical perspective: the score function of a given ensemble h_N^0 can be seen as an estimator of the probability that a single learner predicts the target class. Indeed, when binomial samples are observed i.i.d, the sample mean is an unbiased minimum-variance estimator of the success rate p of the underlying Bernoulli distribution. To hedge against the noise of this estimator when the ensemble size is finite, we introduce the Robust SAA based on building a confidence interval around the estimated threshold $p_{N,\alpha}^*$ as if it were estimated from i.i.d. observations.

Confidence intervals of the success rate of binomial distributions have been studied extensively. In particular, the Agresti-Coull (AC) confidence intervals [1] achieve good coverage of the true success rate in finite samples [6]. The Robust SAA thus uses the threshold:

$$h_N^0(x,\boldsymbol{\xi}^0) \geq \rho_{N,\alpha,\beta}^*, \tag{8}$$

where $\rho^*_{N,\alpha,\beta} = p^*_{N,\alpha} + z_\beta \sqrt{\rho_{AC}(1 - \rho_{AC})/N}$ with z_β being the quantile of the standard normal distribution at $\beta/2$, and $\rho_{AC} = (N \cdot p^*_{N,\alpha} + 2)/(N + 4)$. The confidence level $\beta \in [0,1]$ is a hyperparameter that adjusts the conservativeness of the solution. As β increases, the robustness of the counterfactual increases and so does the distance to the initial observation. Thus, β depends on the data set and needs to be tuned according to the desired robustness level. This approach is always more conservative than the Direct SAA since $\beta = 0$ recovers the Direct SAA.

4 Robustness Guarantees for Ensembles of Convex Learners

Our approach can be supported by statistical guarantees by leveraging the theory on sample-average approximations developed in the stochastic optimization literature. In particular, when the ensemble is made of convex base learners, the Direct SAA is asymptotically consistent.

Proposition 4 (Asymptotic consistency). *As the size of an ensemble of convex base learners increases, the solution of the Direct SAA method converges almost surely to the minimum-cost robust counterfactual explanation.*

When the base learners are convex functions of the features, the condition $t(x;\xi) = 1$ inside the probabilistic constraint of Proposition 1 is also convex. Hence, the proof follows from [30, Chapter 5]. Proposition 4 holds for several classes of randomized ensembles. By definition, it holds for ensembles made of input-convex neural networks [2].

We can show that the result also holds for random forests made of decision trees that use a single decision split (also called stumps). These simplified forests have been studied for instance by [7] to show how bagging reduces variance in random forests. The class prediction of tree stumps can be formulated as a convex constraint for any realization of the uncertain training procedure. Let $\mathcal{X} \subseteq [0,1]^d$ and $t(x,\xi)$ be a decision stump. The robustness constraint can be expressed as a convex constraint as:

$$\mathbb{P}(t(x,\xi) = 1) = \mathbb{P}\left(A(\xi)^\top x - b(\xi) \leq 0\right), \qquad (9)$$

where $A(\xi) \in \{-1,0,1\}^d$ and $b(\xi) \in [-1,1]$. The vector $A(\xi)$ is such that $a_j(\xi) = 0$ if the stump does not split on feature j, $a_j(\xi) = 1$ if it splits on feature j and the left leaf node has class 1 and $a_j(\xi) = -1$ if it splits on feature j and the right leaf node has class 1. The split threshold $b(\xi)$ is positive if the left leaf node has class 1 and negative otherwise. Thus, the function on the left-hand side of the probabilistic Constraint (9) is convex in x.

These results on asymptotic consistency are notable since they hold for a wide variety of randomized ensembles. They do not require any simplification of the training procedure of the base learners, as is common for instance when studying the theoretical properties of random forests [4]. Finally, it is interesting to observe that increasing the size of the ensemble has two effects: (1) it decreases the robustness threshold $p_{N,\alpha}^*$ as shown in Lemma 2, and (2) it increases the accuracy of the Direct-SAA method according to Proposition 4.

Finite-Sample Guarantees. In certain cases, it is not possible to train additional learners when determining counterfactual explanations, such as when only the initial ensemble T^0 is available. In this case, we are interested in finite-sample guarantees on the robustness of explanations. Finite-sample bounds on the quality of the Direct-SAA method are given by [24], but require stringent assumptions on the feature space $\mathcal{X}$ and do not hold if the feature vector contains both continuous and discrete features. We present finite-sample bounds based on a second approximation technique called the convex approximation.

The convex approximation of the probabilistic robustness condition in Eq. (4) results in the following set of constraints:

$$t^0(x, \xi_i^0) = 1, \forall i \in \{1, \ldots, N\}. \tag{10}$$

Thus, the convex approximation is equivalent to taking $\tau(N, \alpha) = 1$ in Eq. (5). Finite-sample bounds on the probability of finding a robust solution by solving the convex approximated model can be obtained. [9] provide a key result when the feasible set $\mathcal{X}$ is convex (for instance, when all features are continuous). [12] generalize it to decision variables that take both continuous and discrete values. We can apply the latter result to obtain the following bound.

Proposition 5 (Finite-sample guarantees). *Given an ensemble of convex learners of size N and a feature vector x with k continuous features and $d - k$ discrete features, if a solution to the convex approximated problem exists, the probability that it is a robust counterfactual explanation is at least (1-δ) with $\delta = \exp\left[\left(2^{d-k}(k+1) - 1\right)(\log(1/\alpha) + \alpha) - (\alpha/2)N\right]$.*

The proof follows directly from [12] when the base learners are convex in x. Note that δ decreases exponentially as the number of trees increases.

5   Experimental Results

We conduct extensive experiments to (i) evaluate the robustness of our proposed approaches, (ii) understand why counterfactual explanations have varying robustness on different data sets, and (iii) demonstrate that our methods provide the best trade-off between robustness and distance. We focus our experiments on random forests, which are arguably the most common randomized ensembles used in practice and remain one of the state-of-the-art classifiers for tabular data [4,14]. We generate counterfactual explanations using the state-of-the-art formulation of [26] based on integer programming. An integer programming method has the advantage of being exact: it returns the counterfactual explanation with minimal distance to the original observation. Hence, it eliminates confounding factors in the analyses due to the choice of a specific heuristic solution.

The simulations are implemented in Python 3.9 using scikit-learn v.1.0.2 to train the random forests. Gurobi 9.5 is used to solve all integer programming models. All experiments are run on an Intel(R) Core(TM) i7-11800H processor at 2.30 Ghz using 16 GB of RAM. The data sets have been pre-processed following [26] to ignore missing values and to take into account feature actionability. Details on the datasets, implementation of our methods and benchmarks, as well as additional results are provided in the online appendix. The code to reproduce all results and figures in this paper is publicly available at the online repository https://github.com/alexforel/RobustCF4RF under an MIT license.

Simulation Setting. We use the standard procedure of scikit-learn to generate random forests of $N = 100$ trees (default value of scikit-learn) with a maximum depth of 4. In each simulation, five samples are selected randomly to serve as new observations for which to derive counterfactuals. A first forest T^0 is trained on the remaining $(n-5)$ points. A second forest T^1 is trained on the same data to asses if the counterfactual explanations are valid. We repeat this procedure forty times.

We implement the Direct-SAA and Robust-SAA methods with $\beta \in \{0.05, 0.1\}$ and vary the tolerance parameter α between 0.5 and 0.01. The key performance indicators are the distance between the initial observation and the counterfactual and their validity, measured as the percentage of counterfactuals that are valid to the test classifier. All distances are measured following a feature-weighted l_1 norm. We use the l_1 norm since it tends to create sparse explanations, that is, explanations with a limited number of changed features. To balance the cost of feature changes between continuous and non-continuous features, we reduce the weight of changes on non-continuous features by a factor $1/4$.

Benchmarks. Existing works on counterfactual explanations do not study the algorithmic uncertainty caused by a randomized training procedure. Hence, there is no directly related benchmark. The closest approach from the existing literature to our setting is arguably the one from [13]. Their algorithm (RobX) aims to find explanations of tree ensembles that are robust to evolving data sets and hyperparameters. Even though they study a different type of robustness, we evaluate the robustness of their explanations to algorithmic uncertainty. RobX is based on iteratively perturbating a naive explanation by moving it toward a neighbor with high stability, in the sense that the prediction model is locally constant. Since it is unclear how this method can be applied with binary, discrete, or categorical features, we only implement it for the SPAMBASE dataset, which has only continuous features.

We further include three benchmarks: (1) the naive approach that uses the constraint $h_N(x) \geq 1/2$, (2) a plausibility-based benchmark that uses isolation forests [22,26], and (3) a plausibility-based benchmark that uses the local outlier factor as [18]. Plausibility methods encourage the explanation to be close to the training data distribution. We include these benchmarks to investigate whether producing explanations that are close to the data distribution is sufficient to ensure robustness to algorithmic uncertainty. The contamination parameter of the isolation forest method is varied as $c \in \{0.05, 0.1, 0.2, 0.3, 0.4, 0.5\}$. Similarly, the weight of the local outlier factor penalty term is varied as $\lambda \in \{1e^{-3}, 1e^{-2}, 1e^{-1}, 1, 1e^1, 1e^2\}$ following [18]. These two hyperparameters control the degree to which the obtained explanations are close to the training data.

Example of Explanations with Increasing Robustness. We illustrate the generated counterfactual explanations with varying robustness targets for the GERMAN CREDIT data set in Fig. 2. The feature values of an initial observation x_{n+1} are shown on the bottom row as a heatmap. Each row then shows the changes to the initial feature values as the target robustness level $(1 - \alpha)$ increases. Positive changes to a feature are shown in blue and negative changes are shown in red. In this example, the "Age" feature can only increase. The row with $(1 - \alpha) = 0.5$ thus shows a naive explanation, whereas the top row with $(1 - \alpha) = 0.99$ shows an explanation with high robustness.

Figure 2 illustrates how the number and magnitude of feature changes vary as $(1 - \alpha)$ increases. It shows that a user can obtain a robust explanation with only a small subset of features changed.

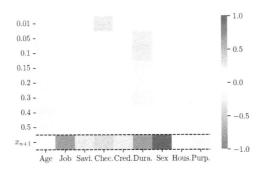

Fig. 2. Initial observation and counterfactual explanations for increasing robustness target $(1 - \alpha)$.

5.1   Achieving Robust Counterfactual Explanations

This section provides our main experimental results. Additional experiments, using larger random forests or an evolving data set, are presented in Appendix C.

Naive Explanations. First, we measure the robustness of naive explanations that ignore algorithmic uncertainty. We present the average validity of the counterfactual explanations in Table 1. Although it depends on the data set, naive explanations are clearly not robust to algorithmic uncertainty. The average validity even falls below 20% for the SPAMBASE dataset.

Table 1. Average naive explanations that remain valid when retraining the ensemble with fixed training data.

Data set	A	C	CC	GC	ON	P	S	SP
Validity [in %]	62	92	27	35	32	80	17	39

Algorithmic Robustness. The validity of the counterfactual explanations generated by the Direct- and Robust-SAA methods is shown in Fig. 3 for varying robustness target levels. The confidence interval of the average validity at the 0.05 level is shown as a shaded area. Figure 3 shows that the Direct-SAA method provides robust counterfactual explanations on all but one data sets. Counterfactual explanations with high robustness can already be found for small robustness targets in two data sets: COMPAS and PHISHING. Conversely, the data set SPAMBASE with $d = 57$ continuous features proves the most difficult. It is only on this data set that the Robust-SAA method is required, yielding robust counterfactuals for moderate and high robustness levels with $\beta = 0.1$ and $\beta = 0.05$, respectively.

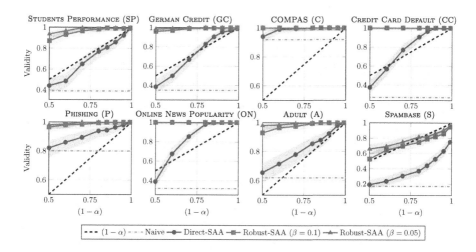

Fig. 3. Validity of robust counterfactuals as a function of the robustness target $(1-\alpha)$.

The same experiment is performed with random forests of size $N = 400$. The results, presented in Appendix C.4, are essentially identical to those in Fig. 3. This confirms that increasing the size of the ensemble does not protect against algorithmic uncertainty if the counterfactual explanation threshold is not chosen appropriately. We also perform experiments when additional data is observed before re-training the random forest. This experiment is presented in Appendix C.5. Again, the results are very similar to those in Fig. 3 and our approaches provide robust explanations on all data sets. This suggests that, for randomized ensembles, the algorithmic uncertainty due to the random training procedure is more critical for the robustness of explanations than the uncertainty caused by observing additional data samples.

Trade-off Between Distance and Robustness. The distance and validity of the counterfactual explanations obtained using our methods and benchmarks are shown in Fig. 4 as a Pareto front. The results show that our methods strictly dominate the plausibility-based benchmarks by consistently providing more robust explanations with lower distance. This means that plausibility and robustness are independent in practice. Even with high contamination parameters, isolation forests do not provide robust explanations although they substantially increase counterfactual distance. Increasing the penalty factor of the local outlier factor (1-LOF) method slightly increases robustness but leads to significantly more distant counterfactual explanations than the ones generated by our robust methods. The low robustness of the naive and plausibility-based benchmarks suggests their counterfactuals are mostly fitting the noise of the random training procedure. The RobX benchmark also does not provide robust explanations. This suggests that the algorithmic uncertainty of model retraining is distinct from the uncertainty of an evolving data set, which is the focus of RobX. Thus, these methods provide neither robust algorithmic recourse nor meaningful explanations of the random forest.

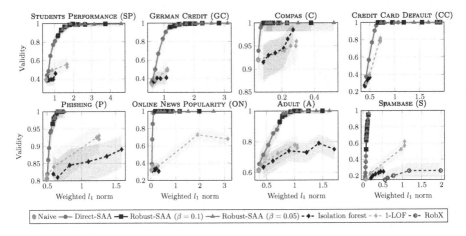

Fig. 4. Trade-off between the distance and robustness of counterfactual explanations.

Figure 4 also highlights that choosing a value for α depends on the preference of the decision-maker. Robust counterfactuals provide more insightful explanations but may be more difficult to act upon due to the increase in the average distance. Still, our results show that a large increase in robustness can be obtained for only a small increase in the counterfactual distance. A satisfying compromise can be found for instance using $\alpha = 0.2$.

5.2   Feature Importance and Robustness

To explain the varying robustness achieved on the different data sets, we analyze two aspects: (i) the average number of features changed in counterfactual explanations, and (ii) the link between the changed features and their predictive importance. Figure 5 shows the average number of changed features as a function of the robustness target level. Data sets that have a few key features with high importance tend to exhibit sparse and robust counterfactual explanations. This is especially true if the features are discrete or categorical as is the case for the COMPAS data set for instance. This sparsity leads to robust counterfactual explanations even when the robustness target level is low. On the contrary, data sets that have many important features have low inherent robustness, especially when these are continuous features as in the case of the SPAMBASE dataset.

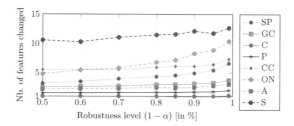

Fig. 5. Average number of features changed for varying robustness targets $(1 - \alpha)$.

The analysis of feature importance is provided in Appendix C.6. It shows that feature changes in counterfactual explanations follow largely their predictive importance: highly predictive features have large changes, whereas unimportant features are mostly unchanged.

6   Conclusions

This work sheds light on a critical, yet overlooked, aspect of interpretability in machine learning: the lack of robustness of existing counterfactual explanation methods with regard to algorithmic uncertainty. We contribute to building more transparent classifiers by providing an effective and easy-to-implement approach to improve robustness. Our method is supported by a probabilistic analysis of randomized ensembles and its value is demonstrated on real-world data. To help practitioners identify situations in which robust counterfactual explanations are essential, we derive valuable insights linking the robustness of counterfactual explanations and the characteristics of the data sets.

References

1. Agresti, A., Coull, B.A.: Approximate is better than "exact" for interval estimation of binomial proportions. Am. Stat. **52**(2), 119–126 (1998)
2. Amos, B., Xu, L., Kolter, J.Z.: Input convex neural networks. In: International Conference on Machine Learning, pp. 146–155. PMLR (2017)

3. Artelt, A., Hammer, B.: Convex density constraints for computing plausible counterfactual explanations. In: Farkaš, I., Masulli, P., Wermter, S. (eds.) ICANN 2020. LNCS, vol. 12396, pp. 353–365. Springer, Cham (2020). https://doi.org/10.1007/978-3-030-61609-0_28

4. Biau, G., Scornet, E.: A random forest guided tour. Test **25**(2), 197–227 (2016)

5. Breiman, L.: Random forests. Mach. Learn. **45**(1), 5–32 (2001)

6. Brown, L.D., Cai, T.T., DasGupta, A.: Interval estimation for a binomial proportion. Stat. Sci. **16**(2), 101–133 (2001)

7. Bühlmann, P., Yu, B.: Analyzing bagging. Ann. Stat. **30**(4), 927–961 (2002)

8. Bui, N., Nguyen, D., Nguyen, V.A.: Counterfactual plans under distributional ambiguity. In: International Conference on Learning Representations (2022)

9. Campi, M.C., Garatti, S.: The exact feasibility of randomized solutions of uncertain convex programs. SIAM J. Optim. **19**(3), 1211–1230 (2008)

10. Charnes, A., Cooper, W.W.: Deterministic equivalents for optimizing and satisficing under chance constraints. Oper. Res. **11**(1), 18–39 (1963)

11. Cui, Z., Chen, W., He, Y., Chen, Y.: Optimal action extraction for random forests and boosted trees. In: Proceedings of the 21th ACM SIGKDD International Conference on Knowledge Discovery and Data Mining, pp. 179–188 (2015)

12. De Loera, J.A., La Haye, R.N., Oliveros, D., Roldán-Pensado, E.: Chance-constrained convex mixed-integer optimization and beyond: two sampling algorithms within S-optimization. J. Convex Anal. **25**(1), 201–218 (2018)

13. Dutta, S., Long, J., Mishra, S., Tilli, C., Magazzeni, D.: Robust counterfactual explanations for tree-based ensembles. In: International Conference on Machine Learning, pp. 5742–5756. PMLR (2022)

14. Grinsztajn, L., Oyallon, E., Varoquaux, G.: Why do tree-based models still outperform deep learning on typical tabular data? In: Thirty-sixth Conference on Neural Information Processing Systems Datasets and Benchmarks Track (2022)

15. Hastie, T., Tibshirani, R., Friedman, J.H.: The Elements of Statistical Learning: Data Mining, Inference, and Prediction, vol. 2. Springer, New York (2009). https://doi.org/10.1007/978-0-387-21606-5

16. Hsu, H., Calmon, F.D.P.: Rashomon capacity: a metric for predictive multiplicity in classification. In: Advances in Neural Information Processing Systems (2022)

17. Ignatiev, A., Narodytska, N., Marques-Silva, J.: On relating explanations and adversarial examples. In: Advances in Neural Information Processing Systems, vol. 32 (2019)

18. Kanamori, K., Takagi, T., Kobayashi, K., Arimura, H.: DACE: distribution-aware counterfactual explanation by mixed-integer linear optimization. In: International Joint Conference on Artificial Intelligence, pp. 2855–2862 (2020)

19. Karimi, A.H., Barthe, G., Schölkopf, B., Valera, I.: A survey of algorithmic recourse: contrastive explanations and consequential recommendations. ACM Comput. Surv. **55**(5), 1–29 (2022)

20. Lakshminarayanan, B., Pritzel, A., Blundell, C.: Simple and scalable predictive uncertainty estimation using deep ensembles. In: Advances in Neural Information Processing Systems, vol. 30 (2017)

21. Laugel, T., Lesot, M.J., Marsala, C., Renard, X., Detyniecki, M.: The dangers of post-hoc interpretability: unjustified counterfactual explanations. In: International Joint Conference on Artificial Intelligence, pp. 2801–2807 (2019)

22. Liu, F.T., Ting, K.M., Zhou, Z.H.: Isolation forest. In: Eighth IEEE International Conference on Data Mining, pp. 413–422. IEEE (2008)

23. Lucic, A., Oosterhuis, H., Haned, H., de Rijke, M.: Focus: Flexible optimizable counterfactual explanations for tree ensembles. In: Proceedings of the AAAI Conference on Artificial Intelligence, vol. 36, no. 5, pp. 5313–5322 (2022)
24. Luedtke, J., Ahmed, S.: A sample approximation approach for optimization with probabilistic constraints. SIAM J. Optim. **19**(2), 674–699 (2008)
25. Pagnoncelli, B.K., Ahmed, S., Shapiro, A.: Sample average approximation method for chance constrained programming: theory and applications. J. Optim. Theory Appl. **142**(2), 399–416 (2009)
26. Parmentier, A., Vidal, T.: Optimal counterfactual explanations in tree ensembles. In: International Conference on Machine Learning, pp. 8422–8431. PMLR (2021)
27. Pawelczyk, M., Agarwal, C., Joshi, S., Upadhyay, S., Lakkaraju, H.: Exploring counterfactual explanations through the lens of adversarial examples: a theoretical and empirical analysis. In: International Conference on Artificial Intelligence and Statistics, pp. 4574–4594. PMLR (2022)
28. Rawal, K., Kamar, E., Lakkaraju, H.: Can i still trust you?: Understanding the impact of distribution shifts on algorithmic recourses. arXiv preprint arXiv:2012.11788 (2020)
29. Sagi, O., Rokach, L.: Ensemble learning: a survey. Wiley Interdisc. Rev. Data Min. Knowl. Disc. **8**(4), e1249 (2018)
30. Shapiro, A., Dentcheva, D., Ruszczynski, A.: Lectures on stochastic programming: modeling and theory. SIAM (2014)
31. Tolomei, G., Silvestri, F., Haines, A., Lalmas, M.: Interpretable predictions of tree-based ensembles via actionable feature tweaking. In: Proceedings of the 23rd ACM SIGKDD International Conference on Knowledge Discovery and Data Mining, pp. 465–474 (2017)
32. Upadhyay, S., Joshi, S., Lakkaraju, H.: Towards robust and reliable algorithmic recourse. In: Advances in Neural Information Processing Systems (2021)
33. Ustun, B., Spangher, A., Liu, Y.: Actionable recourse in linear classification. In: Conference on Fairness, Accountability, and Transparency, pp. 10–19 (2019)
34. Verma, S., Dickerson, J., Hines, K.: Counterfactual explanations for machine learning: a review. arXiv preprint arXiv:2010.10596 (2020)
35. Wachter, S., Mittelstadt, B., Russell, C.: Counterfactual explanations without opening the black box: automated decisions and the GDPR. Harvard J. Law Technol. **31**, 841 (2017)

Certifying MIP-Based Presolve Reductions for 0–1 Integer Linear Programs

Alexander Hoen[1]([✉]) [iD], Andy Oertel[3,4] [iD], Ambros Gleixner[1,2] [iD],
and Jakob Nordström[3,4] [iD]

[1] Zuse Institute Berlin, Takustr. 7, 14195 Berlin, Germany
hoen@zib.de
[2] HTW Berlin, 10313 Berlin, Germany
gleixner@htw-berlin.de
[3] Lund University, Lund, Sweden
andy.oertel@cs.lth.se
[4] University of Copenhagen, Copenhagen, Denmark
jn@di.ku.dk

Abstract. It is well known that reformulating the original problem can be crucial for the performance of mixed-integer programming (MIP) solvers. To ensure correctness, all transformations must preserve the feasibility status and optimal value of the problem, but there is currently no established methodology to express and verify the equivalence of two mixed-integer programs. In this work, we take a first step in this direction by showing how the correctness of MIP presolve reductions on 0–1 integer linear programs can be certified by using (and suitably extending) the VERIPB tool for pseudo-Boolean proof logging. Our experimental evaluation on both decision and optimization instances demonstrates the computational viability of the approach and leads to suggestions for future revisions of the proof format that will help to reduce the verbosity of the certificates and to accelerate the certification and verification process further.

Keywords: Proof logging · Presolving · 0–1 integer linear programming

1 Introduction

Boolean satisfiability solving (SAT) and *mixed-integer programming (MIP)* are two computational paradigms in which surprisingly mature and powerful solvers have been developed over the last decades. Today such solvers are routinely used to solve large-scale problems in practice despite the fact that these problems are *NP*-hard. Both SAT and MIP solvers typically start by trying to simplify the input problem before feeding it to the main solver algorithm, a process known as *presolving* in MIP and *preprocessing* in SAT. This can involve, e.g., fixing

B. Dilkina (Ed.): CPAIOR 2024, LNCS 14742, pp. 310–328, 2024.
https://doi.org/10.1007/978-3-031-60597-0_20

variables to values, strengthening constraints, removing constraints, or adding new constraints to break symmetries. Such techniques are very important for SAT solver performance [6], and for MIP solvers they often play a decisive role in whether a problem instance can be solved or not, regardless of whether the solver uses floating-point [2] or exact rational arithmetic [21].

The impressive performance gains for modern combinatorial solvers come at the price of ever-increasing complexity, which makes these tools very hard to debug. It is well documented that even state-of-the-art solvers in many paradigms, not just SAT and MIP, suffer from errors such as mistakenly claiming infeasibility or optimality, or even returning "solutions" that are infeasible [3,14,27,40,49]. During the last decade, the SAT community has dealt with this problem in a remarkably successful way by requiring that solvers should use *proof logging*, i.e., produce machine-verifiable certificates of correctness for their computations that can be verified by a stand-alone proof checker. A number of proof formats have been developed, such as DRAT [36,37,52], GRIT [18], and LRAT [17], which are used to certify the whole solving process including preprocessing.

Achieving something similar in a general MIP setting is much more challenging, amongst others because of the presence of continuous and general integer variables, which may even have unbounded domains. For numerically exact MIP solvers [15,21,22] the proof format VIPR [12] has been introduced, but it currently only allows verification of feasibility-based reasoning, which must preserve all feasible solutions. In particular, it does not support the verification of dual presolving techniques that may exclude feasible solutions as long as one optimal solution remains. This means that while exact MIP solvers could in principle generate a certificate for the main solving process, such a certificate would only establish correctness under the assumption that all the presolving steps were valid, as, e.g., in [21]. And, unfortunately, the proof logging techniques for SAT preprocessing cannot be used to address this problem, since they can only reason about clausal constraints.

Our contribution in this work is to take a first step towards verification of the full MIP solving process by demonstrating how pseudo-Boolean proof logging with VERIPB can be used to produce certificates of correctness for a wide range of MIP presolving techniques for 0–1 integer linear programs (ILPs). VERIPB is quite a versatile tool in that it has previously been employed for certification of, e.g., advanced SAT solving techniques [8,35], SAT-based optimization (MaxSAT) [4,50], subgraph solving [32,33], and constraint programming [34,42]. However, to the best of our knowledge this is the first time the tool has been used to prove the correctness of reformulations of optimization problems, and this presents new challenges. In particular, the proof system turns out not to be well suited for problem reformulations with frequent changes to the objective function, and therefore we introduce a new rule for objective function updates.

Our computational experiments confirm that this approach to certifying presolve reductions is computationally viable and the overhead for certification aligns with what is known from the literature for certifying problem transformations in other contexts [31]. The analysis of the results reveals new insights into performance bottlenecks, and these insights directly translate to possible revisions of the proof logging format that would be valuable to address in order to decrease the size of the generated proofs and speed up proof verification.

We would like to note that, while our current methods are only applicable to 0–1 ILPs, this covers already a large and important class of MIPs. In particular, there are applications where the exact and verified solution of 0–1 ILPs is highly relevant, see [1,23,46] for some examples.

The rest of this paper is organized as follows. After presenting pseudo-Boolean proof logging and VERIPB in Sect. 2, we demonstrate in Sect. 3 how to produce VERIPB certificates for MIP presolving on 0–1 ILPs. In Sect. 4 we report results of an experimental evaluation, and we conclude in Sect. 5 with a summary and discussion of future work.

2   Pseudo-Boolean Proof Logging with VeriPB

We start by reviewing pseudo-Boolean reasoning in Sect. 2.1, and then explain our extension to deal with objective function updates in Sect. 2.2. In order to make the concept of proof logging more concrete, we conclude this section by providing, in Table 1, a few examples of how the derivation rules explained below are encoded in VERIPB syntax. For space reasons, this list does not include examples of subproofs that may be necessary for some derivations that cannot be proven automatically by VERIPB. Further details on practical aspects and implementation of pseudo-Boolean proof logging can be found in the software repository of VERIPB [30].

2.1   Pseudo-Boolean Reasoning with the Cutting Planes Method

Our treatment of this material will by necessity be somewhat terse—we refer the reader to [9] for more information about the cutting planes method and to [8,31] for detailed information about the VERIPB proof system and format.

We write x to denote a $\{0,1\}$-valued variable and $\bar{x}$ as a shorthand for $1 - x$, and write ℓ to denote such *positive* and *negative literals*, respectively. By a *pseudo-Boolean (PB) constraint* we mean a 0–1 linear inequality $\sum_j a_j \ell_j \geq b$, where when convenient we can assume all literals ℓ_j to refer to distinct variables and all a_j and b to be non-negative (so-called *normalized form*). A *pseudo-Boolean formula* is just another name for a 0–1 integer linear program. For optimization problems we also have an objective function $f = \sum_j c_j x_j$ that should be minimized (and f can be negated to represent a maximization problem).

The foundation of VERIPB is the *cutting planes* proof system [13]. At the start of the proof, the set of *core constraints* $\mathcal{C}$ are initialized as the 0–1 linear inequalities in the problem instance. Any constraints derived as described below are placed in the set of *derived constraints* $\mathcal{D}$, from where they can later be moved to $\mathcal{C}$ (but not vice versa). Loosely speaking, VERIPB proofs maintain the invariant that the optimal value of any solution to $\mathcal{C}$ and to the original input problem is the same. New constraints can be derived from $\mathcal{C} \cup \mathcal{D}$ by performing *addition* of two constraints or *multiplication* of a constraint by a positive integer, and *literal axioms* $\ell \geq 0$ can be used at any time. Additionally, for a constraint $\sum_j a_j \ell_j \geq b$ written in normalized form we can apply *division* by a positive integer d followed by rounding up to obtain $\sum_j \lceil a_j/d \rceil \ell_j \geq \lceil b/d \rceil$, and *saturation* can be applied to yield $\sum_j \min\{a_j, b\} \cdot \ell_j \geq b$.

For a PB constraint $C \doteq \sum_j a_j \ell_j \geq b$ (where we use $\doteq$ to denote syntactic equality), the negation of C is $\neg C \doteq \sum_j a_j \ell_j \leq b - 1$. For a *partial assignment* ρ mapping variables to $\{0, 1\}$, we write $C\restriction_\rho$ for the *restricted constraint* obtained by replacing variables in C assigned by ρ by their values and simplifying the result. We say that C *unit propagates* ℓ *under* ρ if $C\restriction_\rho$ cannot be satisfied unless ℓ is assigned to 1. If unit propagation on all constraints in $\mathcal{C} \cup \mathcal{D} \cup \{\neg C\}$ starting with the empty assignment $\rho = \emptyset$, and extending ρ with new assignments as long as new literals propagate, leads to contradiction in the form of a violated constraint, then we say that C follows by *reverse unit propagation (RUP)* from $\mathcal{C} \cup \mathcal{D}$. Such (efficiently verifiable) RUP steps are allowed in VERIPB proofs when it is convenient to avoid writing out an explicit derivation of C from $\mathcal{C} \cup \mathcal{D}$. We will also write $C\restriction_\omega$ to denote the result of applying to C a *(partial) substitution* ω which can remap variables to other literals in addition to 0 and 1, and we extend this notation to sets in the obvious way by taking unions.

In addition to the cutting planes rules, which can only derive semantically implied constraints, VERIPB has a *redundance-based strengthening rule* that can derive a non-implied constraint C as long as this does not change the feasibility or optimal value of the problem. Formally, C can be derived from $\mathcal{C} \cup \mathcal{D}$ using this rule by exhibiting in the proof a *witness substitution* ω together with subproofs

$$\mathcal{C} \cup \mathcal{D} \cup \{\neg C\} \vdash (\mathcal{C} \cup \mathcal{D} \cup \{C\})\restriction_\omega \cup \{f \geq f\restriction_\omega\}, \qquad (1)$$

of all constraints on the right-hand side from the premises on the left-hand side using the derivation rules above. Intuitively, what (1) shows is that if α is any assignment that satisfies $\mathcal{C} \cup \mathcal{D}$ but violates C, then $\alpha \circ \omega$ satisfies $\mathcal{C} \cup \mathcal{D} \cup \{C\}$ and yields at least as good a value for the objective function f.

During presolving, constraints in the input formula can be deleted or replaced by other constraints, and the proof needs to establish that such modifications are correct. While deletions from the derived set $\mathcal{D}$ are always in order, removing a constraint from the core set $\mathcal{C}$ could potentially introduce spurious solutions. Therefore, deleting a constraint C from $\mathcal{C}$ can only be done by the *checked deletion rule*, which requires to show that C could be rederived from $\mathcal{C} \setminus \{C\}$ by redundance-based strengthening (see [8] for a more detailed explanation).

2.2 A New Rule for Objective Function Updates

When variables are fixed or identified during the presolving process, the objective function f can be modified to a function f'. This modified objective f' can then be used in other presolver reasoning. This scenario arises also in, e.g., MaxSAT solving, and can be dealt with by deriving two PB constraints $f \geq f'$ and $f' \geq f$ in the proof, which encodes that the old and new objective are equal [4].

Table 1. Examples of basic derivation rules in VERIPB syntax. Here, (id) refers to the constraint ID assigned by VERIPB.

Rule	Syntax	Explanation
cutting planes in reverse Polish notation	`pol x1 4 +`	add $x_1 \geq 0$ and (4)
	`pol 3 2 d`	divides (3) by 2
	`pol 1 2 * ~x1 +`	multiplies (1) by 2 and adds $\overline{x}_1 \geq 0$
redundance-based strengthening	`red +1 x1 >= 1; x1 1`	verifies $x_1 \geq 1$ with $\omega = \{x_1 \mapsto 1\}$
	`red +1 x1 +1 x2 >= 1; x1 x2 x2 x1`	verifies $x_1 + x_2 \geq 1$ with $\omega = \{x_1 \mapsto x_2, x_2 \mapsto x_1\}$
RUP	`rup +1 x1 +1 x2 >= 1;`	verifies $x_1 + x_2 \geq 1$ with RUP
move to core	`core id 3`	moves (3) to the core constraints
deletion from core	`delc 3`	deletes (3) from the core constraints
objective function update	`obju new +1 x1 +1 x2 1;`	defines $x_1 + x_2 + 1$ as new objective
	`obju diff +1  x1;`	adds $\overline{x}_1$ to the objective

Whenever the solver argues in terms of f', a telescoping-sum argument with $f' = f$ can be used to justify the same conclusion in terms of the old objective.

However, if the presolver changes f to f' and then uses reasoning that needs to be certified by redundance-based strengthening, then tricky problems can arise. One of the required proof goals in (1) is that the witness ω cannot worsen the objective. If ω does not mention variables in f', then this is obvious to the presolver—ω has no effect on the objective—but if ω assigns variables in the original objective f, then one still needs to derive $f \geq f\lceil_\omega$ in the formal proof, which can be challenging. While this can often be done by enlarging the witness ω to include earlier variable fixings and identifications, the extra bookkeeping required for this quickly becomes a major headache, and results in the proof deviating further and further from the actual presolver reasoning that the proof logging is meant to certify.

For this reason, a better solution is to introduce a new *objective function update rule* that formally replaces f by a new objective f', so that all future reasoning about the objective can focus on f' and ignore f. Such a rule needs to be designed with care, so that the optimal value of the problem is preserved. Due to space constraints we cannot provide a formal proof here, but recall that intuitively we maintain the invariant for the core set $\mathcal{C}$ that it has the same optimal value as the original problem. In agreement with this, the formal requirement for updating the objective from f to f' is to present in the proof log derivations of the two constraints $f \geq f'$ and $f' \geq f$ from the core set $\mathcal{C}$ only.

3   Certifying Presolve Reductions

We now describe how feasibility- and optimality-based presolving reductions can be certified by using VERIPB proof logging enhanced with the new objective function update rule described in Sect. 2.2 above. We distinguish between *primal* and *dual* reductions, where primal reductions strengthen the problem formulation by tightening the convex hull of the problem and preserve all feasible solutions, and dual reductions may additionally remove feasible solutions using optimality-based arguments. More precisely, *weak* dual reductions preserve all optimal solutions, but may remove suboptimal solutions. *Strong* dual reductions may remove also optimal solutions as long as at least one optimal solution is preserved in the reduced problem. Our selection of methods is motivated by the recent MIP solver implementation described in [44]. Before explaining the individual presolving techniques and their certification, we introduce a few general techniques that are needed for the certification of several presolving methods.

3.1   General Techniques

Substitution. In order to reduce the number of variables, constraints, and non-zero coefficients in the constraints, many presolving techniques first try to identify an equality $E \doteq x_k = \sum_{j \neq k} \alpha_j x_j + \beta$ with $\alpha_j, \beta \in \mathbb{Q}$. Subsequently, all occurrences of x_k in the objective and constraints besides E are substituted by the affine expression on the right-hand side and x_k is removed from the problem. The simplest case when x_k is fixed to zero or one, i.e., when $\beta \in \{0,1\}$ and all $\alpha_j = 0$, is straightforward to handle by deriving a new lower or upper bound on x_k. During presolving, every fixed variable is removed from the model. In the cases where some $\alpha_j \neq 0$, first the equation is expressed as a pair of constraints $E_\geq \wedge E_\leq$ and then the variable is removed by aggregation as follows.

Aggregation. In order to substitute variables or reduce the number of non-zero coefficients, certain presolving techniques add a scaled equality $s \cdot E \doteq s \cdot E_\geq \wedge s \cdot E_\leq$, $s \in \mathbb{Q}$, to a given constraint D. We call this an *aggregation*. Since VERIPB certificates expect inequalities with integer coefficients, s is split into two integer scaling factors $s_E, s_D \in \mathbb{Z}$ with $s = s_E/s_D$. In the certificate, the aggregation is expressed as a newly derived constraint

$$D_{new} \doteq \begin{cases} |s_E| \cdot E_\geq + |s_D| \cdot D & \text{if } \frac{s_D}{s_E} > 0 \\ |s_E| \cdot E_\leq + |s_D| \cdot D & \text{otherwise .} \end{cases}$$

Note that the presolving algorithm may decide to keep working with the constraint $(1/s_D)D_{new}$ internally. In this case, it must store the scaling factor s_D in order to correctly translate between its own state and the state in the certificate; this happens in the implementation used in Sect. 4.

Checked Deletion. The derivation of a new constraint D_{new} can render a previous constraint D redundant. A typical example is the case of substituting a variable above. In a (pre)solver, the previous constraint is overwritten, and in order to keep the constraint database in the proof aligned with the solver, one may want to delete the previous constraint from the proof. In order to check the deletion of D, a subproof is required that proves its redundancy. In most cases, this subproof contains the "inverted" derivation of D_{new}. As an example, consider an aggregation $D_{new} \doteq D + E_\leq$ with an equality $E \doteq E_\leq \wedge E_\geq$. In this case, the subproof for the checked deletion is $D_{new} + E_\geq$. Unless stated otherwise, the new constraints are moved to the core and redundant constraints are always removed by inverting the derivation of the constraint that replaces them.

3.2   Primal Reductions

Primal reductions can be certified purely by implicational reasoning.

Bound Strengthening. This preprocessor [24,47] tries to tighten the variable domains by iteratively applying well-known *constraint propagation* to all variables in the linear constraints. Each reduced variable domain is communicated to the affected constraints and may trigger further domain changes. This process is continued until no further domain reductions happen or the problem becomes infeasible due to empty domains. Specifically, for an inequality constraint

$$\sum_{j \in N} a_j x_j \geq b \tag{2}$$

with $a_k \neq 0$, we first underestimate $a_k x_k$ via

$$a_k x_k \geq b - \sum_{j \neq k} a_j x_j \geq b - \sum_{j \neq k, a_j > 0} a_j .$$

If $a_k > 0$, this yields the lower bound

$$x_k \geq \left\lceil \left(b - \sum_{j \neq k, a_j > 0} a_j \right) / a_k \right\rceil , \tag{3}$$

and if $a_k < 0$ we can obtain an analogous upper bound on x_k.

The bound change can be proven either by RUP, or more explicitly by stating the additions and division needed to form (3) from (2) and the bound constraints. We analyze the effect of both variants in Sect. 4.4.

Parallel Rows. Two constraints C_j and C_k are parallel if a scalar $\lambda \in \mathbb{R}^+$ exists with $\lambda(a_{j1}, \ldots, a_{jn}, b_j) = (a_{k1}, \ldots, a_{kn}, b_k)$. Hence, one of these constraints is redundant and can be removed from the model [2,26]. The subproof for deleting the redundant rows must contain the remaining parallel row and λ to prove the redundancy. For a fractional λ the two constraint are scaled to ensure integer coefficients in the certificate.

Probing. The general idea of *probing* [1,47] is to tentatively fix a variable x_j to 0 or 1 and then apply constraint propagation to the resulting model. Suppose x_k is an arbitrary variable with $k \neq i$, then we can learn fixings or implications in the following cases:

1. If $x_j = 0$ implies $x_k = 1$ and $x_j = 1$ implies $x_k = 0$ we can add the constraint $x_j = 1 - x_k$. Analogously, we can derive $x_k = x_j$ in the case that $x_j = 0$ implies $x_k = 0$ and $x_j = 1$ implies $x_k = 1$.
2. If $x_j = 0$ propagates to infeasibility we can fix $x_j = 1$. Analogously, if $x_j = 1$ propagates to infeasibility we can fix $x_j = 0$.
3. If $x_j = 0$ implies $x_k = 0$ and $x_j = 1$ implies $x_k = 0$ we can fix x_k to 0. Analogously, x_k can be fixed to 1 if $x_j = 0$ implies $x_k = 1$ and $x_j = 1$ implies $x_k = 1$.

Cases 1 and 2 can be proven with RUP. To prove correctness of fixing $x_k = 1$ in Case 3 we first derive two new constraints $x_k + x_j \geq 1$ and $x_k - x_j \geq 0$ in the proof log by RUP. Adding these two constraints leads to $x_k \geq 1$. To prove $x_k = 0$ we derive the constraints $x_k + x_j \leq 0$ and $x_k - x_j \leq 0$ leading to $x_k = 0$.

Simple Probing. On equalities with a special structure, a more simplified version of probing called *simple probing* [2, Sect. 3.6] can be applied. Suppose the equation

$$\sum_{j \in N} a_j x_j = b \text{ with } \sum_{j \in N} a_j = 2 \cdot b \text{ and } |a_k| = \sum_{j \in N, a_j > 0} a_j - b$$

holds for a variable x_k with $a_k \neq 0$. Let $\hat{N} = \{p \in N \mid a_p \neq 0\}$. Under these conditions, $x_k = 1$ implies $x_p = 0$ and $x_k = 0$ implies $x_p = 1$ for all $p \in \hat{N}$ with $a_p > 0$. Further, $x_k = 1$ implies $x_p = 1$ and $x_k = 0$ implies $x_p = 0$ for all $p \in \hat{N}$ with $a_p < 0$. These implications can be expressed by the constraints

$$x_k = 1 - x_p \text{ for all } p \in \hat{N} \text{ with } a_p > 0, \tag{4}$$

$$x_k = x_p \text{ for all } p \in \hat{N} \text{ with } a_p < 0. \tag{5}$$

The constraints (4) and (5) can be proven with RUP and used to substitute variables x_p for all $p \in \hat{N}$ from the problem.

Sparsifying the Matrix. The presolving technique sparsify [2,11] tries to reduce the number of non-zero coefficients by adding (multiples of) equalities to other constraints using aggregations. This can be certified as described in Sect. 3.1.

Coefficient Tightening. The goal of this MIP presolving technique, which goes back to [47], is to tighten the LP relaxation, i.e., the relaxation obtained when the integrality requirements are replaced by $x_j \in [0, 1]$. To this end, the coefficients of constraints are modified such that LP relaxation solutions are removed, but all integer feasible solutions are preserved. Suppose we are given a constraint $\sum_{j \in N} a_j x_j \geq b$ with $a_k \geq \varepsilon := a_k - b + \sum_{j \neq k, a_j < 0} a_j > 0$, then the constraint can be strengthened to $(a_k - \varepsilon)x_k + \sum_{j \neq k} a_j x_j \geq b$. The case $a_k < 0$ is handled analogously. This technique is also known as *saturation* in the SAT community [10] and VERIPB provides a dedicated saturation rule that can be used directly for proving the correctness of coefficient tightening. The deletion of the original, weaker constraint can be proven automatically.

GCD-Based Simplification. This presolving technique from [51] uses a divisibility argument to first eliminate variables from a constraint and then tighten its right-hand side. Given $C \doteq \sum_{j \in N} a_j x_j \geq b$ with $|a_1| \geq \cdots \geq |a_n| > 0$. We define the greatest common divisor $g_k = \gcd(a_1, \ldots, a_k)$ as the largest value g such that $a_j/g \in \mathbb{Z}$ for all $j \in \{1, \ldots, k\}$. If for an index k it holds that $b - g_k \cdot \left\lceil \frac{b}{g_k} \right\rceil \geq \sum_{k < j \leq n, a_j > 0} a_j$ and $b - g_k \cdot \left\lceil \frac{b}{g_k} \right\rceil - g_k \leq \sum_{k < j \leq n, a_j < 0} a_j$, then all $a_{k+1}, \ldots, a_n$ can be set to 0. This first step can be certified as *weakening* [41] and VERIPB provides an out-of-the-box verification function for it. Finally, b can be rounded to $g_k \cdot \lceil b/g_k \rceil$. This rounding step can be certified by dividing C with g_k and then multiply it again with g_k.

Substituting Implied Free Variables. A variable x_j is called *implied free* if its lower bound and its upper bound can be derived from the constraints. For example, the constraints $x_1 - x_2 \geq 0$ and $x_2 \geq 0$ imply the lower bound $x_1 \geq 0$. If we have an implied free variable x_j in an equality $E \doteq a_j x_j + \sum_{k \neq j} a_k x_k = b$ with $a_j > 0$, then we can remove x_j from the problem by substituting it with $x_j = (b - \sum_{k \neq j} a_k x_k)/a_j$, see [2] for details.

To apply the substitution in the certificate we use aggregations to remove x_j from all constraints and the objective function update to remove x_j from the objective. If coefficients c_j/a_j or a_k/a_j are non-integer, then the resulting constraints are scaled as described in Sect. 3.1. To prove the deletion of E, we derive two constraints by adding $x_j \geq 0$ and $1 \geq x_j$ to E each, which results in

$$b \geq \sum_{k \neq j} a_k x_k \ \land \ \sum_{k \neq j} a_k x_k \geq b - a_j . \tag{6}$$

Then the deletion of $E_\geq$ can be certified by a witness $\omega = \{x_j \mapsto 1\}$. The constraint simplifies to (6) and is therefore fulfilled. Analogously, we use the witness $\omega = \{x_j \mapsto 0\}$ to certify the deletion of $E_\leq$. Finally, to delete the constraints in (6) we generate a subproof that shows that negation of the auxiliary constraints in (6) leads to $x_j \notin \{0,1\}$. This is a contradiction to the implied variable bounds $0 \leq x_j \leq 1$. Since these bounds are still present through the implying constraints, we can add these implying constraints to (6) in the subproof to arrive at a contradiction.

Singleton Variables. It is well-known that variables that appear only in one inequality constraint or equality can be removed from the problem [2, Sect. 5.2]. This can be certified by applying one of the following primal or dual strategies in this order: First, try to apply duality-based fixing, see Sect. 3.3; second, an implied free singleton variable can be substituted as explained above; otherwise, the singleton variable can be treated as a *slack variable*: substitute the variable in the objective, then relax the equality as in (6), and delete the original constraint.

3.3 Dual Reductions

Dual reductions remove solutions while preserving at least one optimal solution. Hence, to prove the correctness of dual reductions we need to involve the redundance-based strengthening rule of VERIPB. For each derived constraint C we only explain how to prove $f \geq f\lceil_\omega$ (subject to the negation $\neg C$); the proof goals for $C\lceil_\omega$ can be derived in a very similar fashion.

Duality-Based Fixing. This presolving step described in [2, Sect. 4.2] counts the *down-* and *up-lock* of a variable. A down-lock on variable x_j is a negative coefficient, an up-lock on variable x_j is a positive coefficient (for $\geq$ constraints). If x_j has no down-locks and $c_j \leq 0$, it can be fixed to zero; if x_j has no up-locks and $c_j \geq 0$, it can be fixed to one. These reductions can be certified with redundance-based strengthening using the witness $\omega = \{x_j \mapsto v\}$, where v is the fixing value. The proof goal for $f \geq f\restriction_\omega$ is equivalent to $c_j x_j \geq c_j v$, which is fulfilled by the conditions of duality-based fixing.

Dominated Variables. A variable x_j is said to *dominate* another variable x_k [2, 25], in notation $x_j \succ x_k$, if

$$c_j \leq c_k \ \wedge \ a_{ij} \geq a_{ik} \ \text{for all} \ i \in \{1, \dots, m\}, \tag{7}$$

where a_{ij} and a_{ik} are the coefficients of variable x_j and x_k, respectively, in the i-th constraint. Variable x_j is then favored over x_k since x_j contributes less to the objective function, but more to the feasibility of the constraints. For every domination $x_j \succ x_k$, a constraint $C \doteq x_j \geq x_k$ can be introduced. This constraint can be certified by redundance-based strengthening with the witness $\omega = \{x_k \mapsto x_j, x_j \mapsto x_k\}$. The proof goal for $f \geq f\restriction_\omega$ is equivalent to

$$c_j x_j + c_k x_k \geq c_j x_k + c_k x_j. \tag{8}$$

The negated constraint $\neg C \doteq x_j < x_k$ leads to $x_k = 1$ and $x_j = 0$. Substituting these values in (8) leads to $c_k \geq c_j$, which follows directly from Condition (7).

Dominated Variables Advanced. For an implied free variable we can drop the variable bounds and pretend the variable is unbounded. This allows for additional fixings in the following cases of dominated variables:

(a) If the upper bound of x_j is implied and $x_j \succ x_k$, then $x_k = 0$.
(b) If the lower bound of x_k is implied and $x_j \succ x_k$, then $x_j = 1$.
(c) If the upper bound of x_j is implied and $x_j \succ -x_k$, then $x_k = 1$.
(d) If the lower bound of x_j is implied and $-x_j \succ x_k$, then $x_j = 0$.

We use redundance-based strengthening with witness $\omega = \{x_k \mapsto 0\}$ to prove the correctness of (a) as follows. If the upper bound of x_j is implied, this means there exists a constraint with $a_{ij} < 0$ such that $x_j \leq \left\lfloor \frac{b_\ell - \sum_{\ell \neq j, a_{i\ell} > 0} a_{i\ell}}{a_{ij}} \right\rfloor = 1$. Due to Condition (7), it must hold that $0 > a_{ij} \geq a_{ik}$, and the constraint $x_j + x_k \leq 1$ can be derived. Hence, negating and propagating $C \doteq x_k = 0$ with RUP leads to contradiction, which proves the validity of C. Case (b) can be handled analogously using the witness $\omega = \{x_k \mapsto 1\}$. To derive $C \doteq x_k = 1$ in (c) we use redundance-based strengthening with witness $\omega = \{x_k \mapsto 1, x_j \mapsto 1\}$. Then, the proof goal for $f \geq f\restriction_\omega$ is $c_j \cdot x_j + c_k \cdot x_k \geq c_j + c_k$. After propagating $\neg C$, this becomes equivalent to $c_j \leq -c_k$, which is true by Condition (7). Case (d) can be handled analogously using the witness $\omega = \{x_k \mapsto 0, x_j \mapsto 0\}$.

3.4   Example

We conclude this section with an example of a small certificate for the substitution of an implied free variable in Fig. 1, also available with a more detailed description at the software repository of PAPILO [38]. Consider the 0–1 ILP

$$\min x_1 + x_2 \ \text{s.t.} \ x_1 + x_2 - x_3 - x_4 = 1, \tag{9}$$

$$-x_1 + x_5 \geq 0, \tag{10}$$

```
* generates ID 4:          * generates ID 6:          delc 2 ; x1 -> 0
pol 1 ~x1 + ;              pol 3 1 + ;               delc 1 ; x1 -> 1
core id 4                  core id 6                 delc 5
* generates ID 5:          delc 3 ;   ; begin        delc 4 ;   ; begin
pol 2 x1 + ;                  pol 6 2 +                 pol 6 -1 +
core id 5                  end                       end
                           obju new +1 x3 +1 x4 1 ;
```

Fig. 1. A VERIPB certificate to substitute an implied free variable x_1.

in which the lower bound of x_1 is implied by (9) and the upper bound of x_1 is implied by (10). Hence, x_1 is implied free and we can use (9) to substitute it.

In the left section of Fig. 1 we first derive the two auxiliary constraints

$$0 \leq x_2 - x_3 - x_4 \leq 1, \tag{11}$$

which receives the constraint IDs 4 and 5 and are moved to the core. Note that the equality in (9) is split into two inequalities with IDs 1 and 2. In the middle section, we first remove x_1 from (10) by aggregation with (9), perform checked deletion, then remove x_1 from the objective (automatically proven by VERIPB). Last, in the right section, we delete the equality in (9) used for the substitution and the auxiliary constraints in (11) and arrive at the reformulated problem min $x_3 + x_4 + 1$ s.t. $x_2 - x_3 - x_4 + x_5 \geq 1$. From here, we could continue to derive $x_2 = 1$ by duality-based fixing, since x_2 has zero up-locks and objective coefficient zero. This displays the importance of the objective update, as without it x_2 would still contribute to the objective with a positive coefficient, and this would prohibit duality-based fixing to 1.

4   Computational Study

In this section we quantify the cost of *certifying* presolve reductions in a state-of-the-art implementation for MIP-based presolve (Sect. 4.2) and the cost of *verifying* the resulting certificates (Sect. 4.3). In Sect. 4.4, we analyze the impact of certifying constraint propagation by RUP or by an explicit cutting planes proof.

4.1   Experimental Setup

For generating the presolve certificates we use the solver-independent presolve library PAPILO [44], which provides a large set of MIP and LP techniques from the literature, described in Sect. 3. Additionally, it accelerates the search for pre-solving reductions by parallelization, encapsulating each reduction in a so-called transaction to avoid expensive synchronization [28]. Logging the certificate, how-ever, is performed sequentially while evaluating the transactions.

We base our experiments on models from the Pseudo-Boolean Competition 2016 [45] including 1398 linear small integer decision and 532 linear small inte-ger optimization instances of the competitions PB10, PB11, PB12, PB15, and PB16 and 295 decision and 145 optimization instances from MIPLIB 2017 [29] in the OPB translation [19], excluding 10 large-scale instances[1] for which PAPILO reaches the memory limit. This yields a total of 671 optimization and 1681 decision instances. We use PAPILO 2.2.0 [39] running on 6 threads and VERI-PB 2.0 [30]. The experiments are carried out on identical machines with an 11th Gen Intel(R) Core(TM) i5-1145G7 @ 2.60 GHz CPU and 16 GB of memory and are assigned 14,000 MB of memory. The strict time limit for presolve plus certification and verification is three hours. Times (reported in seconds) do not include the time for reading the instance file. For all aggregations, we use the shifted geometric mean with a shift of 1 s.

4.2 Overhead of Proof Logging

In the first experiment, we analyze the overhead of proof logging in PAPILO. The average results are summarized in Table 2, separately over decision (dec) and optimization (opt) instances for PB16 and MIPLIB. Column "relative" indicates the average slow-down incurred by printing the certificate.

The relative overhead of proof logging is less than 6% across all test sets. VERIPB supports two variants to change the objective function. Either printing the entire objective (obju new) or printing only the changes in the objective (obju diff). In our experiments, we only print the changes, since printing the entire objective for each change can lead to a large certificate and overhead, especially for instances with large and dense objective functions. On the PB16 instance NORMALIZED-DATT256, for example, PAPILO finds 135 206 variable fixings. Updating the entire objective function with 262 144 non-zeros for each of these variables leads to a huge certificate of about 138 GB and increases the time from 3.3 s (when printing only the changes) to 6625 s.[2]

For 99% of the instances, we can further observe that the *overhead per applied reduction* is below $0.001 \cdot 10^{-3}$ s over both test sets. This means that the proof logging overhead is not only small on average, but also small per applied reduction on the vast majority of instances. These results show that the overhead scales well with the number of applied reductions and that proof logging remains viable even for instances with many transactions. Here, under applied reductions we subsume all applied transactions and each variable fixing or row deletion in the first model clean-up phase. During model clean-up, PAPILO fixes variables and removes redundant constraints from the problem. While PAPILO technically does not count these reductions as full transactions found during the parallel presolve phase, their certification can incur the same overhead.

4.3 Verification Performance on Presolve Certificates

In this section, we analyze the time to verify the certificates generated by PAPILO. The results are summarized in Table 3. The "verified" column lists

---

[1] NORMALIZED-184, NORMALIZED-PB-SIMP-NONUNIF, A2864-99BLP, IVU06-BIG, IVU59, SUPPORTCASE11, A2864-99BLP.0.S/U, SUPPORTCASE11.0.S/U.

[2] Certificate generated on Intel Xeon Gold 5122 @ 3.60GHz 96 GB with 50,000 MB of memory assigned.

Table 2. Runtime comparison of PAPILO with and without proof logging.

test set	size	default [s]	w/proof log [s]	relative
PB16-dec	1397	0.06	0.06	1.00
MIPLIB-dec	291	0.42	0.43	1.02
PB16-opt	531	0.65	0.66	1.02
MIPLIB-opt	142	0.33	0.35	1.06

Table 3. Time to verify the certificates. VERIPB timeouts are treated with PAR2.

test set	size	verified	PAPILO time [s]		VERIPB time [s]	relative time w.r.t.	
			default	w/proof log		default	w/proof log
PB16-dec	1397	1397	0.06	0.06	0.88	14.67	14.67
MIPLIB-dec	291	267	0.42	0.43	9.64	22.85	22.42
PB16-opt	531	520	0.65	0.66	10.44	16.06	15.82
MIPLIB-opt	142	139	0.33	0.35	5.25	15.91	15.00

the number of instances verified within 3 h. VERIPB timeouts are counted as twice the time limit, i.e., PAR2 score. Similar to Table 2, the "relative" columns report the relative overhead of VERIPB runtime compared to PAPILO.

First note that all certificates are verified by VERIPB (partially on the 38 instances where VERIPB times out). On average, it takes between 14.7 and 22.4 times as much time to verify the certificates than to produce them. Nevertheless, some instances take a longer than average time to verify. Over all test sets, 25% of the instances have an overhead of at least a factor of 193, see also Fig. 2.

To put this result into context, note that presolving amounts more to a transformation than to a (partial) solution of the problem. Each reduction has to be certified and verified while a purely solution-targeted algorithm may be able to skip certifying of a larger part of the findings that are not form a part of the final proof of optimality. Hence, it makes sense to compare the performance of VERIPB on presolve certificates to the overhead for, e.g., for verifying CNF translations [31]. For this study, a similar performance overhead is reported as in Fig. 2.

4.4   Performance Analysis on Constraint Propagation

Finally, we investigate how the performance of VERIPB depends on whether we use RUP (as in Sect. 4.2 and Sect. 4.3) or explicit cutting planes derivations (POL) to certify bound strengthening reductions from constraint propagation. Here, we additionally exclude 9 large-scale instances[3] for which PAPILO

---

[3] NEOS-4754521-AWARAU.0.S, NEOS-827015.0.S/U, NEOS-829552.0.S/U, S100.0.S/U, NORMALIZED-DATT256, S100.

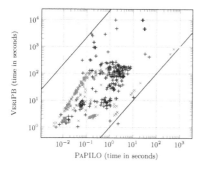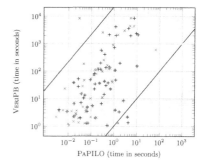

Fig. 2. Running times of VERIPB vs. PAPILO on test sets PB16 (left) and MIPLIB (right), including all instances with more than 1 s in VERIPB and less than 30 min in PAPILO, and excluding timeouts. Green + signs mark optimization and blue × signs mark decision instances. (Color figure online)

reaches the memory limit when certifying with POL. The results are summarized in Table 4. The "verified" column contains the number of instances verified by VERIPB within the time limit. The "time" column reports the time for verification.

Deriving the propagation directly with cutting planes is 3.2% faster on PB16-dec, 2.8% faster on MIPLIB-dec, 13.1% faster on MIPLIB-opt, and 0.7% faster on PB16-opt. On 95% of the decision instances using RUP is at most 9.7% slower. While it is expected that verification is faster when the cutting planes proof is given explicitly, it is surprising that the performance difference between the methods is not more pronounced. This is partly due to the cost of the watched-literal scheme [43,48] used by VERIPB for unit propagation. The overhead of maintaining the watches is present regardless of whether (reverse) unit propagation is used or not. Furthermore, unit propagation is also used for automatically verifying redundance-based strengthening. Together, this limits the potential for runtime savings by providing the explicit cutting planes proof.

Furthermore, providing an explicit cutting planes proof for propagation requires printing the constraint into the certificate. Hence, the certificate size becomes dependent on the number of non-zeros in the constraints leading to propagations. In contrast, the overhead of RUP is constant and much smaller.

All in all, these results suggest to prefer RUP when deriving constraint propagation since it barely impacts the performance of VERIPB and keeps the size of the certificate smaller. The computational cost of RUP could be further reduced by extending it to accept an ordered list of constraints that shall be propagated first, similar as in [16]. Such an extension could also be used for other presolving techniques, in particular probing and simple probing.

Table 4. Comparison of the runtime of VERIPB with RUP and POL over instances with at least 10 propagations.

test set	size	RUP		POL		relative
		verified	time [s]	verified	time [s]	
PB16-dec	284	284	2.21	284	2.14	0.968
MIPLIB-dec	35	31	153.23	31	148.88	0.972
PB16-opt	153	142	28.43	142	28.22	0.993
MIPLIB-opt	16	14	147.11	14	127.83	0.869

5 Conclusion

In this paper we set out to demonstrate how presolve techniques from state-of-the-art MIP solvers can be equipped with certificates in order to verify the equivalence between original and reduced models. Although the pseudo-Boolean proof logging format behind VERIPB [7] was not designed with this purpose in mind, we could show that a limited extension needed for handling updates of the objective function is sufficient to craft a certified presolver for 0–1 ILPs.

However, our experimental study on instances from pseudo-Boolean competitions and MIPLIB also exhibited that the verification of MIP-based presolving can suffer from large and overly verbose certificates. To shrink the proof size we introduced a sparse objective update function but identified further possible improvements. First, a native substitution rule in VERIPB would remove the need for the explicit derivation of new aggregations and the verification of checked deletion as described in Sect. 3.1. For instances where presolving is dominated by substitutions, we estimate that this would reduce certificate sizes by up to 90%, and no more time would be spent on checked deletion for substitutions. Second, augmenting the RUP syntax by the option to specify an ordered list of constraints to propagate first, similarly as in [16], would accelerate RUP, in particular for fast verification of bound strengthenings by constraint propagation.

While VERIPB is currently restricted to operate on integer coefficients only, the certification techniques presented in Sect. 3 do not rely on this assumption and are applicable to general binary programs. It has been shown how to construct VERIPB certificates for bounded integer domains [34, 42], and within the framework of the generalized proof system laid out in [20], our certificates would even translate to continuous and unbounded integer domains. To conclude, we believe our results show convincingly that this type of proof logging techniques is a very promising direction of research also for MIP presolve beyond 0–1 ILPs.

Acknowledgements. The authors wish to acknowledge helpful technical discussions on VERIPB in general and the objective update rule in particular with Bart Bogaerts, Ciaran McCreesh, and Yong Kiam Tan. The work for this article has been partly conducted within the Research Campus MODAL funded by the German Federal Ministry of Education and Research (BMBF grant number 05M14ZAM). Jakob Nordström was supported by the Swedish Research Council grant 2016-00782 and the Independent Research Fund Denmark grant 9040-00389B. Andy Oertel was supported by the Wallenberg AI, Autonomous Systems and Software Program (WASP) funded by the Knut and Alice Wallenberg Foundation. The computational experiments were enabled by resources provided by LUNARC at Lund University.

References

1. Achterberg, T.: Constraint Integer Programming. Doctoral thesis, Technische Universität Berlin, Fakultät II - Mathematik und Naturwissenschaften, Berlin (2007). https://doi.org/10.14279/depositonce-1634
2. Achterberg, T., Bixby, R., Gu, Z., Rothberg, E., Weninger, D.: Presolve reductions in mixed integer programming. INFORMS J. Comput. **32** (2019). https://doi.org/10.1287/ijoc.2018.0857
3. Akgün, Ö., Gent, I.P., Jefferson, C., Miguel, I., Nightingale, P.: Metamorphic testing of constraint solvers. In: Hooker, J. (ed.) CP 2018. LNCS, vol. 11008, pp. 727–736. Springer, Cham (2018). https://doi.org/10.1007/978-3-319-98334-9_46
4. Berg, J., Bogaerts, B., Nordström, J., Oertel, A., Vandesande, D.: Certified core-guided MaxSAT solving. In: Pientka, B., Tinelli, C. (eds.) CADE 2023. LNCS, vol. 14132, pp. 1–22. Springer, Cham (2023)
5. Biere, A., Heule, M.J.H., van Maaren, H., Walsh, T. (eds.): Handbook of Satisfiability, Frontiers in Artificial Intelligence and Applications, vol. 336, 2nd edn. IOS Press, Amsterdam (2021)
6. Biere, A., Järvisalo, M., Kiesl, B.: Preprocessing in SAT solving. In: Biere et al. [5], chap. 9, pp. 391–435
7. Bogaerts, B., Gocht, S., McCreesh, C., Nordström, J.: Certified symmetry and dominance breaking for combinatorial optimisation. In: Proceedings of the 36th AAAI Conference on Artificial Intelligence (AAAI 2022), pp. 3698–3707 (Feb 2022)
8. Bogaerts, B., Gocht, S., McCreesh, C., Nordström, J.: Certified dominance and symmetry breaking for combinatorial optimisation. J. Artif. Intell. Res. **77**, 1539–1589 (2023). preliminary version in AAAI 2022
9. Buss, S.R., Nordström, J.: Proof complexity and SAT solving. In: Biere et al. [5], chap. 7, pp. 233–350
10. Chai, D., Kuehlmann, A.: A fast pseudo-Boolean constraint solver. IEEE Transactions on Computer-Aided Design of Integrated Circuits and Systems **24**(3), 305–317 (2005). preliminary version in *DAC '03*
11. Chang, S.F., McCormick, S.T.: Implementation and computational results for the hierarchical algorithm for making sparse matrices sparser. ACM Trans. Math. Softw. **19**(3), 419-441 (1993). https://doi.org/10.1145/155743.152620
12. Cheung, K.K.H., Gleixner, A.M., Steffy, D.E.: Verifying integer programming results. In: Eisenbrand, F., Koenemann, J. (eds.) IPCO 2017. LNCS, vol. 10328, pp. 148–160. Springer, Cham (2017). https://doi.org/10.1007/978-3-319-59250-3_13
13. Cook, W., Coullard, C.R., Turán, G.: On the complexity of cutting-plane proofs. Discret. Appl. Math. **18**(1), 25–38 (1987)
14. Cook, W., Koch, T., Steffy, D.E., Wolter, K.: A hybrid branch-and-bound approach for exact rational mixed-integer programming. Math. Program. Comput. **5**(3), 305–344 (2013)
15. Cook, W., Koch, T., Steffy, D.E., Wolter, K.: A hybrid branch-and-bound approach for exact rational mixed-integer programming. Math. Program. Comput. **5**(3), 305–344 (2013). https://doi.org/10.1007/s12532-013-0055-6

16. Cruz-Filipe, L., Heule, M.J.H., Hunt, W.A., Kaufmann, M., Schneider-Kamp, P.: Efficient certified rat verification. In: de Moura, L. (ed.) CADE 2017. LNCS, pp. 220–236. Springer, Cham (2017). https://doi.org/10.1007/978-3-319-63046-5_14

17. Cruz-Filipe, L., Heule, M.J.H., Hunt, W.A., Jr., Kaufmann, M., Schneider-Kamp, P.: Efficient certified RAT verification. In: de Moura, L. (ed.) CADE 2017. LNCS, vol. 10395, pp. 220–236. Springer, Cham (2017). https://doi.org/10.1007/978-3-319-63046-5_14

18. Cruz-Filipe, L., Marques-Silva, J.P., Schneider-Kamp, P.: Efficient certified resolution proof checking. In: Legay, A., Margaria, T. (eds.) TACAS 2017. LNCS, vol. 10205, pp. 118–135. Springer, Heidelberg (2017). https://doi.org/10.1007/978-3-662-54577-5_7

19. Devriendt, J.: Miplib 0-1 instances in opb format (2020). https://doi.org/10.5281/zenodo.3870965

20. Doornmalen, J.V., Eifler, L., Gleixner, A., Hojny, C.: A proof system for certifying symmetry and optimality reasoning in integer programming. Technical report 2311.03877, arXiv.org (2023)

21. Eifler, L., Gleixner, A.: A computational status update for exact rational mixed integer programming. Math. Program. (2022). https://doi.org/10.1007/s10107-021-01749-5

22. Eifler, L., Gleixner, A.: Safe and verified gomory mixed integer cuts in a rational MIP framework. SIAM J. Optim. **34**(1), 742–763 (2024). https://doi.org/10.1137/23M156046X

23. Eifler, L., Gleixner, A., Pulaj, J.: A safe computational framework for integer programming applied to chvátal's conjecture. ACM Trans. Math. Softw. **48**(2) (2022). https://doi.org/10.1145/3485630

24. Fügenschuh, A., Martin, A.: Computational integer programming and cutting planes. In: Aardal, K., Nemhauser, G., Weismantel, R. (eds.) Discrete Optimization, Handbooks in Operations Research and Management Science, vol. 12, pp. 69–121. Elsevier (2005). https://doi.org/10.1016/S0927-0507(05)12002-7

25. Gamrath, G., Koch, T., Martin, A., Miltenberger, M., Weninger, D.: Progress in presolving for mixed integer programming. Math. Programm. Comput. **7** (2015). https://doi.org/10.1007/s12532-015-0083-5

26. Gemander, P., Chen, W.K., Weninger, D., Gottwald, L., Gleixner, A.: Two-row and two-column mixed-integer presolve using hashing-based pairing methods. EURO J. Comput. Optim. **8**(3–4), 205–240 (2020). https://doi.org/10.1007/s13675-020-00129-6

27. Gillard, X., Schaus, P., Deville, Y.: SolverCheck: declarative testing of constraints. In: Schiex, T., de Givry, S. (eds.) CP 2019. LNCS, vol. 11802, pp. 565–582. Springer, Cham (2019). https://doi.org/10.1007/978-3-030-30048-7_33

28. Gleixner, A., Gottwald, L., Hoen, A.: PaPILO: a parallel presolving library for integer and linear programming with multiprecision support. INFORMS J. Comput. (2023). https://doi.org/10.1287/ijoc.2022.0171.cd, https://github.com/INFORMSJoC/2022.0171

29. Gleixner, A., et al.: MIPLIB 2017: data-driven compilation of the 6th mixed-integer programming library. Math. Program. Comput. **13**, 443–490 (2021). https://doi.org/10.1007/s12532-020-00194-3

30. Gocht, S., Oertel, A.: Veripb (2023). https://gitlab.com/MIAOresearch/software/VeriPB, githash: dd7aa5a1

31. Gocht, S., Martins, R., Nordström, J., Oertel, A.: Certified CNF translations for pseudo-Boolean solving. In: Proceedings of the 25th International Conference on

Theory and Applications of Satisfiability Testing (SAT 2022). Leibniz International Proceedings in Informatics (LIPIcs), vol. 236, pp. 16:1–16:25 (Aug 2022)

32. Gocht, S., McBride, R., McCreesh, C., Nordström, J., Prosser, P., Trimble, J.: Certifying solvers for clique and maximum common (connected) subgraph problems. In: Simonis, H. (ed.) CP 2020. LNCS, vol. 12333, pp. 338–357. Springer, Cham (2020). https://doi.org/10.1007/978-3-030-58475-7_20

33. Gocht, S., McCreesh, C., Nordström, J.: Subgraph isomorphism meets cutting planes: Solving with certified solutions. In: Proceedings of the 29th International Joint Conference on Artificial Intelligence (IJCAI '20), pp. 1134–1140 (2020)

34. Gocht, S., McCreesh, C., Nordström, J.: An auditable constraint programming solver. In: Proceedings of the 28th International Conference on Principles and Practice of Constraint Programming (CP '22). Leibniz International Proceedings in Informatics (LIPIcs), vol. 235, pp. 25:1–25:18 (2022)

35. Gocht, S., Nordström, J.: Certifying parity reasoning efficiently using pseudo-Boolean proofs. In: Proceedings of the 35th AAAI Conference on Artificial Intelligence (AAAI '21), pp. 3768–3777 (2021)

36. Heule, M.J.H., Hunt Jr., W.A., Wetzler, N.: Trimming while checking clausal proofs. In: Proceedings of the 13th International Conference on Formal Methods in Computer-Aided Design (FMCAD '13), pp. 181–188 (2013)

37. Heule, M.J.H., Hunt, W.A., Wetzler, N.: Verifying refutations with extended resolution. In: Bonacina, M.P. (ed.) CADE 2013. LNCS (LNAI), vol. 7898, pp. 345–359. Springer, Heidelberg (2013). https://doi.org/10.1007/978-3-642-38574-2_24

38. Hoen, A.: Papilo: Parallel presolve integer and linear optimization (2023). https://github.com/scipopt/papilo/tree/develop/check/VeriPB, githash: 5df3dd6d

39. Hoen, A., Gottwald, L.: Papilo: parallel presolve integer and linear optimization (2023). https://github.com/scipopt/papilo, githash: 3b082d4

40. Klotz, E.: Identification, assessment, and correction of ill-conditioning and numerical instability in linear and integer programs. In: Newman, A., Leung, J. (eds.) Bridging Data and Decisions, pp. 54–108. TutORials in Operations Research (2014). https://doi.org/10.1287/educ.2014.0130

41. Le Berre, D., Marquis, P., Wallon, R.: On weakening strategies for PB solvers. In: Pulina, L., Seidl, M. (eds.) SAT 2020. LNCS, pp. 322–331. Springer, Cham (2020). https://doi.org/10.1007/978-3-030-51825-7_23

42. McIlree, M., McCreesh, C.: Proof logging for smart extensional constraints. In: Proceedings of the 29th International Conference on Principles and Practice of Constraint Programming (CP '23). Leibniz International Proceedings in Informatics (LIPIcs), vol. 280, pp. 26:1–26:17 (2023)

43. Moskewicz, M.W., Madigan, C.F., Zhao, Y., Zhang, L., Malik, S.: Chaff: Engineering an efficient SAT solver. In: Proceedings of the 38th Design Automation Conference (DAC '01), pp. 530–535 (2001)

44. PaPILO — parallel presolve for integer and linear optimization. https://github.com/lgottwald/PaPILO

45. Roussel, O.: Pseudo-boolean competition 2016 (2016). http://www.cril.univ-artois.fr/PB16/

46. Sahraoui, Y., Bendotti, P., D'Ambrosio, C.: Real-world hydro-power unit-commitment: dealing with numerical errors and feasibility issues. Energy **184**, 91–104 (2019). https://doi.org/10.1016/j.energy.2017.11.064, shaping research in gas-, heat- and electric- energy infrastructures

47. Savelsbergh, M.: Preprocessing and probing techniques for mixed integer programming problems. ORSA J. Comput. **6** (1994).https://doi.org/10.1287/ijoc.6.4.445

48. Sheini, H.M., Sakallah, K.A.: Pueblo: a hybrid pseudo-Boolean SAT solver. J. Satisfiability, Boolean Model. Comput. **2**(1–4), 165–189 (2006). preliminary version in *DATE '05*

49. Steffy, D.E.: Topics in exact precision mathematical programming. Ph.D. thesis, Georgia Institute of Technology (2011). http://hdl.handle.net/1853/39639

50. Vandesande, D., De Wulf, W., Bogaerts, B.: QMaxSATpb: a certified MaxSAT solver. In: Gottlob, G., Inclezan, D., Maratea, M. (eds.) LPNMR 2022. LNCS, vol. 13416, pp. 429–442. Springer, Cham (2022). https://doi.org/10.1007/978-3-031-15707-3_33

51. Weninger, D.: Solving mixed-integer programs arising in production planning. Phd thesis, Friedrich-Alexander-Universität Erlangen-Nürnberg (2016)

52. Wetzler, N., Heule, M.J.H., Hunt, W.A.: DRAT-trim: efficient checking and trimming using expressive clausal proofs. In: Sinz, C., Egly, U. (eds.) SAT 2014. LNCS, vol. 8561, pp. 422–429. Springer, Cham (2014). https://doi.org/10.1007/978-3-319-09284-3_31

Learning to Solve Job Shop Scheduling Under Uncertainty

Guillaume Infantes[1(✉)], Stéphanie Roussel[2], Pierre Pereira[1], Antoine Jacquet[1], and Emmanuel Benazera[1]

[1] Jolibrain, Toulouse, France
{guillaume.infantes,pierre.pereira,antoine.jacquet,
emmanuel.benazera}@jolibrain.com
[2] ONERA-DTIS, Université de Toulouse, Toulouse, France
stephanie.roussel@onera.fr

Abstract. Job-Shop Scheduling Problem (JSSP) is a combinatorial optimization problem where tasks need to be scheduled on machines in order to minimize criteria such as makespan or delay. To address more realistic scenarios, we associate a probability distribution with the duration of each task. Our objective is to generate a robust schedule, i.e. that minimizes the average makespan. This paper introduces a new approach that leverages Deep Reinforcement Learning (DRL) techniques to search for robust solutions, emphasizing JSSPs with uncertain durations. Key contributions of this research include: (1) advancements in DRL applications to JSSPs, enhancing generalization and scalability, (2) a novel method for addressing JSSPs with uncertain durations. The Wheatley approach, which integrates Graph Neural Networks (GNNs) and DRL, is made publicly available for further research and applications.

Keywords: Job Shop Scheduling Problem · Uncertainty · Graph Neural Network · Deep Reinforcement Learning

1   Introduction

Job-shop scheduling problems (JSSPs) are combinatorial optimization problems that involve assigning tasks to resources (e.g., machines) in a way that minimizes criteria such as makespan, tardiness, or total flow time. While the scheduling of production resources plays an important role in many industries, the JSSPs formulation lacks the handling of uncertainty due to its simplifying assumptions. This leads to several direct practical consequences, such as scheduling from fixed factors which can have significant impact on scheduling performance, and ignoring machine breakdowns or material shortages, leading to poor solutions when faced with real-world uncertainty.

A lot of combinatorial optimization problems are NP-complete, and while there has been a lot of progress in solvers performance [5], classical approaches remain often impractical on large instances. Therefore, approximation and

B. Dilkina (Ed.): CPAIOR 2024, LNCS 14742, pp. 329–345, 2024.
https://doi.org/10.1007/978-3-031-60597-0_21

heuristics-based methods have been proposed [20]; but handling uncertainty within these methods remains challenging. Recent works have considered learning algorithms for such problems and report early advances with deep reinforcement learning (DRL) techniques [6,11,27]. Because it models the world as a runnable environment, and the algorithm learns directly from it, DRL does offer a more natural way to handle uncertainty with JSSPs. As Reinforcement Learning methods are robust to noise, the uncertainty in the problem statement, which is reflected in the learner's environment, is naturally handled by the algorithm.

We present two contributions for tackling JSSP with uncertain durations.

First, this work shows a range of improvements over the DRL and JSSPs literature, from neural network architectures to training hyper-parameters and reward definitions. These directly lead to better generalization and scalability, both to same-size problems and to larger problems.

Second, the proposed method solves JSSPs with uncertain duration, that beats optimal deterministic solutions on expected uncertainty. This is relevant to the general use-case where uncertainty cannot be known in advance and where the best deterministic schedule uses expected uncertainty on tasks duration.

Overall, this leads to a very flexible and efficient approach, capable of naturally handling duration uncertainty, with top results on existing Taillard benchmarks while setting a new benchmark reference for JSSPs with uncertain durations with DRL-based solving approaches. The approach, code-named Wheatley, combines Graph Neural Networks (GNNs) and DRL techniques. The code is made available under an Open Source license at https://github.com/jolibrain/wheatley/.

The paper is organized as follows: related works are introduced in Sect. 2, then the JSSP with uncertainty is formalized as a Markov Decision Process (MDP) in Sect. 3. In Sect. 4, we detail the core technical contributions. Section 5 is dedicated to experiments on both deterministic and stochastic JSSPs. Finally, we conclude and discuss future works in Sect. 6.

2   Related Work

This section provides an overview of techniques developed to address both deterministic and stochastic versions of the Job hop Scheduling Problem [25].

Deterministic JSSPs. Mathematical programming, including techniques such as Constraint Programming (CP) or Integer Linear Programming (ILP), has been favored for solving JSSPs due to its precision and ability to model complex scheduling problems. However, the time and resources required to achieve solutions can be very high for large scenarios [5].

Priority Dispatching Rules (PDRs), are heuristic-based strategies that assign priorities to jobs based on predefined criteria; they make local decisions at each step by picking the highest priority job. Common criteria used in PDRs include the Shortest Processing Time (jobs with the shortest processing time are given priority) and the Earliest Due Date (jobs due the earliest are prioritized). This simplifies the scheduling process, making PDRs particularly useful for real-time or large-scale scenarios where rapid decision-making is essential. However, while PDRs are computationally efficient, they rarely yield the optimal solution. A comprehensive evaluation can be found in [20].

Recently, there has been a surge in machine learning and data-driven approaches to solve JSSPs. Instead of relying on handcrafted heuristics, the Learning to Dispatch strategy (L2D) uses machine learning to emulate successful dispatching strategies, as described in [28]. A significant advantage of this method is its size-agnostic nature, as it uses the disjunctive graph representation of the JSSPs. Graph Neural Networks (GNNs) process these graphs to capture intricate relations between operations and their constraints. Deep Reinforcement Learning (DRL) guides the decision-making process, optimizing scheduling decisions based on the features extracted by the GNNs. In [16], the authors leverage a GNN to convert the JSSP graph into node representations. These representations assist in determining scheduling actions. Proximal Policy Optimization (PPO) is employed as a training method for the GNN-derived node embeddings and the associated policy. This method employs an event-based simulator for the JSSP and directly incorporates times into the states of the base Markov Decision Process. This specificity complicates its adaptation to uncertain scenarios. [22] also uses GNNs and PPO for addressing the flexible job-shop problem where the agent also has to choose machines for tasks; the authors add nodes for machines and use two different types of message-passing. The same problem is addressed using a bipartite graph and custom message passing in [9], with good results. The Reinforced Adaptive Staircase Curriculum Learning (RASCL) approach [8] is a Curriculum Learning method that adjusts difficulty levels during learning by dynamically focusing on challenging instances.

Robustness in JSSPs. Stochastic JSSPs (SJSSP) account for uncertainties in processing times by modeling them as random variables. Some techniques use classic solvers to generate robust solutions by anticipating potential disruptions or modeling worst-case scenarios. For instance, in [13], for a given JSSP, several processing times scenarios are sampled. The objective is therefore to generate a unique schedule that is good for all sampled scenarios. PDRs can also be used but they can lack a global view, as in the deterministic case. Several works address SJSSP through meta-heuristics, as described in [2]. Other techniques involving genetic algorithms and their hybridization are also widely used [3,21]. [1] introduces a way to robustify solution to deterministic relaxation of the SJSSP. [14] presents a both proactive and reactive scheduling: a multi-agent architecture is responsible for the proactive robust scheduling and a repair procedure is involved for machine breakdown and arrival of rush jobs.

Some works address the dynamic variant of SJSSP, in which new jobs can arrive at any time. [26] is a review of such extensions along with corresponding proposed solving methods. Classical approaches involve complex mathematical programming models that do not scale well [15]. Online reactive recovery approaches are a possible solution if no robust solution is available [17]. More recent approaches explore the use of using DRL and GNN [12]. However, to the best of our knowledge, there are no work that use such techniques for the SJSSP.

3 JSSP with Uncertainty as a MDP

In this section, we first recall the JSSP definition. Then, we describe how to represent uncertainty and define the corresponding MDP. We finally present how to use Reinforcement Learning (RL) for solving the MDP.

3.1 Background

A JSSP is defined as a pair $(\mathcal{J}, \mathcal{M})$, where $\mathcal{J}$ is a set of jobs and $\mathcal{M}$ is a set of machines. Each job $J_i \in \mathcal{J}$ must go through n_i machines in $\mathcal{M}$ in a given order $O_{ij} \to ... \to O_{in_i}$, where each element $O_{ij}(1 \leq j \leq n_i)$ is called an operation of J_i. The binary relation $\to$ is a precedence constraint. The size of a JSSP instance is classically denoted as $|\mathcal{J}| \times |\mathcal{M}|$. In the following, the set of all operations is denoted $\mathcal{O}$. To be executed, each operation $O_{ij} \in \mathcal{O}$ requires a unique machine $m_{ij} \in \mathcal{M}$ during a processing time denoted p_{ij} ($p_{ij} \in \mathbb{N}^+$). Each machine can only process one job at a time, and preemption is not allowed.

A *solution* σ of a JSSP instance is a function that assigns a start date S_{ij} to each operation O_{ij} so that precedence between operations of each job are respected and there is no temporal overlap between operations that are performed on the same machine. The completion time of an operation O_{ij}, is $C_{ij} = S_{ij} + p_{ij}$. A solution σ is *optimal* if it minimizes the makespan $C_{max} = max_{O_{ij} \in \mathcal{O}}\{C_{ij}\}$, *i.e.* the maximal completion time of operations.

As described in [18], the disjunctive graph is defined by $\mathcal{G} = (\mathcal{O}, \mathcal{C}, \mathcal{D})$ of a JSSP $(\mathcal{J}, \mathcal{M})$ as:

- $\mathcal{O}$ is the set of vertices, i.e. there is one vertex for each operation $o \in \mathcal{O}$;
- $\mathcal{C}$ is a set of directed arcs representing the precedence constraints between operations of each job (conjunctions);
- $\mathcal{D}$ is a set of edges (disjunctions), each of which connects a pair of operations requiring the same machine for processing.

Figure 1a shows the disjunctive graph of a JSSP with 3 jobs and 3 machines. A *selection* is a state of the graph in which a direction is chosen for some edges in $\mathcal{D}$, denoted $\mathcal{D}^O$. If an edge $(O_{ij}, O_{i'j'})$ in $\mathcal{D}$ becomes oriented (in that order), then it represents that operation O_{ij} is performed before $O_{i'j'}$ on their associated machine. A selection is valid if the set of oriented arcs $(\mathcal{C} \cup \mathcal{D}^O)$ makes the graph acyclic. A solution σ can be defined by a valid selection in which all edges in $\mathcal{D}$ have a direction ($\mathcal{D} = \mathcal{D}^O$) and in which start dates of operations are the earliest possible dates consistent with the selection precedences, as done in a classical Schedule Generation Scheme (SGS). Figure 1b illustrates a valid selection for the toy JSSP instance of Fig. 1a.

3.2 Representing Uncertainty

JSSPs can be easily extended with duration uncertainty as bounds on tasks' duration, and effect uncertainty as failure outcomes of a task. While task failures could be represented using special nodes representing completely different

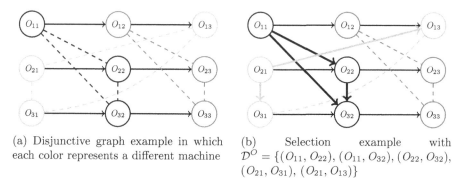

(a) Disjunctive graph example in which each color represents a different machine

(b)   Selection   example   with $\mathcal{D}^O = \{(O_{11}, O_{22}), (O_{11}, O_{32}), (O_{22}, O_{32}), (O_{21}, O_{31}), (O_{21}, O_{13})\}$

Fig. 1. Disjunctive graph representation

outcomes, this would push the boundaries outside JSSP formal capabilities. In this work, failures are handled as retries that consume an uncertain time duration a fixed maximum number of times. In the following, we focus on uncertain task duration as a generic enough scheme to capture relevant uncertainty use-cases.

In the JSSP definition, every operation O_{ij} has a deterministic processing time p_{ij}. We extend this definition by considering that processing times are not known in advance, but that each operation O_{ij} has an associated probability distribution $\mathbb{P}_{ij}$ over its possible duration values. The objective is to minimize the average makespan, that is formally defined by $max_{O_{ij} \in \mathcal{O}} \int_0^{+\infty} (S_{ij} + p_{ij})\, d\mathbb{P}_{ij}$.

3.3   Sequential Decision Making

The scheduling problem boils down to a Markov Decision Process. Inputs of the problem are the the original disjunctive graph $\mathcal{G} = (\mathcal{O}, \mathcal{C}, \mathcal{D})$ and the probability distribution over duration values $\mathbb{P}_{ij}$ associated with each operation O_{ij}.

State. The state s_t at decision step t is defined by:

- the current selection $\mathcal{D}_t^O$,
- the set of already scheduled operations $\mathcal{S}_t$ at step t.

The initial state s_0 is the disjunctive graph representing the original JSSP instance with $\mathcal{D}_0^O = \emptyset$ and $\mathcal{S}_0 = \emptyset$. The terminal state s_T is a complete solution where $\mathcal{D}_T^O = \mathcal{D}$, *i.e.* all disjunctive arcs have been assigned a direction, and $\mathcal{S}_T = \mathcal{O}$, *i.e.* all operations have been scheduled.

Actions. At each step t, candidate actions consist in selecting an operation O_{ij} to put directly after the last scheduled operation on the corresponding machine. This is a simple way to ensure that cycles are never added in $\mathcal{G}$. Intuitively, it consists in choosing an operation to do before all the ones that have not yet been scheduled, and update the current selection accordingly. Furthermore, we force that an operation is a candidate for selection only if its preceding tasks in the same job have been scheduled. As exactly one operation is scheduled at each step, the final state is reached at step $T = card(\mathcal{O})$.

Candidates actions $\mathcal{A}_t$ at step t are formally defined by $\mathcal{A}_t = \{O_{ij} \in \mathcal{O} \mid O_{ij} \notin \mathcal{S}_t \text{ and } \forall j' < j, O_{ij'} \in \mathcal{S}_t\}$.

Transitions. If the chosen action at step t consists in selecting the operation O_{ij} in $\mathcal{A}_t$, then it leads to adding O_{ij} to the scheduled operations set and adding the arc (O_{kl}, O_{ij}) to $\mathcal{D}^O$ for each operation O_{kl} scheduled at step t on the same machine. Formally, this gives:

- $\mathcal{S}_{t+1} = \mathcal{S}_t \cup \{O_{ij}\}$
- $\mathcal{D}^O_{t+1} = \mathcal{D}^O_t \cup \{(O_{kl}, O_{ij}) \in \mathcal{D} \mid O_{kl} \in \mathcal{S}_t \text{ and } m_{kl} = m_{ij}\}$

Reward/Cost. In most approaches to solve MDPs with reinforcement learning, it is preferred to use non-sparse rewards, *i.e.* an informative reward signal at every step. For instance the authors of [28] propose to use the difference of makespan induced by the affectation of the task, as this naturally sums to makespan at the end of the trajectory. With the presence of uncertainty, while we could compute bounds on the makespan in the same way, it is not obvious to aggregate such bounds into a uni-dimensional reward signal.

We use a different approach: we draw durations only when the schedule is complete (at time T) and give it as a cost. Start date of each operation is its earliest possible date considering conjunctive and disjunctive precedence arcs in the solution. All other rewards are null. Formally, it is defined as follows:

- $\forall t < T, r_t = 0$;
- $r_T = max_{O_{ij} \in \mathcal{O}}(S_{ij} + p^{sample}_{ij})$ where $p^{sample}_{ij} \sim \mathbb{P}_{ij}$.

As reinforcement learning aims at minimizing the expectation of costs sum along trajectories, this corresponds to our objective of minimizing the average makespan.

3.4   Solving the MDP with Reinforcement Learning

We use a reinforcement learning setup, where the agent selects tasks to schedule, and gets corresponding partial schedules as observation along with rewards. Using this modeling, effects of actions are deterministic (as they only add edges in the graph), all uncertainty is in the reward value.

The objective is to find a policy that minimizes the average makespan over a set of test problems that are not used during the training phase. To do so, we use a simulator that generates problems that are close to the test problems, and aim at obtaining a policy that minimizes the expectation of makespan along the problems generated by the simulator. Such a policy has to be able to generalize to test problems, *i.e.* give good results without further learning. As the only source of uncertainty is the durations for which parameters only are observable, we want our parametric policy to be able to adapt to these parameters.

Algorithm. In order to learn a policy, we consider a parametric policy and use the Proximal Policy Optimization (PPO) algorithm [19], with action masking [7]. PPO is an on-policy actor-critic RL algorithm. Its current stochastic policy is the actor, while the critic estimates the quality of the current state. More precisely, as shown in Algorithm 1, the algorithm starts by randomly initializing parameters of policy (actor) and value function estimator (critic). Then, for a given number of iterations, it collects trajectory data in the form of a tuple (*observation, action, next observation*) from train problem instances where actions are chosen using current stochastic policy. It then computes makespan values corresponding to train instances by sampling durations and applying chosen order of actions. Using this, it computes returns at every timestep, then advantages (difference of sampled returns to value estimation) using current critic. Observed graphs are then rewired (see Sect. 4.1). The PPO update algorithm itself samples a subset of corresponding observations, actions, advantages, value prediction and returns, then updates the actor parameters (including GNN) using the gradient of the advantages, and updates the critic (including GNN) using gradient of mean square error between the critic value and the observed returns. This PPO update is repeated a small number of times or until the variation in the policy would lead to out-of-distribution critic estimations (because samples are collected using the old policy, i.e. the one before PPO updates).

We then evaluate the current policy (actor) by playing the argmax of the stochastic policy on a given set of validation problems. We repeat these steps until the policy does not seem to improve on validation instances (the N in the external for loop is an upper bound on the number of iterations, which is set to a large value and the for loop is interrupted manually).

4   GNN Implementation: Rewiring, Embedding and Addressing Uncertainty

An overview of the architecture is shown in Fig. 2. The agent takes as an input a partial schedule in the form of a graph, as in Fig. 1b. Several elements, described in this section, are within the actor and allow to choose one action. This action is treated by the simulator to update the schedule graph by adding arcs and simulates the uncertainty when the last state is reached. Note that PPO uses the schedule graph, the action, the reward and the value estimation in order to update the embedders and the GNN.

4.1   Graph Rewiring and Embedding

The representation above allows to model partial schedules (where some conflicts are not resolved) as disjunctive graph representation. Generally speaking, Message-Passing Graph Neural Networks (MP-GNN) use the graph structure as a computational lattice, meaning that information has to follow the graph adjacencies and only them. We thus have to make the difference between the input

---

Algorithm 1: General algorithm

---

1 Generate validation instances, compute heuristic and ortools performance on these instances
    `// actor is ~ current policy` π_θ
2 Init actor
    `// critic is ~ value function estimator`
3 Init critic
4 **for** $i = 1, 2, \ldots N$ **do**
    `// Collect dataset`
5   Generate train instances
6   Collect trials data $\mathcal{D}_i = ((s_t, a_t, r_t, s_{t+1}), ...)$ using current actor
7   For each trial : sample makespan using system simulator (= final cost)
8   Compute returns on trials
9   Compute advantages on trials using current critic
10   Rewire graphs in trial data
    `// PPO update algorithm`
11   **repeat**
12   Sample a minibatch of n data points over shuffled collected data
13   Update actor over the minibatch data towards advantage maximization
14   Update critic by MSE regression
15   **until** *max number of iterations or too large KL-divergence between current and updated policy*
16   Evaluate current policy (actor) on validation instances

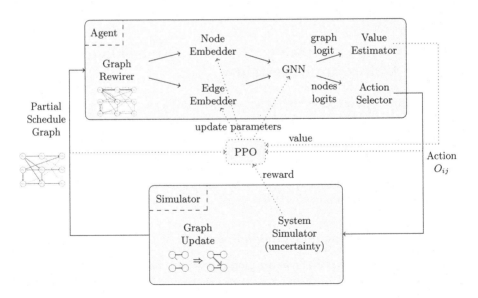

Fig. 2. General Architecture

graph and the graph used by the MP-GNN. This is known as "graph rewiring" in the MP-GNN literature. In our case, if we use only precedencies as adjacencies, this would mean the we explicitly forbid information to go from future tasks to present choice of dispatch, which is definitely not what we want: we want the agent to choose task to dispatch based on effects on future conflicts, meaning that we want information go from future to present task.

Precedences. In order to have a rewired graph as small as possible, we remove from $\mathcal{D}^O$ all edges that are not necessary to obtain the complete order. For instance, we remove from Fig. 1b, the edge (O_{11}, O_{32}). Links are then added in the rewired graph in both directions for every precedence, with different types for precedence and reverse-precedence edges. This enables learned operators to differentiate between chronological and reverse-chronological links and allows the network to pass information in a forward and backward way, depending on what is found useful during learning phase.

Conflicts. Remains the challenge of allowing message circulation between tasks sharing the same machine in the GNN. Two options are possible: 1) adding a node representing a machine with links to tasks using the machine and edges in both directions (from tasks to this machine node and in the opposite direction), or 2) directly connecting tasks that share a machine, resulting in a clique per machine in the message-passing graph. In this paper, we choose the second approach as it showed better results than the first one in preliminary experiments (Fig. 3).

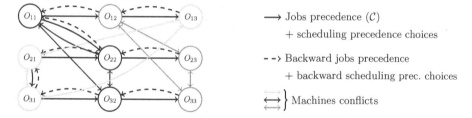

Fig. 3. Rewired graph example with precedences, backward precedences and conflicts as cliques. Each type of arc on the right has its own encoding. Operations O_{11}, O_{21}, O_{22} and O_{31} have here been scheduled in this order.

Edge Attributes. They are used to give explicitly a different type to edges, allowing the network to learn to pass different messages for reverse precedence, precedence and conflicts links. This helps the GNN to effectively handle interactions between tasks of different jobs that share machines.

Node Attributes. We define for node n_{ij} associated with task O_{ij} the following attributes: a boolean A_{ij} indicating if the corresponding task has already been scheduled (affected); a boolean Sel_{ij} indicating if the node is selectable and the machine identifier M_{ij}. We also give parameters of probabilistic distribution of tasks durations, and corresponding task completion time distribution parameters. Task completion times are initialized as if there were no conflicts, i.e. using only durations of task and previous one from same job. There are updated once tasks are affected, considering conflicts with previously affected tasks.

Graph Pooling. For the GNN to give a global summary of nodes, there are two options: either a global isotropic operator on nodes like mean, maximum or sum (or any combination); or a special node that is connected to every task nodes. The latter case is equivalent to learning a custom pooling operator. In this work, following preliminary experimental results, we have chosen the maximum isotropic operator.

Output of the GNN. The message-passing GNN yields a value for every node and a global value for the graph (see Graph Pooling). As nodes represent tasks, these values can be directly used as values for later action selection. In our implementation, we also concatenate the global value to every node value.

Ability to Deal with Different Problem Sizes. The GNN outputs a logit for each node, and there is a one-to-one mapping between nodes and actions, whatever the number of nodes/actions. Internally, the message passing scheme collects messages from all neighbors, making the whole pipeline agnostic to the number of nodes. Learning best actions boils down to node regression, with target values being given by the reinforcement learning loop. This still needs some careful implementation with respect to data structures and batching, but the direct mapping from nodes to actions allows to deal with different problem sizes.

4.2   Handling Uncertainty

On the observation/agent side, we have durations defined with $\mathbb{P}_{ij}$. From these durations distributions, we can compute approximate distributions of tasks completion time simply by propagating completion time parameters recursively upon the precedence graph, whenever precedences are added. The true real duration of the full schedule is computed only once the complete schedule is known based on all $\mathbb{P}_{ij}$. It is then passed as a cost signal. As the RL algorithm naturally handles uncertainty of MDPs, it learns to evaluate partial schedule quality based on expectation of costs, which is exactly our objective.

4.3   Implementation Details

Connecting to PPO. In most generic PPO implementation, the actor (policy) consists of a feature extractor whose structure depends on the data type of the observation, followed by a MLP with a output dimension matching the number of actions. Same holds for the critic (value estimator), with the difference that the output dimension is 1. Some layers can be shared (the feature extractor and first layers of the MLPs). In our case, we do not want to use such generic structure because we have a one-to-one matching from the number of nodes of the observation to the actions. We thus always keep the number of nodes as a dimension of the data tensors.

Graph Embedder. The graph embedder builds the rewired graph by adding edges as stated in Sect. 4.1. It embeds node attributes using a learnable MLP, and edge attributes (here type of edge only) using a learnable embedding. The output dimension of embeddings is an open hyper-parameter *hidden_ dim*, we found a size of 64 being good in our experiments.

Message-Passing GNN As a message passing GNN, we use *EGATConv* from the DGL library [24], which enriches GATv2 graph convolutions [4] with edges attributes. We used 4 attention heads, leading to output of size 4× *hidden_ dim*. This dimension is reduced to *hidden_ dim* using learnable MLPs, before being passed to next layer (in the spirit of feed-forward networks used in transformers). This output of a layer can be summed with the input of the layer, using residual connections. For most of our experiments, we used 10 such layers.

Action Selection. Action selection aims at giving action probabilities given values (logits) output from the GNN. We can either use logits output by the last layer, or use a concatenation of logits output from every layer. We furthermore concatenate the global graph logits of every layer, leading to a data size of $((n\_layers + 1) \times hidden\_ dim) \times 2$ per node. This dimension is reduced to 1 using a learnable linear combination (minimal case of a MLP; we did not find using a full MLP to be useful). Finally, a distribution is built upon these logits by normalizing them, taking into account action masks at this point. As node numbers correspond to action numbers, we directly have action identifier when drawing a value from the distribution.

Normalization. Along all neural network components, we did not find any kind of normalization to be useful. On the opposite hand, durations are normalized in the [0,1] range.

5   Experiments

5.1   Uncertainty Modeling

The framework presented in this paper could accommodate to any kind of duration probability distribution. The main parameters of this distribution belong to the node features and must therefore been described formally.

In order to deal with duration uncertainties, for each operation O_{ij}, we use a triangular distribution with 3 parameters min_{ij}^p, max_{ij}^p, $mode_{ij}^p$, as this is often used in the context of manufacturing processes.

We also have in the simulator the real processing time of a task, denoted $real_{ij}^p$, which is observed by the agent in the final state. With such definition, tasks completion times can respectively be represented with their min, max and mode times as follows: $min_{ij}^C = S_{ij}+min_{ij}^p$, $max_{ij}^C = S_{ij}+max_{ij}^p$, $mode_{ij}^C = S_{ij}+$

$mode_{ij}^p$. Task completion time parameters are updated when adding precedences to the graph (min, max, mode start times are computed based on precedence relations). We give both duration distribution parameters and task completion times distribution parameters as node attributes. The real task completion times $real_{ij}^C = S_{ij} + real_{ij}^p$ can be computed only during the simulation of a complete schedule, giving the real makespan $M_{real} = max_{ij}(real_{ij}^C)$, used as a cost given to the learning agent.

5.2   Benchmarks and Baselines

Benchmarks. Our approach has been tested on instances generated using Tail-lard rules [23]: durations are uniformly drawn in [1,99], and machine affectation is randomly chosen. For *stochastic* instances, this duration corresponds to $mode_{ij}^p$. Minimum and maximum value are uniformly drawn in [$0.95 \times mode_{ij}^p, 1.1 \times mode_{ij}^p$], meaning that tasks can take at most 5% less time and at most 10% more time than mode value.

Baselines. We compare several approaches:

- we first train Wheatley on instances of various sizes. More precisely, W-nxm denotes our approach tested on instances of size $n \times m$, with $(n, m) \in \{(6, 6), (10, 10), (15, 15)\}$;
- for deterministic instances, we can compare with L2D using values reported in the associated paper [28];
- we test several popular Priority Dispatch Rules [20], namely Most Operations Remaining (MOPNR), Shortest Processing Time (SPT), Most Work Remaining (MWKR) and Minimum Ratio of Flow Due Date to Most Work Remaining (FDD/WKR). When computing a schedule, these rules use the mode duration of operations. Next, we retrieve the operations sequence scheduled for each machine and run this sequence with the real operation duration, as done in Schedule Generation Scheme (SGS);
- we use the CP-SAT solver of OR-Tools, denoted *OR-Tools* in deterministic instances test. For stochastic instances, we use it with mode durations and with real instances, respectively denoted *OR-Tools mode* and *OR-Tools real*. As for PDRs, we retrieve the order on each machine and use SGS;
- for stochastic instances, we implement the approach proposed in [13], here denoted *CP-stoc*, that consists in finding the schedule that minimize the average makespan over a given number of sampled instances. We found using 50 samples was a very good compromise between solution quality and computation time (100 gives not much improvement and needs too much time). It is implemented with CP Optimizer 22.10 through docplex [10].

Classical techniques like *CP-stoc* and *OR-Tools/CP-sat* are anytime algorithms that need to compute solution for every problem, while our approach uses a large offline training time and the resulting agent only takes a small inference time for every problem. We decide to give 3 min to classical techniques, as

Table 1. Epoch number for best model

	W-6x6	W-10x10	W-15x15
deterministic	962	542	519
stochastic	712	714	434

Table 2. Comparison of Wheatley wrt training instance sizes.

Evaluation	Deterministic			Stochastic		
	W-6x6	W-10x10	W-15x15	W-6x6	W-10x10	W-15x15
6×6	**508**	521	521	**700**	714	715
10×10	927	**890**	915	1269	**1217**	1232
15×15	1557	**1388**	1392	2297	**1889**	**1889**
20×15	1798	**1583**	1622	2585	**2181**	2188
20×20	2314	1959	**1888**	3632	2643	**2608**

they tend to give very quickly very good solutions, including for large problems, but generally need up to hours to find optimal solution as soon as the problem size becomes large. On the opposite hand, Wheatley takes from 1 h to a few days of training (depending on the problem size), but has a fixed inference time that can become very small when correctly optimized (linear in the number of tasks). The number of iterations to reach the best model is given in Table 1.

5.3 Results

Wheatley Baselines. We first compare the three Wheatley baselines together: we have tested them on small instances, both deterministic and stochastic.

Table 2 presents results obtained for deterministic and stochastic instances on Taillard problems of several sizes. For each size $n \times m$, we have generated 100 instances and, for the stochastic evaluation, we have then sampled one duration scenario for each instance. We then compute the average makespan for each set of instances sizes. Results show that W-10x10 is a good compromise, both for deterministic and stochastic problems. Therefore, in the following, we only present results associated with this approach.

Deterministic JSSP. We compare W-10x10 with baselines presented previously for the deterministic case. In Table 3, we present the average makespan and the average gap[1] obtained for all instances of each category size. Note that we do not present results obtained for each PDR but only the best result one.

Results show that OR-Tools outperforms the other approaches for these sizes, but W-10x10 manages to get close results, even for large instances. More precisely, in comparison with L2D, which is also an approach based on DRL and GNN that was developed for solving JSSPs [28], W-10x10 returns better schedules in average. W-10x10 competes with the best PDR, which is mostly MOPNR

---

[1] Gap for an approach a is equal to $100 \cdot \frac{makespan(a) - makespan^{best}}{makespan^{best}}$.

Table 3. Results on *deterministic* Taillard instances

Evaluation	W-10x10	L2D	Best PDR	OR-Tools
6 × 6	521 (7.4)	571 (17.7)	545 (12.4)	**485 (0)**
10 × 10	890 (9.6)	993 (22.3)	948 (16.8)	**812 (0)**
15 × 15	1389 (17.2)	1501 (26.7)	1419 (19.8)	**1185 (0)**
20 × 15	1583 (16.9)	-	1642 (21.3)	**1354 (0)**
20 × 20	1959 (24.9)	2026 (29.2)	1870 (19.3)	**1568 (0)**
30 × 10	1829 (5.5)	-	1878 (8.9)	**1725 (0)**
30 × 15	2043 (14.5)	-	2092 (17.3)	**1784 (0)**
30 × 20	2377 (22.0)	-	2331 (19.7)	**1948 (0)**
50 × 15	3060 (8.3)	-	3079 (9.0)	**2825 (0)**
50 × 20	3322 (14.9)	-	3295 (14.0)	**2891 (0)**
60 × 10	3357 (1.7)	-	3376 (2.3)	**3301 (0)**
100 × 20	5886 (6.9)	-	5786 (5.1)	**5507 (0)**

Table 4. Results on *stochastic* Taillard instances

Evaluation	W-10x10	Wd-10x10	MOPNR	CP-stoc	OR-Tools	
					mode	real
6 × 6	714 (16.3)	817 (33.1)	699 (13.8)	**669** (9.0)	728 (18.6)	*614 (0)*
10 × 10	1217 (21.5)	1464 (46.1)	1252 (25.0)	**1177 (17.5)**	1262 (25.9)	*1002 (0)*
15 × 15	1889 (29.3)	2406 (64.7)	1988 (36.1)	**1872 (28.1)**	1925 (31.8)	*1461 (0)*
20 × 15	**2181 (30.5)**	2729 (63.3)	2314 (38.5)	2222 (33.0)	2244 (34.3)	*1571 (0)*
20 × 20	2643 (36.4)	3511 (81.2)	2708 (40.0)	**2631 (35.8)**	2619 (35.1)	*1938 (0)*
30 × 10	**2425 (14.1)**	3511 (65.2)	2532 (19.1)	2476 (16.5)	2598 (22.2)	*2126 (0)*
30 × 15	**2792 (26.7)**	3251 (47.5)	2964 (34.5)	2892 (31.2)	2943 (33.5)	*2204 (0)*
30 × 20	**3305 (36.9)**	4186 (73.3)	3390 (40.4)	3355 (39.0)	3299 (36.6)	*2415 (0)*
50 × 15	**4043 (16.5)**	4413 (27.1)	4262 (22.8)	4239 (22.1)	4435 (27.7)	*3472 (0)*
50 × 20	**4520 (26.8)**	5351 (50.1)	4679 (31.2)	4682 (31.3)	4758 (33.4)	*3566 (0)*
60 × 10	**4315 (6.3)**	4475 (10.2)	4451 (9.6)	4442 (9.4)	4579 (12.8)	*4061 (0)*
100 × 20	**7591 (11.8)**	8377 (23.3)	7956 (17.1)	8203 (20.8)	8188 (20.5)	*6793 (0)*

in the case of instances larger than 20 × 20. These results show that Wheatley is able to learn task selection strategies that generalize to much larger problems.

Stochastic JSSP. Table 4 shows results obtained for stochastic problems. Note that the solver *OR-Tools real* is perfect, in the sense that it works with real operations duration values, which is unknown for other approaches at the scheduling time. Therefore, the makespan value computed by *OR-Tools real* is much lower than that of other approaches. Results show that the closest to *OR-Tools real* is *CP-stoc* for small problem sizes. In fact, despite the 50 scenarios it works with, it manages to find a good average makespan. However, when the instances size increases, W-10x10 clearly outperforms other approaches. We also present the results for the deterministic of version of Wheatley run on modes as Wd-10x10. This shows that Wheatley is able to successfully generalize on larger problems.

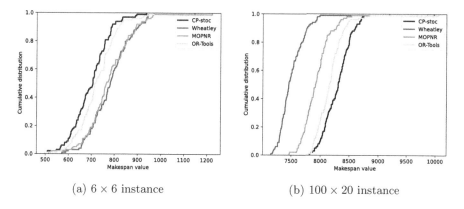

(a) 6 × 6 instance   (b) 100 × 20 instance

Fig. 4. Cumulative makespan of W-10x10, *CP-stoc*, *MONPR* and *OR-Tools mode* for 100 duration scenarios.

In order to further compare the approaches in terms of scatter of the results, we have sampled 100 duration scenarios for one problem of size 6 × 6 and 100 scenarios for one problem of size 100 × 20. Cumulative makespan are presented on Fig. 4. It shows that results presented on Table 4 are representative of several scenarios. In fact, for the 6 × 6 problem, *CP-stoc* returns the lowest makespans, then *OR-Tools*, and *MOPNR* and W-10x10 equivalently (Fig. 4a). That order is completely reversed in the case of the 100 × 20 problem, in which W-10x10 returns the best results (Fig. 4b).

6   Conclusion and Future Works

This paper presents Wheatley, a novel approach for solving JSSPs with uncertain operations duration. It combines Graph Neural Networks and Deep Reinforcement Learning techniques in order to learn a policy that iteratively selects the next operation to execute on each machine. The policy is updated during the training phase through PPO. Results show that Wheatley is competitive in the case of deterministic JSSPs and outperforms other approaches for stochastic JSSPs. Moreover, Wheatley is able to generalize to larger instances.

This work could be extended in several directions. First, it would be possible to extend the experiments with other JSSPs data and particularly instances coming from the industry. It would also be interesting to study the effect of pre-training the policy before running PPO. Finally, we are convinced that the GNN and DRL could be applied to other scheduling problems, such as the Resource-Constrained Project Scheduling Problem, in which handling uncertainty is essential in an industrial context.

References

1. Beck, J., Wilson, N.: Proactive algorithms for scheduling with probabilistic durations. In: IJCAI International Joint Conference on Artificial Intelligence, pp. 1201–1206 (2005)
2. Bianchi, L., Dorigo, M., Gambardella, L.M., Gutjahr, W.J.: A survey on metaheuristics for stochastic combinatorial optimization. Nat. Comput. **8**, 239–287 (2009)
3. Boukedroun, M., Duvivier, D., Ait-el Cadi, A., Poirriez, V., Abbas, M.: A hybrid genetic algorithm for stochastic job-shop scheduling problems. RAIRO: Operations Research (2804-7303) **57**(4) (2023)
4. Brody, S., Alon, U., Yahav, E.: How attentive are graph attention networks? CoRR **abs/2105.14491** (2021). https://arxiv.org/abs/2105.14491
5. Da Col, G., Teppan, E.C.: Industrial-size job shop scheduling with constraint programming. Oper. Res. Perspect. **9**, 100249 (2022). https://doi.org/10.1016/j.orp.2022.100249, https://www.sciencedirect.com/science/article/pii/S2214716022000215
6. Han, B.A., Yang, J.J.: Research on adaptive job shop scheduling problems based on dueling double DQN. IEEE Access **8**, 186474–186495 (2020). https://doi.org/10.1109/ACCESS.2020.3029868
7. Huang, S., Ontañón, S.: A closer look at invalid action masking in policy gradient algorithms. In: The International FLAIRS Conference Proceedings, vol. 35 (2022). https://doi.org/10.32473/flairs.v35i.130584
8. Iklassov, Z., Medvedev, D., de Retana, R.S.O., Takác, M.: On the study of curriculum learning for inferring dispatching policies on the job shop scheduling. In: Proceedings of the Thirty-Second International Joint Conference on Artificial Intelligence, IJCAI 2023, 19th-25th August 2023, Macao, SAR, China, pp. 5350–5358. ijcai.org (2023). https://doi.org/10.24963/ijcai.2023/594
9. Kwon, Y.D., Choo, J., Yoon, I., Park, M., Park, D., Gwon, Y.: Matrix encoding networks for neural combinatorial optimization. In: 35th Conference on Neural Information Processing Systems (NEURIPS) (2021)
10. Laborie, P., Rogerie, J., Shaw, P., Vilím, P.: IBM ILOG CP optimizer for scheduling: 20+ years of scheduling with constraints at IBM/ILOG. Constraints **23**, 210–250 (2018)
11. Liu, C.L., Chang, C.C., Tseng, C.J.: Actor-critic deep reinforcement learning for solving job shop scheduling problems. IEEE Access **8**, 71752–71762 (2020). https://doi.org/10.1109/ACCESS.2020.2987820
12. Liu, C.L., Huang, T.H.: Dynamic job-shop scheduling problems using graph neural network and deep reinforcement learning. IEEE Trans. Syst. Man, Cybern. Syst. (2023)
13. LocalSolver: Stochastic job shop shceduling problems (2023). https://www.localsolver.com/docs/last/exampletour/stochastic-job-shop-scheduling-problem.html
14. Lou, P., Liu, Q., Zhou, Z., Wang, H., Sun, S.X.: Multi-agent-based proactive-reactive scheduling for a job shop. Int. J. Adv. Manuf. Technol. **59**, 311–324 (2012). https://link.springer.com/article/10.1007/s00170-011-3482-4
15. Luh, P., Chen, D., Thakur, L.: An effective approach for job-shop scheduling with uncertain processing requirements. IEEE Trans. Robot. Autom. **15**(2), 328–339 (1999). https://doi.org/10.1109/70.760354

16. Park, J., Chun, J., Kim, S.H., Kim, Y., Park, J.: Learning to schedule job-shop problems: representation and policy learning using graph neural network and reinforcement learning. Int. J. Prod. Res. **59**(11), 3360–3377 (2021). https://doi.org/10.1080/00207543.2020.1870013

17. Raheja, A.S., Subramaniam, V.: Reactive recovery of job shop schedules - a review. Int. J. Adv. Manuf. Technol. **19**, 756–763 (2002). https://link.springer.com/article/10.1007/s001700200087

18. Roy, B., Sussmann, B.: Les problemes d'ordonnancement avec contraintes disjonctives. Note ds **9** (1964)

19. Schulman, J., Wolski, F., Dhariwal, P., Radford, A., Klimov, O.: Proximal policy optimization algorithms (2017)

20. Sels, V., Gheysen, N., Vanhoucke, M.: A comparison of priority rules for the job shop scheduling problem under different flow time-and tardiness-related objective functions. Int. J. Prod. Res. **50**(15), 4255–4270 (2012)

21. Shen, J., Zhu, Y.: Chance-constrained model for uncertain job shop scheduling problem. Soft. Comput. **20**, 2383–2391 (2016). https://doi.org/10.1007/s00500-015-1647-z

22. Song, W., Chen, X., Li, Q., Cao, Z.: Flexible job-shop scheduling via graph neural network and deep reinforcement learning. IEEE Trans. Industr. Inf. **19**(2), 1600–1610 (2023). https://doi.org/10.1109/TII.2022.3189725

23. Taillard, E.: Benchmarks for basic scheduling problems. Eur. J. Oper. Res. **64**(2), 278–285 (1993)

24. Wang, M., et al.: Deep graph library: towards efficient and scalable deep learning on graphs. CoRR **abs/1909.01315** (2019), http://arxiv.org/abs/1909.01315

25. Xiong, H., Shi, S., Ren, D., Hu, J.: A survey of job shop scheduling problem: the types and models. Comput. Oper. Res. **142**, 105731 (2022). https://doi.org/10.1016/j.cor.2022.105731, https://www.sciencedirect.com/science/article/pii/S0305054822000338

26. Xiong, H., Shi, S., Ren, D., Hu, J.: A survey of job shop scheduling problem: the types and models. Comput. Oper. Res. **142** (2022). https://doi.org/10.1016/j.cor.2022.105731, https://www.sciencedirect.com/science/article/pii/S0305054822000338

27. Zeng, Y., Liao, Z., Dai, Y., Wang, R., Li, X., Yuan, B.: Hybrid intelligence for dynamic job-shop scheduling with deep reinforcement learning and attention mechanism (2022)

28. Zhang, C., Song, W., Cao, Z., Zhang, J., Tan, P.S., Chi, X.: Learning to dispatch for job shop scheduling via deep reinforcement learning. In: Larochelle, H., Ranzato, M., Hadsell, R., Balcan, M.F., Lin, H. (eds.) Advances in Neural Information Processing Systems, vol. 33, pp. 1621–1632. Curran Associates, Inc. (2020). https://proceedings.neurips.cc/paper/2020/file/11958dfee29b6709f48a9ba0387a2431-Paper.pdf

Author Index

Printed in the United States
by Baker & Taylor Publisher Services